T0298350

ELEMENTS OF ELECTRICITY and MAGNETISM

Theory and Applications

ELEMENTS OF ELECTRICITY and MAGNETISM

Theory and Applications

Dr. R.K. Tyagi
Professor of Physics
Government PG College,
New Tehri, Uttarakhand

CRC Press is an imprint of the
Taylor & Francis Group, an **informa** business

First published 2025
by CRC Press
4 Park Square, Milton Park, Abingdon, Oxon, OX14 4RN

and by CRC Press
2385 NW Executive Center Drive, Suite 320, Boca Raton FL 33431

CRC Press is an imprint of Informa UK Limited

British Library Cataloguing-in-Publication Data
A catalogue record for this book is available from the British Library

Print edition not for sale in South Asia (India, Sri Lanka, Nepal, Bangladesh, Pakistan or Bhutan)

ISBN13: 9781032867755 (hbk)
ISBN13: 9781032867779 (pbk)
ISBN13: 9781003529125 (ebk)

DOI: 10.4324/9781003529125

Typeset in Times New Roman
by Manakin Press, Delhi

PREFACE

The contents of this book is somewhat different from the other. The basic principles are defined to make this book very remarkable. The book covers two semesters or one year course and I hope that the materials covered in the present book will be beneficial for those students who specially have interest in studying the electricity and magnetism. Moreover, if the students have more difficulty in solving the problems using first principle and they stand in such a situation where they are unable to treat the problems, this book is designed keeping all such problems in mind. At the end of each chapter, the solved problems are given which boosts the students for understanding the basic concepts. It is obvious that the present book is intended first year undergraduate students with the elementary knowledge of the mathematical analysis. The many important illustrative examples are worked out using simple mathematics.

Chapter-1 provides and idea of the charge and Coulomb's Law. In the Chapter–2 and 3, the electric field is introduced with some mathematical techniques and Gauss's theorem is derived in different way. However, some other special topics are given in detail. The electric potential is described in the Chapter-4. The applications of the electric potential play an important role in stimulating and empowering the students with intuitive feel. Chapter-5 describes the concepts of the capacitors and dielectrics. It provides an information, "How is energy stored in the electric field?". Moreover, how does the electric field change when the matter is placed in the electric field. The Chapter-6 deals with the resistance and concerned laws. On the other hand, the solutions of the Laplace and Poisson equations are discussed in the Chapter-7. No section is omitted from the chapter. The concept of magnetic field in detail is given in the Chapter-8. Biot-Savart Law, Ampere's Law and Faraday Law have been discussed with suitable applications. Moreover, paramagnetism, diamagnetism and ferromagnetism and their properties are also given in detail.

In Chapter-9, Alternating currents are encountered and contains a fairly detailed treatment of the necessary and desired circuits containing different passive elements. In the Chapter-10, much of the material on and no topic is avoided but treated with all trappings. Maxwell's equations and proceeding via electromagnetic wave equation are described in the Chapter-11. This chapter is also devoted to the section of scalar and vector potentials.

Author

Detailed Contents

1 CHAPTER

Electric Charge and Coulomb's Law

1.1. ELECTRIC CHARGE

In our universe, there are many particles which are the sources of the electromagnetic radiation. Many of them have definite amount of charge which creates the electromagnetic field. In the atom, the negatively charged electrons move around the positively charged nucleus, Although, nucleus consists of neutrons and protons where the neutrons do not have any charge, while the protons have positive charge. In this way, we can say that the charge is a fundamental property of the particle and it is bound to the matter. If a rod of amber is rubbed with fur, it acquires the property of attracting the small pieces of the papers. In this way, we can say that the some electrons are rubbed off the fur that is why, it acquires the property of the attraction. Further more, if a glass rod is rubbed with silk cloth, the glass rod becomes positively charged.

1.1.1. Kinds of the Electric Charge

There are two kinds of the electric charge, one is positive charge and another is negative charge. The positive charge is said to be the source of the electric field and the negative charge is the sink of the electric field. Thus, we write

(1) Positive charge: a source of the electric field.

(2) Negative charge: a sink of the electric field.

It is observed that the like charges repel each other and unlike charges attract each other.

1.1.2. The Unit of Charge

We know that the smallest charge exists in the nature is the charge of an electron. The magnitude of an electronic charge is 1.60219×10^{-19} coulomb. Thus, we can write,

$$\boxed{e = 1.6 \times 10^{-19} \text{ coulomb}} \quad ...(1.1)$$

where coulomb (C) is the unit of the electric charge. The charge is an invariant quantity. The invariant means that its value does not depend on the choice of the frame of reference and the laws of physics do not deteriorate their form when a positive charge is replaced by a negative charge. Thus, we may say that the charge of all matters is quantized and it is equal to the integral multiple of the electronic charge e, that is,

$$\boxed{q = ne} \quad ...(1.2)$$

1.1.3. Conservation of the Charge

The law of nature is that the charge remains conserved every where. If we consider a closed system consisting of a large number of particles, this closed system must obey the law of the electric charge conservation. The conserve means that the electronic charge can not be created or destroyed. However, the charge can be transferred from one body to another. Moreover, the sum of all positive and the negative charges remains conserved, that is,

$$\boxed{\sum_n q_n = \text{constant}} \quad ...(1.3)$$

1.2. ELECTRIC CONDUCTORS AND INSULATORS

The electrical conduction is defined as the motion of the electrons in a substance. All substances are not good conductors of the electricity. The electrical conductors are those having large numbers of free electrons for carrying the current and the conductors have very low electrical resistance. On the other hand, the insulators are those having no free electrons for the conduction of the electricity. In the insulators, all electrons are tightly bound to the nucleus. Thus, the insulators have a very high resistance. The examples of the insulators are mica, rubber, wood and glass. More-over, when a glass rod is rubbed with fur, only the rubbed portion of the rod becomes charged and this induced charge is known as static or stationary charge. This static charge can not move to another part of the rod. This is because that the glass rod is an insulator. However, if we take a metallic rod, the induced charge spread over the whole region of the rod. This metallic rod cannot attract the piece of papers, thus, as a result it can be said that the conductors can not be charged.

Experiment: Fig. 1.1 shows a phenomenon of the electrostatic induction. It shows, how a conductor does accumulate the charge without touching it. Now, take a negatively charged rubber rod R and the rod R is brought near to the conductor AB, the negatively charged particles (electrons) will face a repulsive force due to the presence of the electrons at the end of the rubber rod. As a result, some of the electrons move away from the end A and thus, the positive charges are accumulated at the end A of the rod AB. This accumulation of the positive charges are due to the migration of the electrons from the edge A. Now, when the end B of the conductor AB is connected to the earth as shown in fig. 1.1, all the electrons are repelled to the earth and as a result, the positive charges are left on the other end.

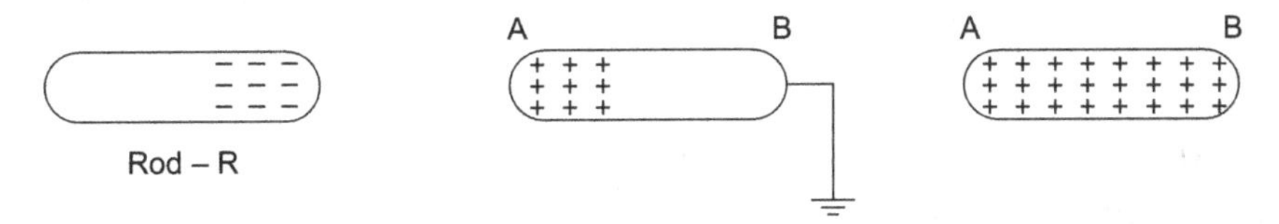

Fig. 1.1. The phenomenon of electrostatic induction

Moreover, if the conductor AB is grounded, all the positive charges have been distributed uniformly over the whole surface of the conductor AB.

1.3. COULOMB'S LAW OF FORCE

In 1784, a French scientist Charles Coulomb measured the electric force between the two charge bodies using torsion balance. This electric interaction between two charged bodies is observed as a coulomb's law, which states that the electric interaction between two static point charges is proportion to the product of the magnitude of two charges and to the inverse of the square of the distance between them. Note that this force is directed along the line joining the two charges. The force between two like charges is repulsive and the force between two unlike charges is attractive in nature. Moreover, the force between any two charges is independent of the presence of the other charges. Suppose that the two point charges q_1 and q_2 are separated by a distance r as shown in fig. 1.2.

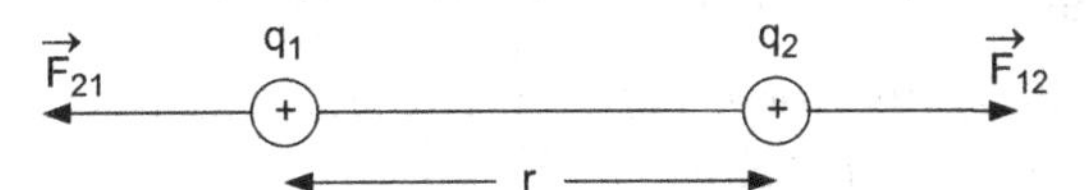

Fig. 1.2. Illustration of the coulomb's Law

According to coulomb's law, the force exerted by q_1 on q_2 is,

$$F \propto \frac{q_1 q_2}{r^2}$$

or
$$F = k\frac{q_1 q_2}{r^2}\hat{r} \qquad ...(1.4)$$

where $\hat{r}$ is the unit vector along the distance r.

If we denote the force exerted by the charge q_1 on the charge q_2 as $\vec{F}_{12}$, then

$$\vec{F}_{12} = k\frac{q_1 q_2}{r^2}\hat{r} \quad ...(1.5)$$

Therefore, The Newton's third law gives.

$$\vec{F}_{12} = -\vec{F}_{12} \quad ...(1.6)$$

where k is the constant of proportionality.

In M.K.S. system of units, The force $\vec{F}_{12}$ is taken in Newton, r in meters and the charge in coulomb.

Thus, k is given by

$$k = \frac{1}{4\pi \epsilon_0} = 9 \times 10^9 \text{ Nm}^2\text{c}^{-2} \quad ...(1.7)$$

where ϵ_0 is called the permittivity of the free space. The quantitative value of ϵ_0 is given by

$$\epsilon_0 = 8.85 \times 10^{-12} \text{ C}^2\text{N}^{-1}\text{m}^{-2} \quad ...(1.8)$$

Thus, the Eq. (1.5) takes a standard form as

$$\boxed{\vec{F}_{12} = \frac{1}{4\pi \epsilon_0}\frac{q_1 q_2}{r^2}\hat{r}} \quad ...(1.9)$$

Furthermore, consider a situation if charges q_1 and q_2 are opposite in nature, Then, the force between them is attractive as illustrated in the fig. 1.3.

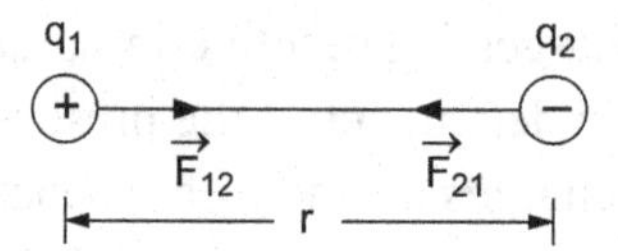

Fig. 1.3. Force between two opposite charges

Example 1.1. Compute the force between two equal charges $q_1 = q_2 = 1C$, separated by a distance $r = 1$ metre.

Solution: According to coulomb's Law

$$F = \frac{1}{4\pi \epsilon_0}\frac{q_1 q_2}{r^2}$$

$$= 9 \times 10^9 \cdot \frac{1C \times 1C}{1\text{m}}$$

or $$F = 9 \times 10^9 \text{ Newton}$$

F is repulsive in nature.

In the example 1.1, the force between two equal charges, separated by a distance 1.0 m apart, comes out 9×10^9 N. From this argument, we can define one coulomb as "one coulomb is that charge which repels or attracts the similar charge placed 1.0 m apart in the vacuum, with a force of 9×10^9 N.

Units of Charge: In S.I. units, the unit of charge is coulomb and in C.G.S. system of units, it is stat coulomb.

$$1 \text{ C} = 3 \times 10^9 \text{ stat coulomb}$$
$$= 3 \times 10^9 \text{ e.s.u.}$$

1.4. LIMITATIONS OF COULOMB'S LAW

We have following limitations on coulomb's law.

(1) The coulomb's law can be applied to any pair of the point charges.

(2) The system of charges must be in stationary position.

1.5. SUPERPOSITION PRINCIPLE

Since coulomb's law determines the force, when two point charges are situated at a distance r apart, it may also be applied to a system consisting of large numbers of charges. Then, the net force on any charge is the vector sum of the forces due to rest charges. This is known as principle of superposition. If we consider a system of three charges q_1, q_2 and q_3 as shown in fig. 1.4, the resultant force on the charge q_1 will be the vector sum of the forces due to charges q_2 and q_3. Thus,

$$\vec{F}_1 = \vec{F}_{12} + \vec{F}_{13} \quad \text{...(1.10)}$$

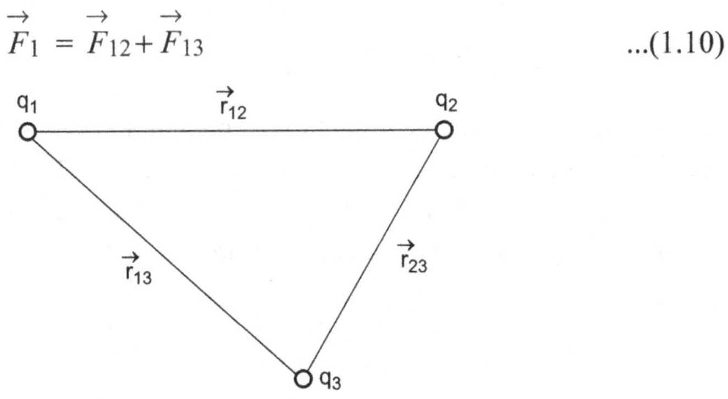

Fig. 1.4. System of three point charges

Moreover, for the system of N charges, the net force experienced by the charge q_1 will be

$$\vec{F}_1 = \vec{F}_{12} + \vec{F}_{13} + \vec{F}_{14} + \ldots + \vec{F}_{1N}$$

$$= \sum_N \vec{F}_{1N} = \frac{1}{4\pi \in_o} \sum_{N=2}^{N} \frac{q_1 q_N}{r_{1N}^2} \hat{r}_{1N} \quad \ldots(1.11)$$

Example 1.2. A charge q is divided into two parts q_1, and $q - q_1$. If these charges are placed at a distance x for which they experience maximum coulomb repulsion, Find the relation between q and q_1.

Solution: The force between q_1 and $q - q_1$ is

$$F = \frac{1}{4\pi \in_0} \frac{q_1(q - q_1)}{x^2}$$

For the force to be maximum, we have

$$\frac{dF}{dq_1} = 0$$

or $$\frac{1}{4\pi \in_o x^2} \frac{d}{dq_1}(qq_1 - q_1^2) = 0$$

or $$q_1 = \frac{q}{2}$$

and $$\frac{d^2F}{dq_1^2} = -2(-\text{ve})$$

The F will be maximum if $q = 2q_1$

Example 1.3. Two identical metallic charged balls, each of mass m, are suspended at a common point c by threads of negligible mass and length l as shown in fig. 1.5. Each ball carries a charge q, so that the balls repel each other. Show that the system comes in equilibrium at a distance

$$x = \left(\frac{q^2 l}{2\pi \in_0 \text{mg}}\right)^{1/3}$$

$$q^2 = 16\pi \in_o \text{mg}\, l^2 \sin^2\theta \tan\theta\,.$$

Solution: Let θ be the angle made by the thread to the vertical at the position of the equilibrium,

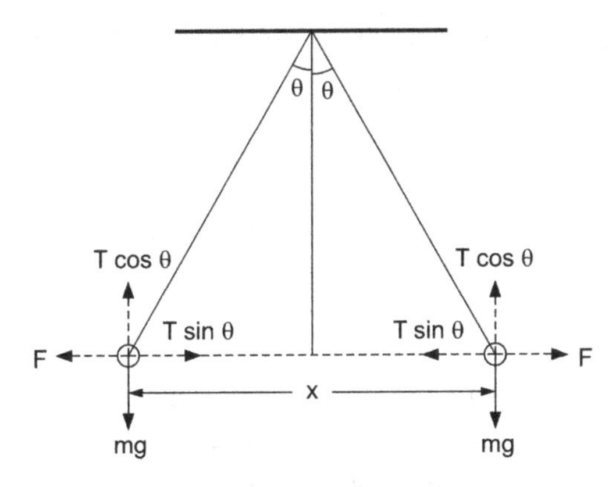

Fig. 1.5. Two charged balls.

$$F = \frac{1}{4\pi \epsilon_0} \frac{q^2}{x^2}$$

$$\therefore \qquad T \sin \theta = F = \frac{1}{4\pi \epsilon_0} \frac{q^2}{x^2}$$

and $\qquad T \cos \theta = \text{mg}$

on dividing the equations, we get, for small θ,

$$\sin \theta = \tan \theta = \frac{x}{2l} = \frac{1}{4\pi \epsilon_0} \frac{q^2}{x^2 \text{mg}}$$

Thus, $\qquad x = \left(\frac{q^2 l}{2\pi \epsilon_0 \text{ mg}} \right)^{1/3}$

Since, $\qquad \tan \theta = \frac{q^2}{4\pi \epsilon_0 . x^2 \text{ mg}}$

$$\therefore \qquad q^2 = 4\pi \epsilon_0 \text{ mg } x_2 \tan \theta$$

for small θ, $\qquad x = 2l \sin \theta$

Thus, $\qquad q^2 = 16\pi \epsilon_0 \, mg \, l^2 \sin^2 \theta \tan \theta.$

1.6. PERMITTIVITY AND DIELECTRIC CONSTANTS

When two charges are placed at a some distance apart in a medium, the force between these two charges is affected, due to the medium. Thus, the permittivity of the medium is defined as the property of the medium which determines the forces between two point charges. Consider the two point charges q_1 and q_2

separated by a distance r. In the medium, the force between two point charge is given by

$$\vec{F}_m = \frac{1}{4\pi \epsilon} \frac{q_1 q_2}{r^2} \hat{r} \quad ...(1.12)$$

On the other hand, if two charges are placed in a vacuum, the force will be

$$\vec{F} = \frac{1}{4\pi \epsilon_0} \frac{q_1 q_2}{r^2} \hat{r} \quad ...(1.13)$$

This force may be attractive or repulsive, it depends on the nature of the charges. In the Eq. (1.12) the subscript m represents the force between two charges q_1 and q_2 in the medium and ϵ is the absolute permittivity of the medium. On dividing Eq. (1.13) by the Eq. (1.12), we get,

$$\frac{F}{F_m} = \frac{\epsilon}{\epsilon_0} \quad ...(1.14)$$

This ratio ϵ/ϵ_o is called the relative permittivity of the medium, Thus, we write

$$\boxed{\epsilon_r = \frac{\epsilon}{\epsilon_0}} \quad ...(1.15)$$

Hence, the relative permittivity of the medium is defined as the ratio of the absolute permittivity of the medium to the permittivity of the free space.

Example 1.4. Consider three charges q_1, $-q_2$ and $-q_3$, where the charge q_1 is situated at origin as shown in fig. 1.6. The distances $r_{12} = a$ and $r_{13} = b$. Compute the net force on the charge q_1 exerted by q_2 and q_3.

Solution: We have

$$\vec{F}_1 = \vec{F}_{12} + \vec{F}_{13}$$

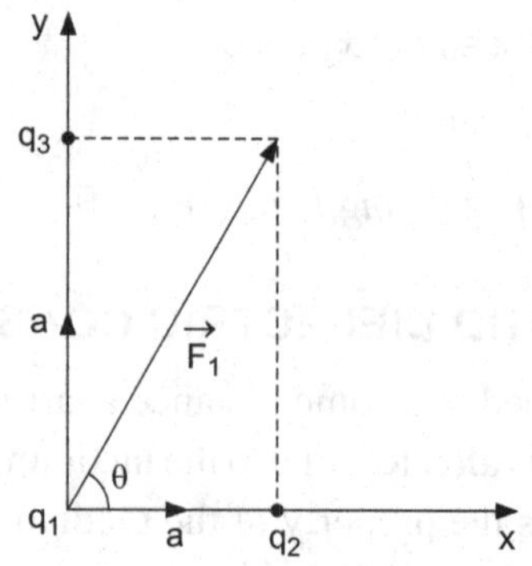

Fig. 1.6. Force between charges.

Now, $$\vec{F}_{12} = \frac{1}{4\pi \in_0} \frac{q_1 q_2}{r_{12}^2} \hat{r}_{12}$$

$$= \frac{1}{4\pi \in_0} \frac{q_1 q_2}{b^2} \hat{i}$$

and $$\vec{F}_{13} = \frac{1}{4\pi \in_0} \frac{q_1 q_3}{a^2} \hat{j}$$

The force on the charge q_1 will be

$$|F_1| = \sqrt{F_{12}^2 + F_{13}^2}$$

The direction of the force $\vec{F}_1$ with x-axis is given by

$$\tan\theta = \frac{F_{13}}{F_{12}}$$

or $$\theta = \tan^{-1}\left(\frac{F_{13}}{F_{12}}\right)$$

Example 1.5. Two charges of Q coulombs each are placed at two opposite corners of a square. What should be the value of the additional charges-q placed at each of the other two corners that will reduce the resultant force on each of the charges Q to zero.

Solution: Consider a square of side a as shown in fig. 1.7.

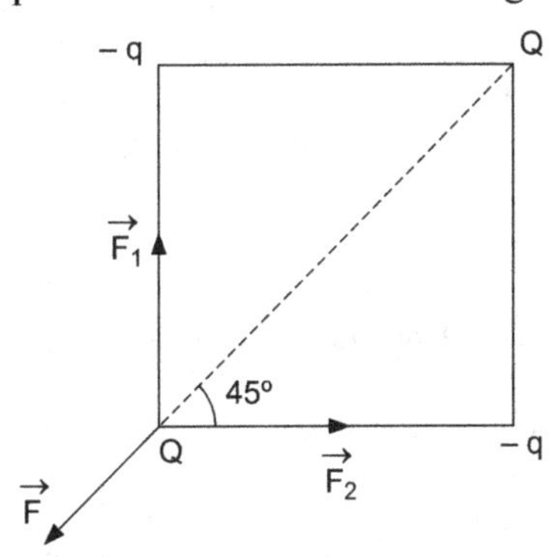

Fig. 1.7. System of charges.

$\therefore$ $$\vec{F}_1 = \vec{F}_1 \cos 45 + \vec{F}_2 \cos 45$$

$$= 2F_1 \cos 45$$

$\therefore$ $$F_1 = \frac{1}{4\pi \in_0} \frac{q \cdot Q}{a^2}$$

and
$$F = \frac{1}{4\pi\epsilon_0} \cdot \frac{Q^2}{2a^2}$$

Thus,
$$\frac{1}{4\pi\epsilon_0} \cdot \frac{Q^2}{2a^2} = \frac{\sqrt{2}}{4\pi\epsilon_0} \frac{qQ}{a^2}$$

Thus,
$$Q = 2\sqrt{2}\, q$$

Example 1.6. The three point charges are placed at the equal distance along the straight line as shown in fig. 1.8. What should be the magnitude of the charge Q in order to make the net force on the charge at the origin to be zero.

Solution:

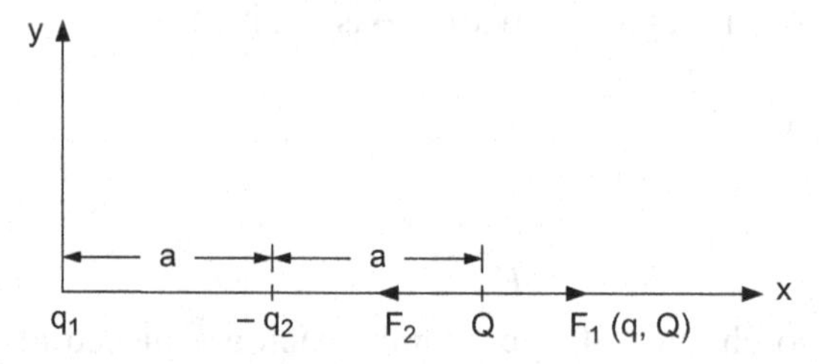

Fig. 1. 8. Three point charges.

we have two forces, $F_1(q_1, Q)$ and $F_2(-q_2, Q)$

Thus,
$$F_1 = F_2$$

$$\frac{1}{4\pi\epsilon_0} \frac{q_1 Q}{(2a)^2} = \frac{1}{4\pi\epsilon_0} \frac{q_2 Q}{a^2}$$

Thus, we get,
$$q_1 = 4q_2$$

EXERCISES

1.1. What is the electric charge? A body has the charge of 1μC, what does it mean.

1.2. Find the number of the electrons in

(a) a charge of 1C.

(b) a charge of 1μC.

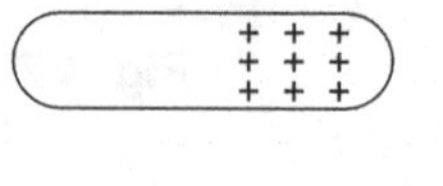

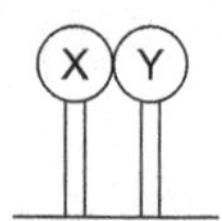

Fig. 1.9. Conductors.

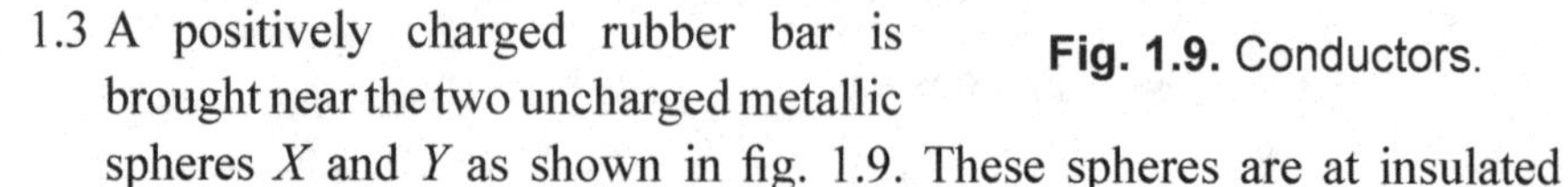

1.3 A positively charged rubber bar is brought near the two uncharged metallic spheres X and Y as shown in fig. 1.9. These spheres are at insulated stands.

(a) what happens, if the bar is brought near to the sphere *X*.

(b) What happens, when the bar is removed.

(c) When bar is removed and the spheres *X* and *Y* are separated.

1.4. An isolated metallic sphere is given a positive charge. Does it mass increases, decreases or remains same.

1.5. Two similar charges, 1 cm distance apart experience a force of 1.0×10^{-2} N. Compute the magnitude of each charge.

1.6. Two charges of –4μC and 6μC are separated by a distance of 1.0 cm. Compute the force between them.

1.7. Four equal point charges are placed at the four corners of a square of the side *L*. Find the magnitude of the charge placed at the centre of the square for which the system should be in equilibrium.

1.8. Two positive point charges are placed at a distance *d* apart. The sum of the both charges is *Q*. Compute the values of the charges if coulomb force between them is maximum.

1.9. Three point charges, each having magnitude *q*, are placed at the vertices of an equilateral triangle. Find the force on a charge *Q* placed at the centre of the triangle.

1.10. The three charges q_1 = 5μC, q_2 = –3μC and q_3 = 7μC are placed in a triangle as shown in fig. 1.10. Compute the net force on the charge q_3 if $a = 3$ cm, and $b = 4$ cm.

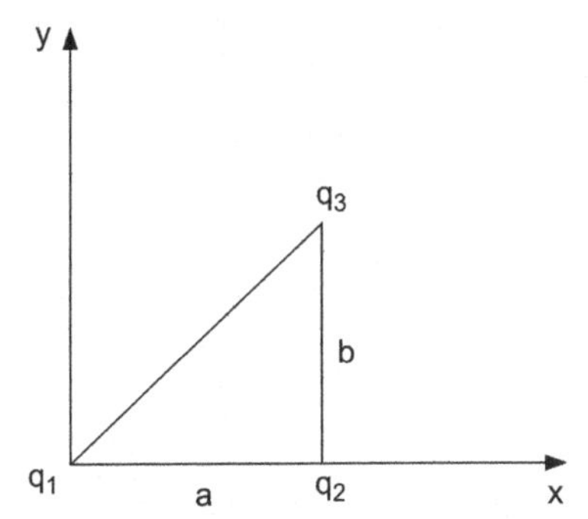

Fig. 1.10. Charge system.

1.11. A thin metallic ring of the radius a has the charge *q*. What should be the tension in the wire, if the charge *Q* is placed at the centre of the ring.

■ ■ ■

Fig. 1.10 Charge system.

2 The Electrostatic Field

CHAPTER

The gravitational force is a common force that acts between two distant masses. Similarly, the coulomb force acts between two stationary charges. The limitation of the coulomb force is that it acts between two point charge. If there is a distribution a charges, then the concept of the electric field is a powerful tool to handle such a system. A stationary charge acts an electrostatic force on the other charges in the region of space. It mean that a single charge creates an electric field in the space near by it. In this chapter, we shall discuss the electric field and its applications.

2.1. THE ELECTRIC FIELD

Since the static charge produces an electric field surrounding it, thus, the electric field at a given point in the region of space may be defined as the force experienced by a positive test charge in that region of space. The concept of the positive test charge is very interesting that it is a tool to defect whether field is present or not in the region of space. The magnitude of the test charge in negligibly small. The electric field is a vector field because it measures the force. We know that a charge q produces an electric field, then a positive test charge q_0 is placed at a point where the electric field is to be determined. As a result, the test charge q_0 experiences a force due to the interaction of the electric field produced by the charge q. Thus, the force on the test charge q_0 will be

$$\vec{F} = \frac{1}{4\pi\epsilon_o} \frac{q \cdot q_o}{r^2} \hat{r} \qquad ...(2.1)$$

where r is a distance of a point from the charge q at which the electric field is to be computed.

Thus, the electric field strength or electric field intensity is defined as,

$$\boxed{\vec{E} = \lim_{q_o \to o} \frac{\vec{F}}{q_o}} \qquad ...(2.2)$$

Hence,

$$\vec{E} = \frac{1}{4\pi\epsilon_o} \frac{q}{r^2} \hat{r} \qquad ...(2.3)$$

Thus, the electric field intensity at a point in the region of space is defined as the force per unit test charge. However, it is a field surrounding the charge q. Moreover, a test charge is taken into the electric field to measure the strength and direction of the electric field. The presence of the test charge q_0 does not affect the original electric field. To understand this point consider a point charge q which generates an electric field $\vec{E}$ directed away from the charge q, as shown in fig. 2.1. Now, to compute the electric field at a point P at a distance r from the point charge q,

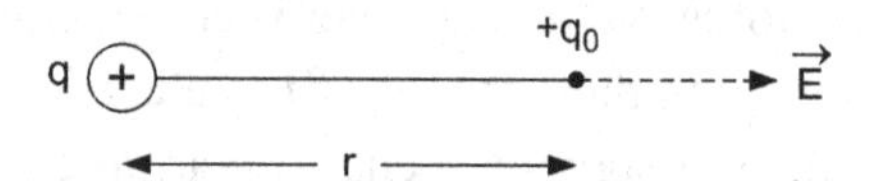

Fig. 2.1. A test charge q_o is placed at a distance r from the positive point charge q.

on placing the positive test charge q_0, the electric field at a point P is given by

$$\vec{E} = \frac{\vec{F}}{q_o} = \frac{1}{4\pi\epsilon_o} \frac{q}{r^2} \hat{r} \qquad ...(2.4)$$

where $\hat{r}$ is the unit vector in the direction of $\vec{r}$. Since two positive charges repel each other, the direction of the force $\vec{F}$ will be the direction of the electric field. The positive test charge experiences the different forces at the different points in the region of space. Thus, the electric field varies from the point to point in the region. As a result, one may say that there are many electric fields due to the several charges in the region of space. Using super-position principle, the total electric field in the space is equal to the vector sum of the electric fields due to the individual charges. Hence,

$$\vec{E} = \vec{E_1} + \vec{E_2} + \vec{E_3} + ...$$

or

$$\boxed{\vec{E} = \sum_{n=1}^{\infty} \vec{E_n}} \qquad ...(2.5)$$

In general, the electric field at any point is given by

$$\vec{E} = \frac{1}{4\pi\epsilon_o} \sum_i \frac{q_i}{r_i^2} \hat{r} \qquad ...(2.6)$$

The unit of the electric field E is newton/coulomb (N/C).

2.2. ELECTRIC FIELD DUE TO UNIFORM CHARGE DISTRIBUTION

Suppose that the charges are distributed in the region, such a distribution of charges is known as continuous charge distribution. We can, now, compute the electric field intensity by taking a charge element dq. Thus, the electric field strength at a point P at a distance r from the charge element dq is given by

$$\vec{dE} = \frac{1}{4\pi\epsilon_o}\frac{dq}{r^2}\hat{r} \qquad ...(2.7)$$

where $\hat{r}$ is unit vector along $\vec{r}$. Now, according to principle of superposition, the net electric field intensity $\vec{E}$ is the vector sum of all such electric fields produced by the individual elements. Thus,

$$\vec{E} = \sum_i dE_i \qquad ...(2.8)$$

$$= \frac{1}{4\pi\epsilon_o}\sum_i \frac{dq_i}{r_i^2}\hat{r}_i$$

or

$$\boxed{\vec{E} = \frac{1}{4\pi\epsilon_o}\int \frac{dq}{r^2}\hat{r}} \qquad ...(2.9)$$

Example 2.1. The four point charges $q_1 = 1\text{C}$, $q_2 = 2\text{C}$, $q_3 = 3\text{C}$ and $q_4 = -2C$ are placed at the four corners of a square of side 2 m. Compute the electric field strength at the centre of the square.

Solution: The four charges q_1, q_2, q_3 and q_2 are situated at the four corners of the square of side 2 m as shown in fig. 2.2.

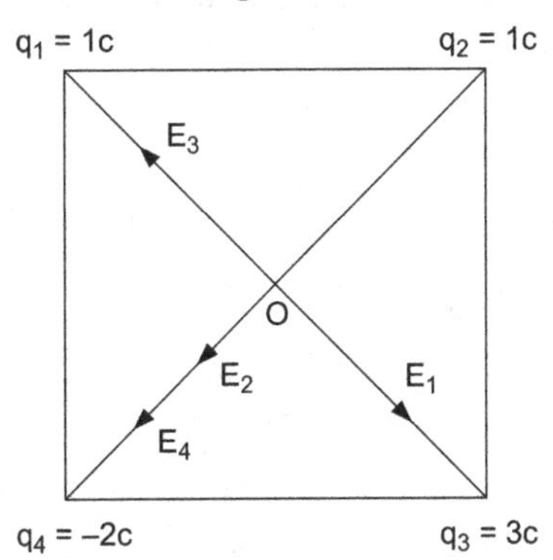

Fig. 2.2. Four charges on square.

$$E_1 = \frac{1}{4\pi\epsilon_o}\frac{q_1}{r_1^2} = 9\times10^9\times\frac{1}{(\sqrt{2})^2} = 4.5\times10^9 \text{ N/C}$$

$$E_2 = \frac{1}{4\pi\epsilon_o}\frac{q_2}{r_2^2} = 9\times10^9\times\frac{2}{2} = 9.0\times10^9 \text{ N/C}$$

$$E_3 = \frac{1}{4\pi\epsilon_o}\frac{q_3}{r_3^2} = 9\times10^9\times\frac{3}{2} = 13.5\times10^9 \text{ N/C}$$

and $$E_4 = \frac{1}{4\pi\epsilon_o}\frac{q_4}{r_4^2} = 9\times10^9\times\frac{2}{2} = 9\times10^9 \text{ N/C}$$

Thus, $E'_1 = E_3 - E_1$ (since E_1 and E_3 are opposite)

$= 9 \times 10^9$ N/C

and $E'_2 = E_2 + E_4$ (since E_2 and E_4 are in same directions)

$= 18.0 \times 10^9$ N/C

The net electric field at the centre 'O' will be

$$E = \sqrt{E_1'^2 + E_2'^2}$$

$$= 15.58 \times 10^9 \text{ N/C}$$

2.3. MOTION OF A CHARGED PARTICLE IN THE UNIFORM ELECTRIC FIELD

There are two cases for the motion of the charged particle in the uniform electric field, viz,

(a) When the charged particle is moving parallel to the electric field,

(b) When the charged particle is moving perpendicular to the electric field.

(a) Consider a charged particle, having a charge q, moving between two parallel plates as shown in fig. 2.3. These plates consist of opposite charges.

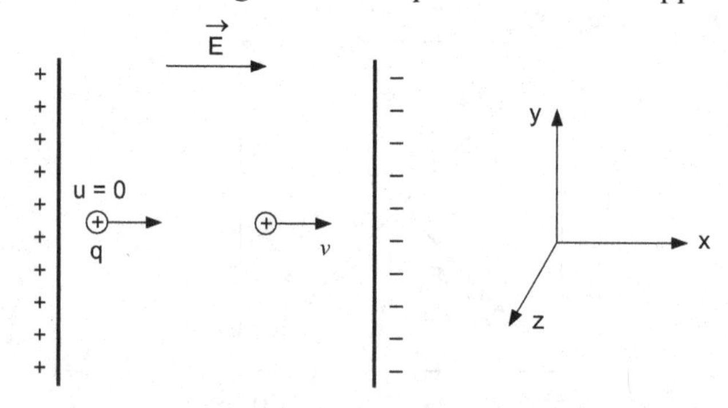

Fig. 2.3. A positive charge is moving parallel to the electric field.

When the charge q moves in a uniform electric field, it experiences an electrostatic force,

$$\vec{F} = q\vec{E} \qquad ...(2.10)$$

Since the field is uniform, the intensity of the electric field, will remain same every where. Suppose that the initial velocity of the charge is zero, the charge will move in a straight line in the direction of the electric field with a constant acceleration.

Thus, the acceleration of the charge q is given by

$$\vec{a} = \frac{\vec{F}}{m}$$

or

$$\vec{a} = \frac{q\vec{E}}{m} \qquad ...(2.11)$$

where m is the mass of the charged particle. Since the electric field E is along x-axis, we have

$$\vec{E} = E_x\hat{i} \qquad ...(2.12)$$

Substituting the value of $\vec{E}$ from the Eq. (2.12) in the Eq. (2.11), we get

$$\boxed{\vec{a} = \frac{qE_x}{m}\hat{i}} \qquad ...(2.13)$$

The positive charge q experiences a force when placed initially between the plates, and it starts moving towards the negatively charged plates. The velocity gained by the charged particle at the time t may be obtained as

$$\vec{v} = \vec{v} + \vec{a}t \qquad ...(2.14)$$

or

$$\vec{v} = \vec{a}t$$

$$\boxed{v = \frac{qE_xt}{m}} \qquad ...(2.15)$$

If x is the distance travelled by the charge q in the time t,

$$x = ut + \frac{1}{2}at^2 \qquad ...(2.16)$$

$$= \frac{1}{2}at$$

or

$$\boxed{x = \frac{1}{2}\frac{qE_x}{m}t^2} \qquad ...(2.17)$$

Moreover, if v is the final speed before it strikes the negative charged plate, Thus,

$$v^2 = 2ax \qquad ...(2.18)$$

$$= \frac{2qE_x x}{m}$$

or

$$\boxed{v = \left(\frac{2qE_x x}{m}\right)^{1/2}} \qquad ...(2.19)$$

Again, the Kinetic energy of the charge q after after attaining the velocity v is given by

$$\text{K.E.} = \frac{1}{2}mv^2$$

$$= \frac{1}{2}m.\frac{2q.E_x.x}{m}$$

$$= qE_x.x$$

$$\boxed{\text{K.E.} = Fx} \qquad ...(2.20)$$

The Eq. (2.20) is the direct consequence of the work-energy theorem.

(b) **When the motion of the charged particle is perpendicular to the electric field.** Suppose that a particle of mass m, charge q enters into an electric field with the initial velocity velocity v_0 as shown in fig. 2.4. The electric field E is uniform and the particle experiences an upward force,

$$\vec{F} = q\vec{E} \qquad ...(2.21)$$

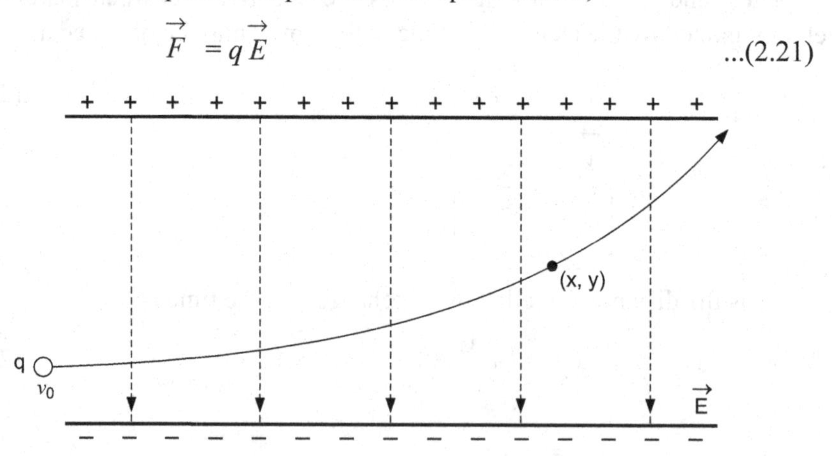

Fig. 2.4. Motion of a charge q in a uniform electric field.

The electric field is uniform every where, the acceleration of the electron will be

$$\vec{a} = \frac{\vec{F}}{m}$$

$$a = \frac{qE}{m} \quad ...(2.22)$$

Let $(x\ y)$ be the position coordinate of the particle at any time t, thus,

$$x = v_o t \quad ...(2.23)$$

and

$$y = \frac{1}{2}at^2 \quad ...(2.24)$$

Substituting the value of a from the Eq. (2.22) in the Eq. (2.24), we get

$$y = \frac{1}{2}\frac{qE}{m}t^2 \quad ...(2.25)$$

Now, eliminating t from the Eq. (2.25) by using the Eq. (2.23), we get

$$y = \frac{1}{2}\left(\frac{qE}{mv_0^2}\right)x^2 \quad ...(2.26)$$

Since the kinetic energy of the particle is

$$K = \frac{1}{2}mv_0^2,$$

thus,

$$\boxed{y = \left(\frac{qE}{4K}\right)x^2} \quad ...(2.27)$$

The Eq. (2.27) shows that the trajectory of the particle between the plates is a parabola.

Example 2.2. An electron is released between the plates of a cathode ray oscilloscope and there is a uniform electric field of 2×10^5 N/C. If the initial kinetic energy of the electron is 1 KeV, find the deflection of the electron if it enters perpendicular to the field. The length of assembly is 4 cm.

Solution: Using Eq. (2.27),

$$y = \frac{qE}{4K}.x^2$$

$$= \frac{1.6\times10^{-19}\times2\times10^5\times4\times4\times10^{-4}}{4\times10^3\times1.6\times10^{-9}}$$

$$= 8 \times 10^{-2} = 8 \times 10^{-2} \text{ m} = 8 \text{ cm}.$$

2.4. ELECTRIC LINES OF FORCE

Michael Faraday introduced the concept of the electric lines of force to visualize the nature of the electric field. The concept of the electric lines of force is very useful when dealing with the spatial electric field problems. It is due to the fact that the force between two point charges depends on the distance between them. The electric lines of force are imaginary lines that predicts the pictorial visualization of the field. Moreover, tangent to these lines of force at any point represents the direction of the electric field at that point.

Properties of the Electric Field Line

We have following properties of the electric field lines,

(1) The electric field lines originate from the positive charge which is called the source of the field lines and end-up on the negative charge which is known as the sink. The electric field lines for the positive and negative charges are shown in fig. 2.5.

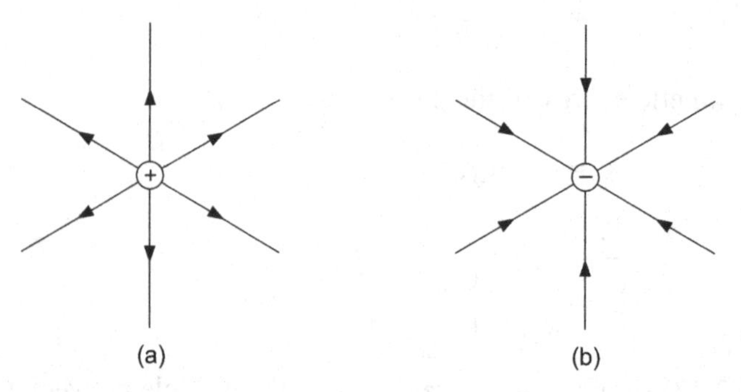

Fig. 2.5. Electric lines of force for (a) positive charge (b) negative charge.

(2) The tangent at any point to the electric lines of force predicts the direction of the electric field at that point.

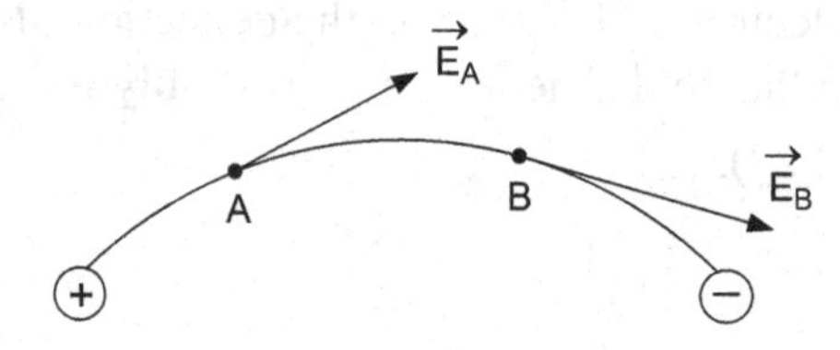

Fig. 2.6. Directions of the electric fields at different points.

In Fig 2.6. $\vec{E}_A$ represents the direction of the electric field at the point A and $\vec{E}_B$ at the point B. The directions of the electric fields at the points A and B are different.

(3) The number of field lines originating from the positive charge and end on negative charge is proportional to the magnitude of the charge. In the fig. 2.7, more and more lines originate from the charge $2q$ than the charge q.

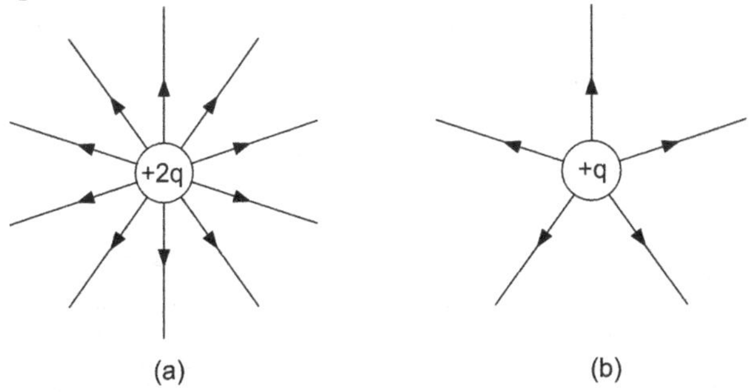

Fig. 2.7. Density of lines depends on the magnitude of charge.

(4) Greater the density of electric lines of force, greater the electric field strength.

(5) Since there are no closed field lines in electrostatics, the electric field is irrotational. That is, the curl of the electric field must be zero.

(6) Two electric field lines never intersect each other. That is, the direction of the electric field is unique. If it is not, then, at the point of intersection, the electric field will have two directions which is not possible.

(7) The electric field lines emerging from a conductor are perpendicular to its surface. To visualize this argument, consider a conducting sphere as shown in fig. 2.8. The total charge reside on the surface of the sphere, and the electric lines of force are emerging, normally, from its conducting surface.

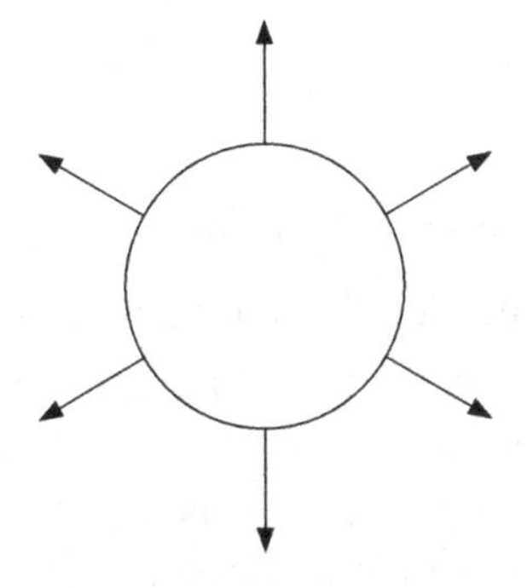

Fig. 2.8. Lines of force from a conductor.

(8) The electric field about a point charge is isotropic. If we take an isolated point charge, the electric field lines emerge in all direction as shown in fig. 2.5.

(9) The electric field lines do not penetrate the conductor, since the electric field inside the conductor is zero. Moreover, the electric field lines for the charges are shown in fig. 2.9.

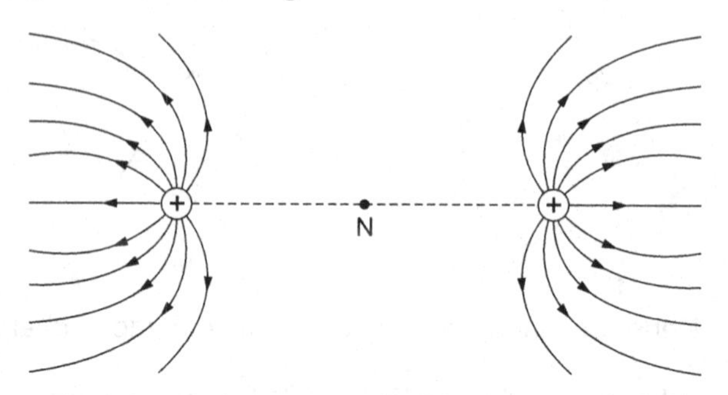

Fig. 2.9. Electric field lines for positive charges.

In this pattern, there is a point, where the electric field $\vec{E} = 0$. This point is known as null point and no electric line of force will pass through it.

If the electric field lines are curved, the electric field does not follow the field lines. Fig. 2.10, shows that the direction of the electric field lines are radially outward for the +ve charge and radially inward for the negative charge.

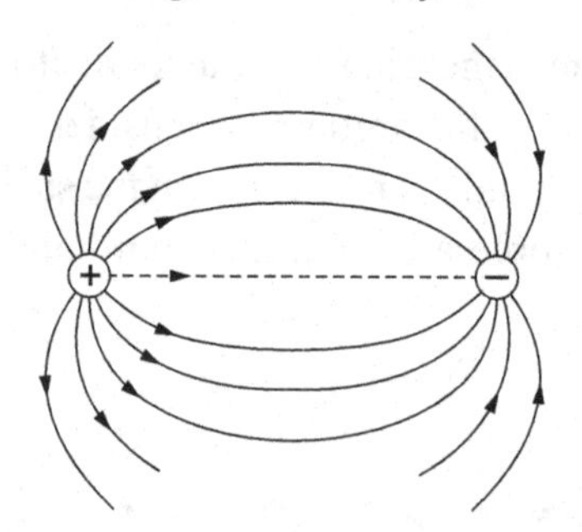

Fig. 2.10. Source and sink for field lines.

It can also be seen that the electric field lines are symmetrical.

Example 2.3. Compute the electric field strength due to a point charge using the concept of electric field lines.

Solution: Suppose that a positive charge is situated at a point O, and we compute the electric field strength at a distance r from the point O, as shown in fig. 2.11.

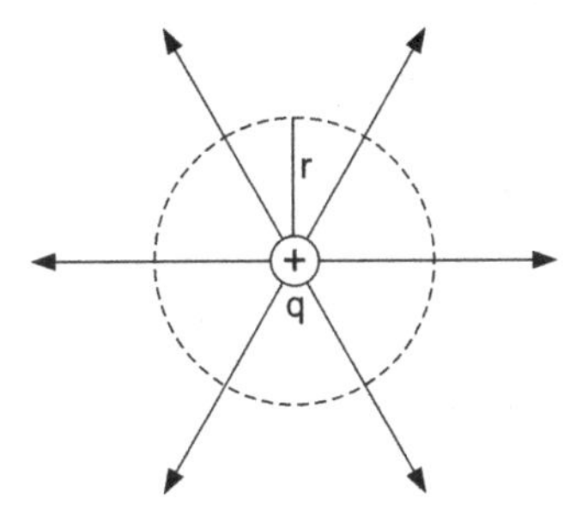

Fig. 2.11. Field lines from a charge.

Thus, sketch a spherical surface of the radius r around the charge $+q$. Since the spherical surface is uniform, the electric lines of force pass through a spherical surface normally. Thus, the surface area of the sphere,

$$s = 4\pi r^2$$

The charge density
$$\sigma = \frac{\text{Charge}}{\text{Area}}$$
$$= \frac{q}{4\pi r^2}$$

the electric field intensity at a point on the spherical surface will be

$$E = \frac{\sigma}{\epsilon_o}$$
$$= \frac{1}{\epsilon_o}\left(\frac{q}{4\pi r^2}\right)$$
$$E = \frac{q}{4\pi \epsilon_o r^2}$$

2.5. CHARGE DENSITIES

If the source of the electric field is the continuous charge distribution, the electric field may be calculated by considering a small charge element and integrated it over the region of the space. Since, large number of charges are distributed uniformly over the region, as shown in fig. 2.12, the integration of the electric field intensity due to the small charge element gives the total electric field produced by the uniform charge distribution in the region of the space. We have the following charge densities, viz,

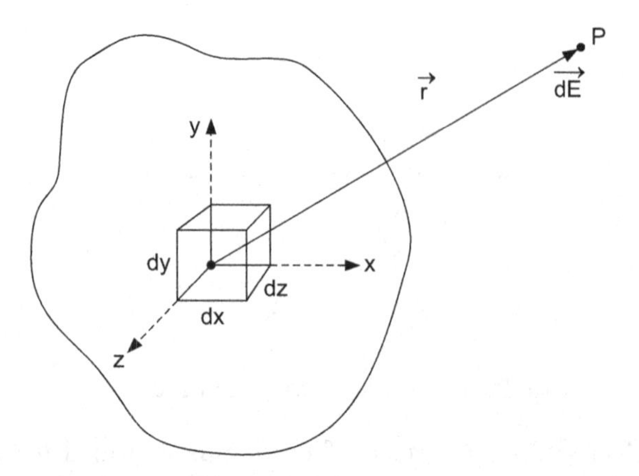

Fig. 2.12. Charge element inside a uniform charge distribution region of space.

(a) Volume charge density

(b) Surface charge density

(c) Line charge density.

Now, we shall discuss all the charge densities.

(a) **Volume Charge Density:** Consider a system where the charge is uniformly distributed and the volume of the system is divided into large number of small elements. Let us consider such an element of volume $dx\, dy\, dz$ and consisting of the charge dq. If the charge dq is positive, the electric field at a point P due to this element points away. The volume charge density $\rho(v)$ is given by

$$\rho(v) = \frac{\text{charge of the small element}}{\text{volume}}$$

$$= \frac{dq}{dx\, dy\, dz}$$

or $$dq = \rho(v)\, dy\, dy\, dz \qquad ...(2.28)$$

Thus, the total charge of the system is given by

$$q = \int_v dq \qquad ...(2.29)$$

$$= \int_v \rho(v)\, dx\, dy\, dz$$

or $$\boxed{q = \int_v \rho(v)\, dV} \qquad ...(2.30)$$

where $dv = dx\, dy\, dz$ is the volume of the element. Here, ρ is a scalar function.

(b) **Surface Charge Density:** If the charge is distributed uniformly over a surface of the area S in the region of space, we must define a surface charge density over a small surface area $\vec{dS}$. The surface charge density is denoted by σ. Thus, we write,

$$\sigma = \frac{dq}{dS} \qquad ...(2.31)$$

or

$$dq = \sigma\, dS \qquad ...(2.32)$$

The net charge may be computed as

$$\boxed{q = \int_S \sigma\, \vec{dS}} \qquad ...(2.33)$$

(c) **Line Charge Density:** If the charge distribution is along a line (one dimension), we must define a line charge density and it is represent by λ.

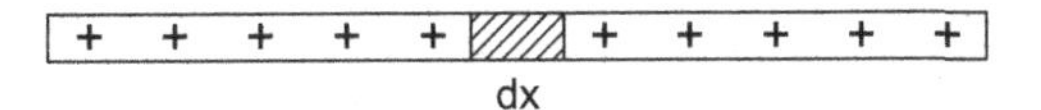

Fig. 2.13. Line element

Let us consider a line of length l consisting of a positive charge distribution uniformly over the length l. Let dx be a line element having charge dq, thus, the line charge density

$$\lambda = \frac{dq}{dx} \qquad ...(2.34)$$

net charge of the line will be

$$q = \int dq$$

$$\boxed{q = \int \lambda\, dx} \qquad ...(2.35)$$

2.6. THE ELECTRIC DIPOLE

An electric dipole is an arrangement of two equal and opposite charges separated by a small distance as shown in fig. 2.14.

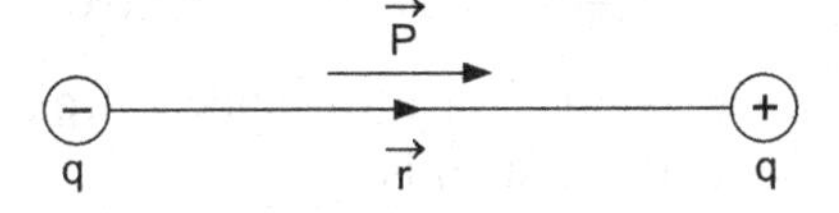

Fig. 2.14. An electric dipole

Suppose that two charge $-q$ and $+q$ are situated at a distance $\vec{r}$, the electric dipole moment $\vec{p}$ is defined as

$$\boxed{\vec{p} = q\vec{r}} \quad ...(2.36)$$

The electric dipole moment is a vector quantity and its magnitude $|\vec{p}|$ is equal to the product of the magnitude of either charge and the distance of separation. **The direction of an electric dipole is taken from negative charge to the positive charge**. The line joining the two charges is called the axis of the dipole. In M.K.S units, the unit of dipole is coulomb metre. For a system of charges, the electric dipole moment is given by

$$\vec{p} = \sum_i q_i r_i \quad ...(2.37)$$

where $\sum_i$ is the summation over all dipole moments. All the polar molecules like, HCl, HBr, CO and H_2O etc possess the electric dipole moment. For example we take HCl molecule as

$$H^+ \text{——} Cl^-$$

In HCl molecule the positive and negative ions (charges) are separated by the certain distance, thus, it possesses the electric dipole moment. On the other hand, the non-polar molecules do not possess the electric dipole moment because the positive and negative charges are not separated for these molecules. Since two point charges are connected, the electric field due to a dipole may be computed in a simple way and without using integration.

2.7. THE ELECTRIC FIELD DUE TO AN ELECTRIC DIPOLE AT A POINT ALONG ITS AXIS

Suppose that an electric dipole is lying along the x-axis as shown in fig. 2.15.

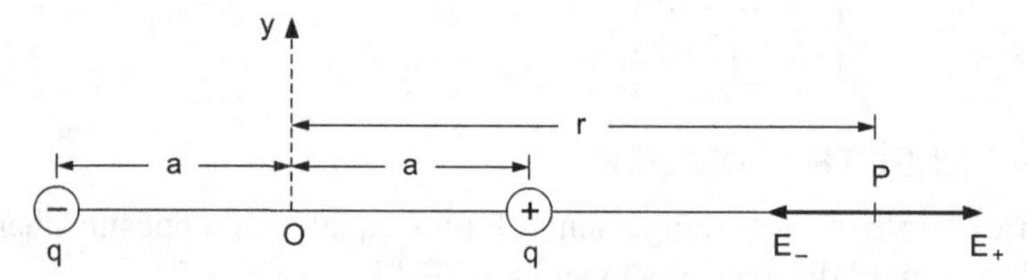

Fig. 2.15. Electric field due to dipole.

Let us consider an axial point P at a distance r from the centre of the dipole. The dipole moment of the system is

$$\vec{p} = 2qa\hat{i} \quad ...(2.38)$$

where $2a$ is the separation of the charges. If E_- and E_+ are the electric fields at a point P due to charges $-q$ and $+q$ respectively, Thus,

$$E_{-} = \frac{1}{4\pi\epsilon_0}\frac{q}{(r+a)^2}(-\hat{i}) \qquad ...(2.39)$$

and

$$E_{+} = \frac{1}{4\pi\epsilon_0}\frac{q}{(r+a)^2}(+\hat{i}) \qquad ...(2.40)$$

Then, the resultant electric field at P will be

$$E = E_{+} + E_{-}$$

$$= \frac{q}{4\pi\epsilon_0}\left[\frac{1}{(r-a)^2} - \frac{1}{(r+a)^2}\right]\hat{i}$$

or

$$E = \frac{2q}{4\pi\epsilon_0}\frac{2ar\hat{i}}{(r^2-a^2)^2} \qquad ...(2.41)$$

since,

$$p = 2aq,$$

$$E = \frac{2q}{4\pi\epsilon_0}\frac{2pr\hat{i}}{(r^2-a^2)^2}$$

if we take $r >> a$,

$$\boxed{E = \frac{1}{4\pi\epsilon_0}\frac{2p}{r^3}\hat{r}} \qquad ...(2.42)$$

The direction of the electric field is along $(+x)$ and p is $(-x)$.

2.8. THE ELECTRIC FIELD AT A POINT ON THE PERPENDICULAR BISECTOR OF THE DIPOLE AXIS

Suppose that the two charges are separated by a distance $2a$ constituting a dipole of the dipole moment

$$\vec{p} = 2aq\,\hat{i} \qquad ...(2.43)$$

Consider a point P at a distance y on the perpendicular bisector of the dipole axis as shown in fig. 2.16.

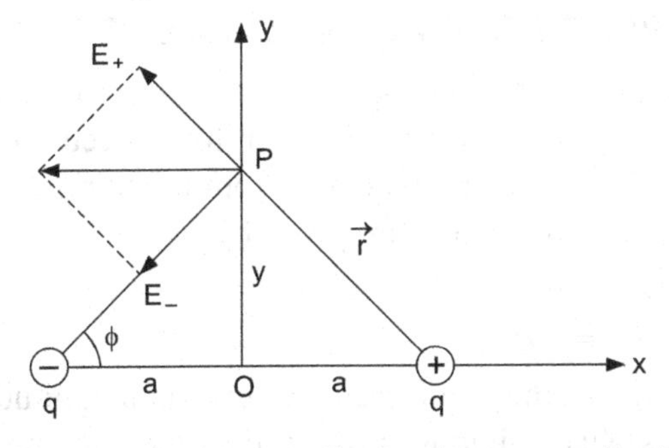

Fig. 2.16. Electric field due to dipole at a perpendicular bisector.

The electric field at a point P due to $+q$ charge is,

$$\vec{E}_+ = \frac{1}{4\pi\epsilon_0}\frac{q}{r^2}\hat{r}_+ \quad ...(2.44)$$

where $\hat{r}_+$ is a unit vector, given by

$$\hat{r}_+ = \cos\phi\,\hat{i} + \sin\phi\,\hat{j} \quad ...(2.45)$$

Again, the electric field due to $-q$ charge is given by

$$\vec{E}_- = \frac{1}{4\pi\epsilon_0}\frac{q}{r^2}\hat{r} \quad ...(2.46)$$

where $$\hat{r}_- = -\cos\phi\,\hat{i} - \sin\phi\,\hat{j} \quad ...(2.47)$$

The resultant electric field at the point P is

$$\vec{E} = \vec{E}_+ + \vec{E}_- \quad ...(2.48)$$

$$= \frac{2q}{4\pi\epsilon_0}\frac{\cos\phi}{r^2}(-\hat{i}) \quad ...(2.49)$$

since $$\cos\phi = \frac{a}{r}$$

Thus, $$\vec{E} = \frac{2aq}{4\pi\epsilon_0 r^3}(-\hat{i})$$

or $$\vec{E} = \frac{1}{4\pi\epsilon_0}\frac{\vec{p}}{r^3}(-\hat{i}) \quad ...(2.50)$$

Here, it can be observed that the electric field points in $-x$ direction, and falls off as $\frac{1}{r^3}$. Moreover, the electric field due to a dipole along its axis is twice the electric field at a point lying on perpendicular bisector.

2.9. AN ELECTRIC DIPOLE IN UNIFORM ELECTRIC FIELD

An electric dipole consists of two charges of equal magnitude and at a distance $2a$ apart. Now consider an electric dipole in a uniform electric field at an angle θ as shown in fig. 2.17. The electric field acts from left to right or along x-axis. Thus, the forces acting on the dipole are given by

$$F_+ = qE \quad ...(2.51)$$

and $$F_- = -qE$$

where $|F_+| = |F_-|$, that is, the two forces are equal in magnitude but pointing in opposite directions. Since, field is uniform, the dipole moment is given by

$$p = 2aq \quad ...(2.52)$$

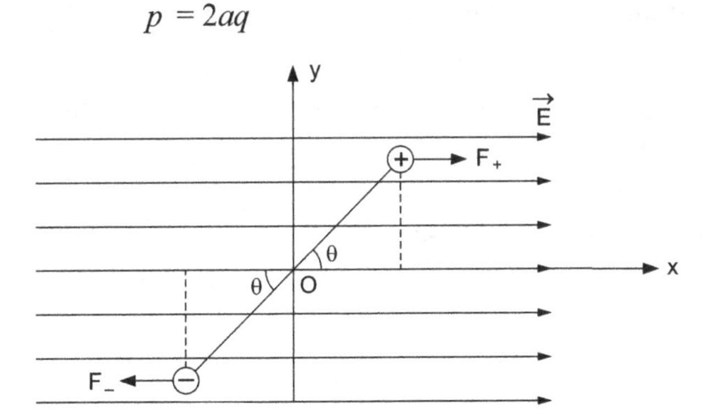

Fig. 2.17. An electric dipole in a uniform electric field.

Here, two equal and opposite forces are acting on the dipole, net force on the dipole will be

$$F_{net} = F_+ + F_-$$
$$= qE - qE = 0 \quad ...(2.53)$$

Thus, the net force on the dipole is zero. However, two forces acting in opposite directions produce a torque which rotates the electric dipole in clockwise direction, as a result of this, the electric dipole becomes parallel to the electric field. Now, the torque on the dipole is, then,

$$\tau = F_+ \,.\, a \sin \theta + F_- \,.\, a \sin \theta$$

$\therefore$

$$F = F_+ = F_- = qE$$

Thus,

$$\tau = 2aF \sin \theta \quad ...(2.54)$$

or

$$\tau = 2a\, qE \sin \theta \quad ...(2.55)$$

Since, $p = 2aq$, we have

$$\boxed{\tau = PE \sin \theta} \quad ...(2.56)$$

In vector form, it is written as

$$\boxed{\vec{\tau} = \vec{p} \times \vec{E}} \quad ...(2.57)$$

Now, we have two cases.

(1) The torque τ will be maximum, when $\theta = 90$

$$\tau = PE$$

That is, the torque acting on the dipole will be maximum if the electric dipole is along *y*-axis, a direction perpendicular to the electric field.

(2) The torque τ is minimum if $\theta = 0$

$$\tau = 0$$

if the dipole is along the electric field direction, the torque acting on the dipole becomes zero.

Example 2.4. An electric dipole consists of two equal and opposite charges of magnitude 2μC separated by a distance of 1.0 cm. If the dipole is placed at an angle of 30° in a uniform electric field of 10^5 N/C, calculate the,

(a) dipole moment.

(b) torque on the dipole.

Solution:

$$q = 2 \times 10^{-6} \text{ C}$$
$$2a = 1.0 \times 10^{-2} \text{ m}$$
$$E = 10^5 \text{ N/C}$$
$$\theta = 30°.$$

(a) Dipole moment

$$p = 2aq$$
$$= 1.0 \times 10^{-2} \times 2 \times 10^{-6}$$
$$p = 2 \times 10^{-8} \text{ C-m}$$

(b) Torque

$$\tau = pE \sin\theta$$
$$= 2 \times 10^{-8} \times 10^5 \times \sin 30°$$
$$= 1.0 \times 10^{-3} \text{ N.m}$$

2.10. AN ELECTRIC DIPOLE IN A NON-UNIFORM ELECTRIC FIELD

Consider an electric dipole in a non-uniform electric field as shown in fig. 2.18.

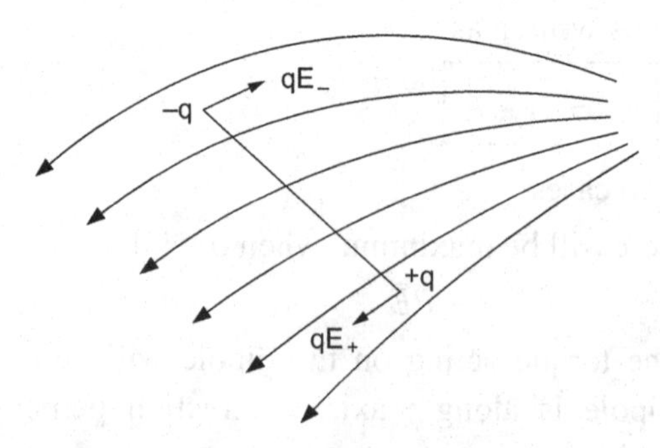

Fig. 2.18. Dipole in a non-uniform electric field.

The dipole moment is given by

$$p = 2aq \qquad \text{...(2.58)}$$

where $2a$ is the distance between the two charges. Since the dipole is in a non-uniform electric field, a net force acts on the dipole. As a result, a net torque would act on the dipole which depends on the orientation of the dipole in the electric field. Now, suppose that E_- and E_+ are the electric fields at the charges $-q$ and $+q$ respectively, where E_+ and E_- are given by

$$E_+ = E(Z + 2a)$$

$$E_- = E(Z) \qquad \text{...(2.59)}$$

Expanding $E(Z + 2a)$ using Taylor series, we get

$$E(Z+2a) = E(Z) + 2a\frac{dE}{dZ} + \frac{(2a)^2}{2}\frac{d^2E}{dZ^2} + \ldots \qquad \text{...(2.60)}$$

neglecting higher order terms, we write

$$E(Z+2a) = E(Z) + 2a\frac{dE}{dZ} \qquad \text{...(2.61)}$$

Here, we have assumed that the dipole is lying in z-direction and the electric field varies along the z-direction. Due to the variation in the electric field, the dipole consists of two types of motion, linear motion and rotation about its axis. Now, we compute the net force acting on the dipole as,

$$F = qE \qquad \text{...(2.62)}$$

or

$$F = q(E_+ - E_-) \qquad \text{...(2.63)}$$

Now, substituting the values of E_+ and E_- from the Eqs (2.59) and (2.61) in the Eq (2.63) we get,

$$F = q\left[E(z) + 2a\frac{dE}{dZ} - E(z)\right]$$

$$F = 2aq\frac{dE}{dZ} \qquad \text{...(2.64)}$$

Since, $p = 2aq$, we can write the Eq. (2.64) as

$$F = p\frac{dE}{dZ} \qquad \text{...(2.65)}$$

The Eq. (2.65) represents the net force acting on a dipole in the non-uniform electric field.

Example 2.5. An electric dipole is placed in a non-uniform electric field as shown in fig. 2.19.

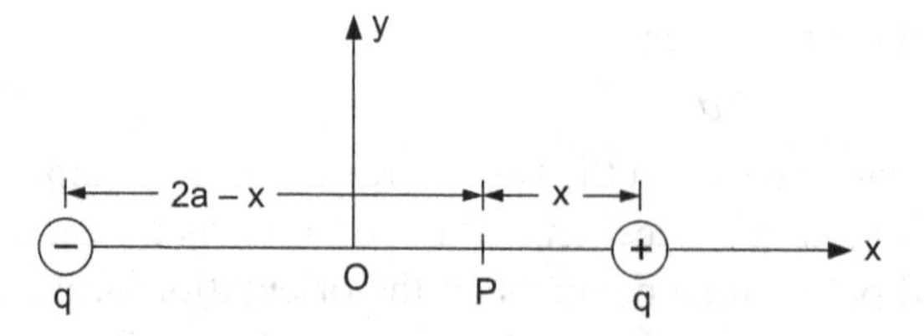

Fig. 2.19. Dipole in non-uniform field.

Derive an expression for the electric field gradient at a point P from the centre O. Compute also the field gradient at $x = a$.

Solution: The electric field for the system is given by

$$E = \frac{1}{4\pi\epsilon_0}\left[\frac{1}{x^2} - \frac{1}{(2a-x)^2}\right]$$

$$\left(\frac{\partial E}{\partial x}\right) = -\frac{2q}{4\pi\epsilon_0}\left[\frac{1}{x^2} + \frac{1}{(2a-x)^3}\right]$$

$$\left(\frac{\partial E}{\partial x}\right)_{x=a} = \frac{1}{4\pi\epsilon_0}\cdot\frac{4aq}{a^4}$$

$$\left(\frac{\partial E}{\partial x}\right)_{x=a} = \frac{1}{4\pi\epsilon_0}\cdot\frac{2p}{a^4}$$

2.11. POTENTIAL ENERGY OF DIPOLE IN AN ELECTRIC FIELD

Suppose that a dipole is placed in the uniform electric field, a couple acts on the dipole which tends to rotate the dipole. Thus, a work is done by the field to rotate the dipole and this work in rotating the dipole is stored as the potential energy. If the dipole is rotated through a small angle $d\theta$, the work done will be

$$dw = \tau\, d\theta$$

or

$$dw = pE \sin\theta\, d\theta \qquad ...(2.66)$$

Now, the net amount of the work done by the electric field to rotate the dipole from $\theta = \theta_o$ to $\theta = \theta$ is given by

$$w = \int dw$$

$$= \int_{\theta=\theta_0}^{\theta} \tau\, d\theta$$

$$= \int_{\theta_0}^{\theta} pE \sin\theta\, d\theta$$

or $$\boxed{U = -pE(\cos\theta - \cos\theta_o)} \quad ...(2.67)$$

This expression shows that a negative work is done by the electric field.

If initially, the electric dipole is at right angle to the electric field, we have $\theta_o = 90°$. From the Eq. (2.67), we may write the potential energy of the dipole as

$$U = -pE\cos\theta \quad ...(2.68)$$

or $$\boxed{U = -\vec{p}.\vec{E}} \quad ...(2.69)$$

Example 2.6. In the example 2.4, compute the work done by the electric field in rotating the dipole from $\theta = 0°$ to $\theta = 180°$.

Solution: The work done by the electric field is

$$w = -pE(\cos\theta - \cos\theta_o)$$
$$= -2 \times 10^{-8} \times 10^5 \times (\cos 180 - \cos o)$$
$$w = 4 \times 10^{-3}\ \text{J}$$

Example 2.7. Find the electric field at the centre of a hemisphere of the radius a and of uniform charge density σ.

Solution: Consider a hemi-sphere of the radius a as shown in fig. 2.20.

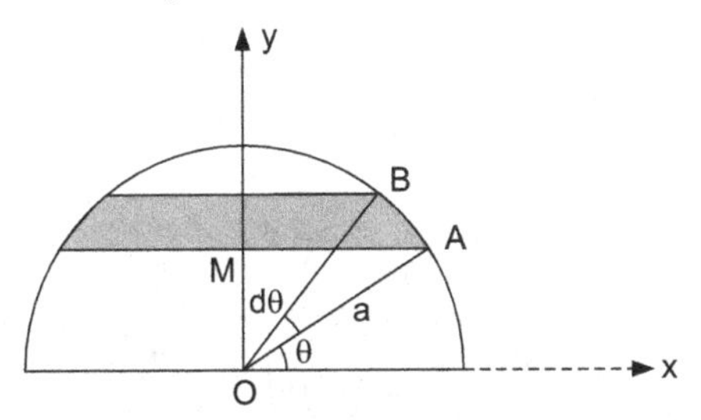

Fig. 2.20. Electric field due to hemi-sphere

Let us consider a ring on the hemi-sphere of the radius AM, where $AM = a\cos\theta$ and $OM = a\sin\theta$

Now, the thickness of the ring $AB = ad\theta$,

and $$\text{area} = 2\pi(a\cos\theta)\, ad\theta$$
$$= 2\pi a^2 \cos\theta\, d\theta$$

Thus, The charge on the ring is given by

$$dq = \sigma\ \text{area}$$
$$= 2\pi a^2 \sigma \cos\theta\, d\theta$$

Now, the electric field at O due to the ring is given by

$$dE = \frac{dq}{4\pi\epsilon_0}\frac{a\sin\theta}{[a^2\cos^2\theta + a^2\sin^2\theta]^{3/2}}$$

$$= \frac{\sigma . 2\pi a^2 \cos\theta \, d\theta}{4\pi \in_0} \frac{a \sin\theta}{a^3}$$

$$= \frac{\sigma}{2 \in_0} \cos\theta \sin\theta \, d\theta$$

The net electric field at a point O due to hemi-sphere will be

$$E = \int_0^{\pi/2} dE$$

$$= \frac{\sigma}{2 \in_0} \int_0^{\pi/2} \sin\theta \cos\theta \, d\theta$$

$$E = \frac{\sigma}{4 \in_0}$$

Example 2.8. A uniform charged rod of length L is bent into a semicircle of radius r as shown in fig 2.21. Find the magnitude and the direction of the electric field at the centre of the semi-circle.

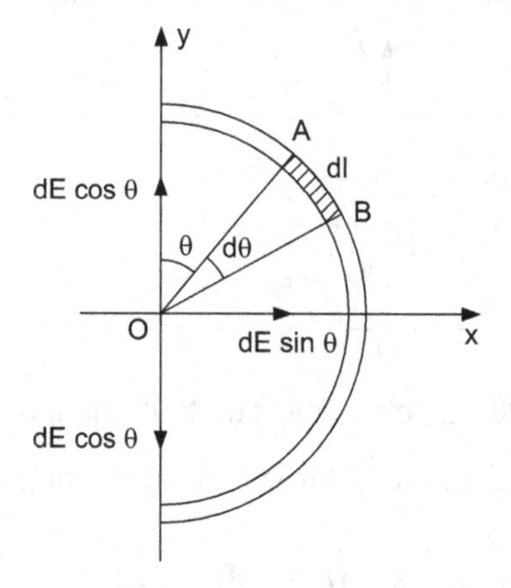

Fig. 2.21. Charged semicircular rod.

Let λ be the line charge density and given by

$$\lambda = \frac{q}{L}$$

where q is the charge on the rod. Suppose that dl is the charge element whose charge is given by

$$dq = \lambda \, dl$$

$$= \lambda . r d\theta$$

From the fig 2.21, it is clear that the y component of the electric field is zero and the x-component of the electric field is then,

$$E_x = \int dE \sin\theta = \int \frac{1}{4\pi\epsilon_0} \cdot \frac{dq \sin\theta}{r^2}$$

$$= \frac{1}{4\pi\epsilon_0} \frac{\lambda}{r} \int_0^{\pi} \sin\theta \, d\theta$$

$$E_x = \frac{\lambda}{2\pi\epsilon_0 r} = \frac{1}{2\pi\epsilon_0 r} \frac{q}{L}$$

but, $$r = \frac{L}{\pi}$$

$\therefore$ $$E_x = \frac{q}{2\epsilon_0 L^2}$$

E_x is along $+x$ axis.

EXERCISES

2.1. Find the electric field intensity at a point P due to a point charge q at the origin as shown in fig. 2.22.

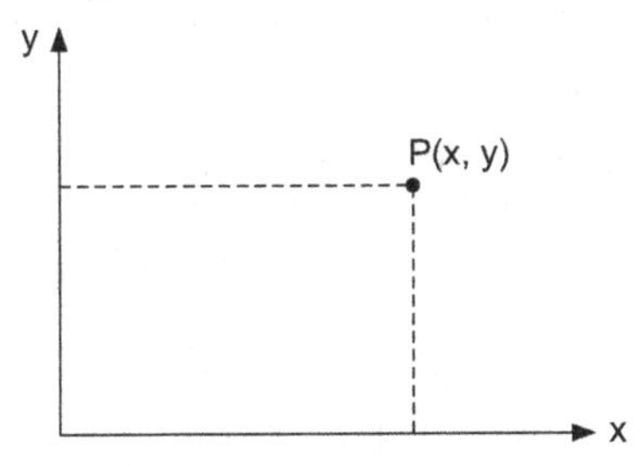

Fig. 2.22. A charge in a system.

Hint: $$\vec{E} = \frac{1}{4\pi\epsilon_0} \frac{q}{r^2} \hat{r}, \ \vec{r} = x\hat{i} + y\hat{j}$$

2.2. Define the electric dipole moment and obtain an expression for the torque when the dipole is placed in the uniform electric field.

2.3. Obtain an expression for the force when an electric dipole is placed in a non-uniform electric field.

2.4. If an electric dipole is placed in a uniform electric field, find the expression for its potential energy.

2.5. When a dipole of moment $p = 2aq$ C-m is placed in an electric field which varies as x^3 along the x-axis. Compute the force acting on the dipole.

2.6. Compute the electric field at a point on the axis perpendicular to disc surface and passing through its centre. The disc has a uniform charge density σ, sketch the electric field also.

2.7. Compute the electric field at a point lying on the axis of the ring of uniform charge density λ.

2.8. The four point charges are placed at the vertices of a square of side a as shown in fig. 2.23. compute

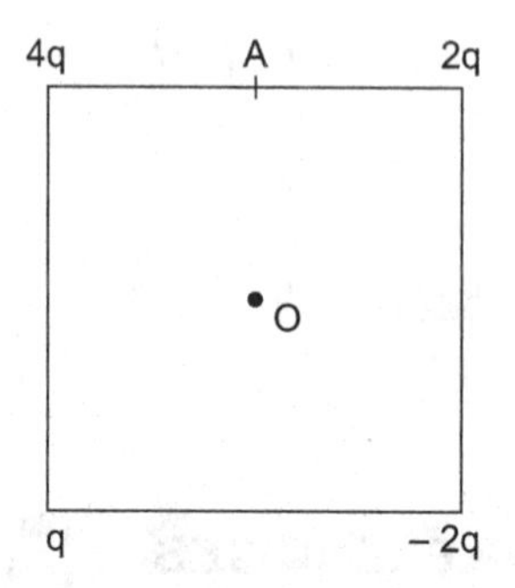

Fig. 2.23. Charges on a square.

(a) The magnitude and direction of the electric field at the centre '*O*'.

(b) The electric field at point *A*.

2.9. Determine the magnitude and the direction of the electric field at the point *P* shown in the fig. 2.24.

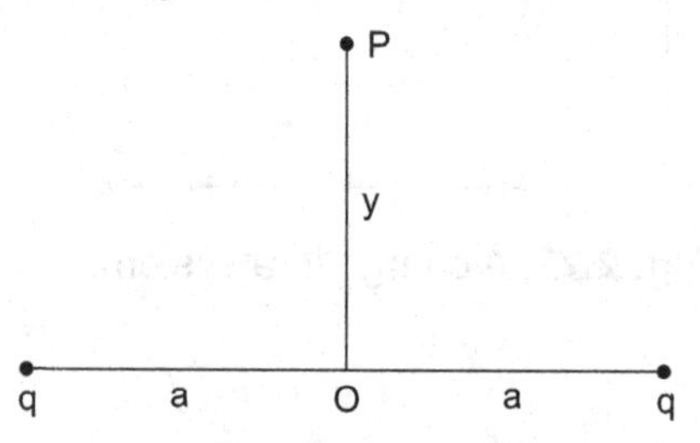

Fig. 2.24. Charge system.

2.10. Three charges, $q_1 = 1C$, $q_2 = 2C$, $q_3 = -2C$ are placed at the vertices of an equilateral triangle of side 2 cm. Compute the electric field at the centre of the triangle.

2.11. A particle of charge 2μC experiences a force of 10^{-5} N downward. Compute the magnitude and direction of the electric field.

■■■

CHAPTER 3

Gauss's Law

In the previous chapters, we have discussed the force between two charges and the applications of the electric field. We have investigated the electric field due to a charge or an assembly of the charges using the concept of positive test charge q_0. In this chapter we shall solve the electrostatic problems using Gauss's law.

3.1. ELECTRIC FLUX

Since the positive and negative charges are the source and sink of the electric lines of force, we can define the density of the electric field lines passing through a given surface. In this way we can develop a relation between the electric field and the electric flux (density of the field lines).

The electric flux is defined as the number of the electric field lines passing through the given surface area. It is denoted by ϕ_e. The electric flux ϕ_e is directly proportional to the electric charge where the electric field lines originate. Suppose that a spherical surface is at some distance apart from a charge q situated at a point 'O' as shown in fig. 3.1.

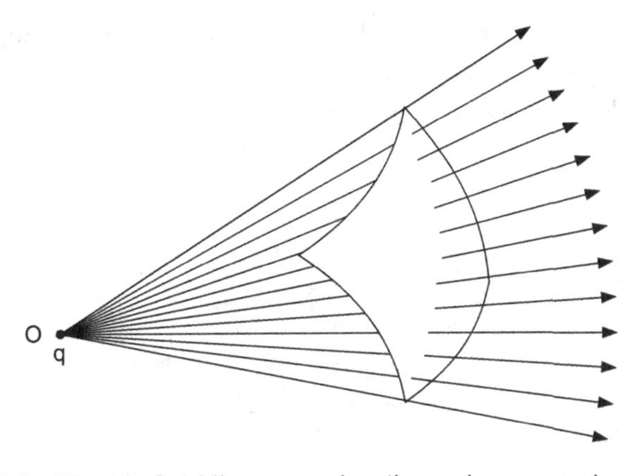

Fig. 3.1. Electric field lines passing through a curved surface.

Thus, the electric flux through the surface is equal to the number of electric field lines passing through the area of the surface. Then, we write

$$\phi_e = \vec{E} \cdot \vec{S} \qquad ...(3.1)$$

where S is the area of the surface. Now, consider a plane surface in a uniform electric field $\vec{E}$ as shown in fig. 3.2.

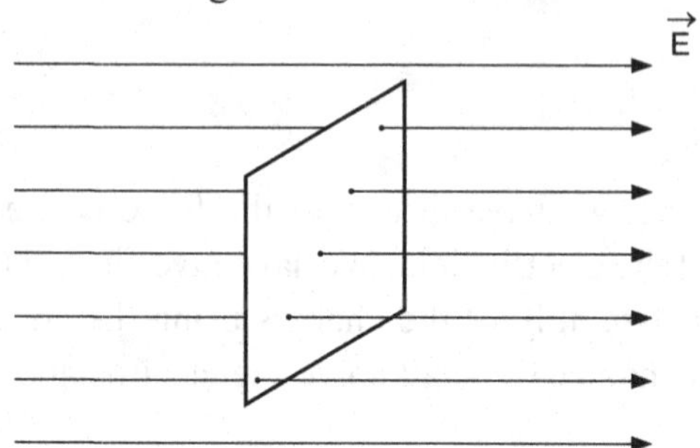

Fig. 3.2. Flux passes through a plane surface.

In this case, the flux through the surface is

$$\phi_e = \vec{E} \cdot \vec{S}$$

Where S denotes the area of the plane surface.

or
$$\phi_e = \vec{E} \cdot \hat{n} S$$
$$= ES \qquad ...(3.2)$$

where $\hat{n}$ is a unit normal vector to the surface $\vec{S}$. Moreover, if the plane surface is parallel to the electric field, no field line passes through the surface and

$$\phi_e = 0$$

Now, consider a case where the electric field lines make an angle θ at the surface. That is, the surface is neither parallel nor perpendicular to the electric field lines. Consider such a arbitrary surface as shown in fig. 3.3.

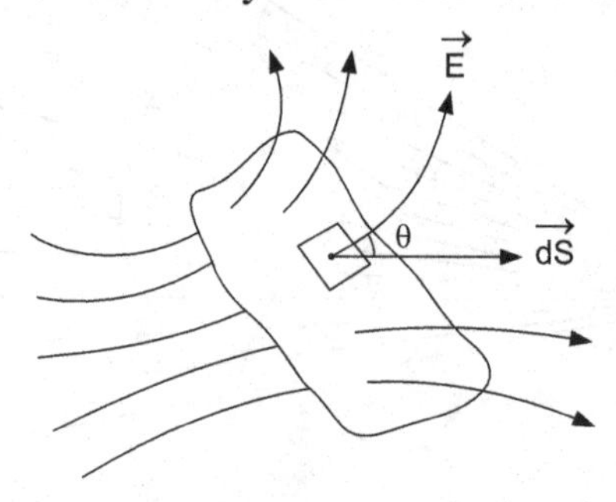

Fig. 3.3. Field lines passing through a closed surface.

Assume that the whole surface is divided into infinitesimally small patches such that for each patch, we have a flat surface. Here, the electric field $\vec{E}$ varies over the curved surface S. Now to determine the electric flux passing through the whole surface S, consider an electric field vector $\vec{E_i}$ at the ith patch whose are vector is $\vec{dS_i}$.

where $\vec{dS_i} = \hat{n}\, dS_i, \hat{n}$ is a unit normal vector. The electric flux passing through the area $\vec{dS_i}$ is given by

$$d\phi_e = \vec{E_i} \cdot \vec{dS_i} \qquad ...(3.3)$$

The dot product of electric field vector $\vec{E_i}$ and surface area vector $\vec{dS_i}$ is purely a scalar quantity.

Thus, the net flux passing through the surface will be,

$$\phi_e = \sum_i \vec{E_i} \cdot \vec{dS_i} \qquad ...(3.4)$$

or

$$\boxed{\phi_e = \int_s \vec{E} \cdot \vec{dS}} \qquad ...(3.5)$$

This surface integration can be evaluated by specifying the surface S. Since the vector $\vec{E}$ makes an angle θ with the surface area vector $\vec{dS}$, we may write the Eq. (3.5) as,

$$\boxed{\phi_e = \int_S E\, dS \cos\theta} \qquad ...(3.6)$$

3.2. CONCEPT OF SOLID ANGLE

To understand the concept of the solid angle, it is important to have a look at the simple angle as shown in fig. 3.4. Let $d\theta$ be the angle made by an arc dl, then

$$\text{Angle} = \frac{\text{Arc}}{\text{Radius}} \qquad ...(3.7)$$

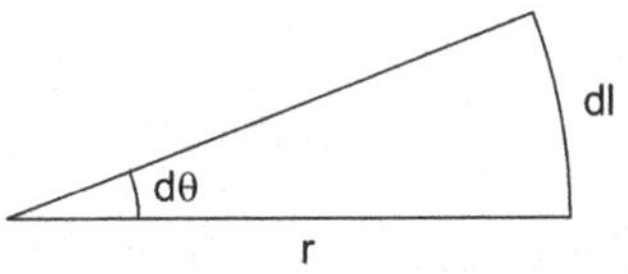

Fig. 3.4. Angle subtended by an arc.

$$\therefore \qquad d\theta = \frac{dl}{r} \qquad ...(3.8)$$

or

$$dl = r\,d\theta \qquad ...(3.9)$$

The ratio $\frac{dl}{r}$ is very meaningful. In a similar way, we can define a solid angle. For this, consider two concentric spheres of radii r_1 and r_2 respectively as shown in fig. 3.5.

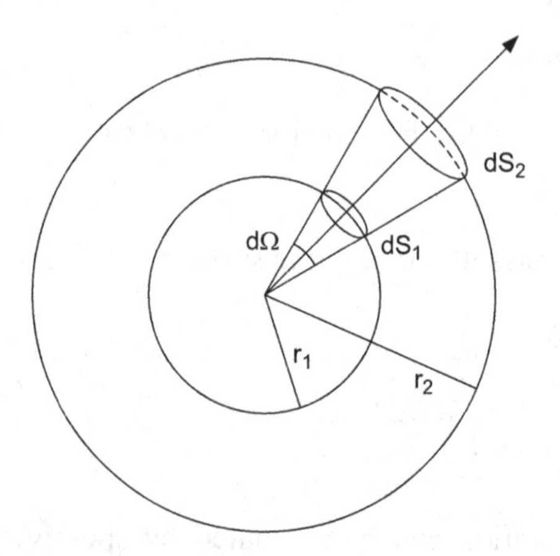

Fig. 3.5. Solid angle

Here, area dS_1 and dS_2 are proportional to r_1^2 and r_2^2 respectively. Thus,

$$\frac{dS_1}{r_1^2} = \frac{dS_2}{r_2^2} = d\Omega \qquad ...(3.10)$$

where dΩ is known as solid angle and is used to define the solid angle of the cone as shown in fig. 3.5. The unit of the solid angle is steradian. The steradian is defined as the solid angle subtended at the centre of a sphere of radius 1.0 meter by an area of 1.0 m^2 lying on the surface of the sphere. It is clear that solid angle increases with the surface area dS. Since the surface area of a sphere of radius r is $4\pi r^2$.

Then, the maximum solid angle is given by

$$\Omega = \frac{4\pi r^2}{r^2} = 4\pi \qquad ...(3.11)$$

To prove that Eq. (3.11). Consider a sphere of radius r as shown in fig. 3.6.

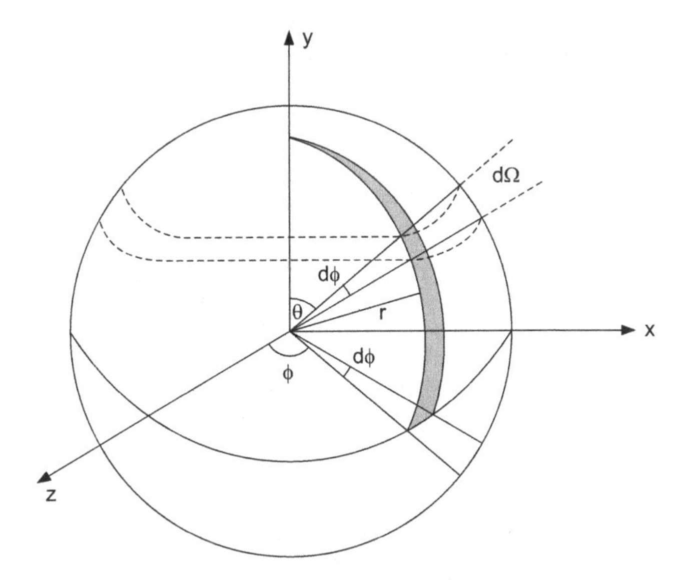

Fig. 3.6. Geometry of solid angle

In spherical coordinates, the surface area is given by

$$dS = (r\,d\theta)\,(r\sin\theta\,d\phi)$$
$$= r^2 \sin\theta\,d\theta\,d\phi \qquad ...(3.12)$$

The elementary solid angle $d\Omega$ is now,

$$d\Omega = \frac{dS}{r^2} = \sin\theta\,d\theta\,d\phi$$

total solid angle $\quad \Omega = \int_S d\Omega$

$$= \int_{\theta=0}^{\pi} \int_{\phi=0}^{2\pi} \sin\theta\,d\theta\,d\phi = 4\pi \qquad ...(3.13)$$

3.3. GAUSS'S LAW OF ELECTROSTATICS

Since electric field can be evaluated algebraically using coulomb's law, Gauss' law provides an easy way for computing the electric field for the uniform charge distribution. K.F. Gauss, a German mathematician derived a relation between the electric field and the charge in a closed system. Actually, this idea was derived from the fluid dynamics. Suppose that a charge q is situated at the centre of a sphere, the electric field lines are perpendicular to the surface and hence the electric field $\vec{E}$, at every point on the surface of the sphere, is perpendicular to the surface as shown in fig. 3.7. Thus, at the surface of the sphere, the electric field is given by

$$\vec{E} = \frac{1}{4\pi\epsilon_0}\frac{q}{r^2}\hat{r} \quad ...(3.14)$$

where r is the radius of the sphere.

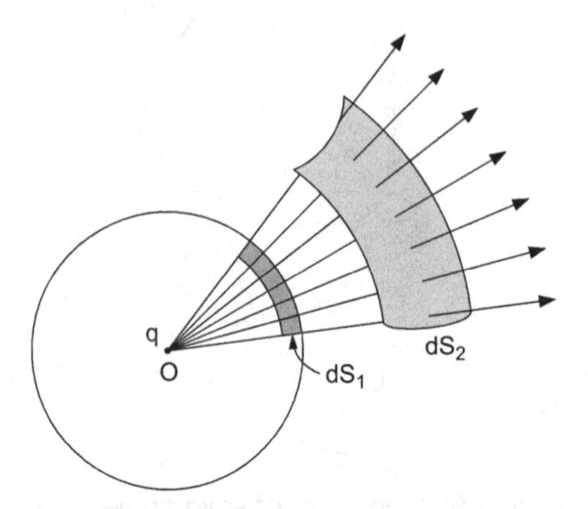

Fig. 3.7. Electric field lines passing through the segments dS_1 and dS_2.

It is clear that the flux passing through the segment dS_1 is equal to the flux through the areal segment dS_2. Thus, we write

$$E_1 dS_1 = E_2\, dS_2 \quad ...(3.15)$$

Since the electric flux depends on the magnitude of the charge, consider the concept of the solid angle to prove the Eq. (3.15). Let $d\Omega$ be the solid angle subtended at the centre of the sphere by the surfaces dS_1 and dS_2 as shown in fig. 3.8.

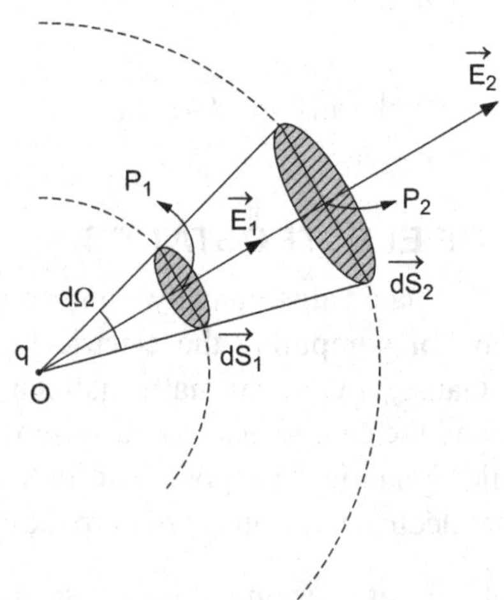

Fig. 3.8. Solid angle subtended by the area elements dS_1 and dS_2.

Thus, we have

$$d\Omega = \frac{dS_1}{r_1^2} = \frac{dS_2}{r_2^2} \quad ...(3.16)$$

where $OP_1 = r_1$ and $OP_2 = r_2$. Thus, the electric field at the point P_1 is

$$E_1 = \frac{1}{4\pi\epsilon_0}\frac{q}{r_1^2} \quad ...(3.17)$$

and the electric field at the point P_2 is given by

$$E_2 = \frac{1}{4\pi\epsilon_0}\frac{q}{r_2^2} \quad ...(3.18)$$

Now, we calculate the electric flux passing through the surface dS_1 is

$$\phi_1 = E_1\, dS_1 = \frac{q}{4\pi\epsilon_0}\frac{dS_1}{r_1^2} \quad ...(3.19)$$

Similarly, electric flux passing through the area dS_2 is given by

$$\phi_2 = E_2 dS_2 = \frac{1}{4\pi\epsilon_0}\frac{dS_2}{r_2^2} \quad ...(3.20)$$

since, we know that

$$\phi_1 = \phi_2 \quad ...(3.21)$$

or

$$E_1\, dS_1 = E_2\, dS_2 \quad ...(3.22)$$

Thus, for an arbitrary surface dS, we write

$$\phi = \int_S \vec{E}\cdot\vec{dS} = \int_S \frac{1}{4\pi\epsilon_0}\frac{q}{r^2}\,dS$$

$$= \frac{1}{4\pi\epsilon_0}\cdot\int_S \frac{dS}{r^2} \quad ...(3.23)$$

Here $\int_S \frac{dS}{r^2}$ is the solid angle $d\Omega$, which has value 4π.

$$\therefore \quad \phi_e = \frac{q}{4\pi\epsilon_0}\int_S d\Omega \quad ...(3.24)$$

$$= \frac{q}{4\pi\epsilon_0}\cdot 4\pi = \frac{q}{\epsilon_0}$$

$$\boxed{\phi_e = \int_S \vec{E}\cdot\vec{dS} = \frac{q}{\epsilon_0}} \quad ...(3.25)$$

Thus, Gauss's Law states that the net flux through any closed surface is proportional to the net charge enclosed by the surface. If there are large number of charges enclosed by the surface, we write,

$$= q_1 + q_2 - q_3 - q_4 \ldots\ldots$$

That is, $\sum_i q_i$ represents the algebraic sum of all the charges which enclosed by the gaussian surface. If we consider a irregular body as shown in fig. 3.9. we may use the concept of solid angle to prove the Gauss's law.

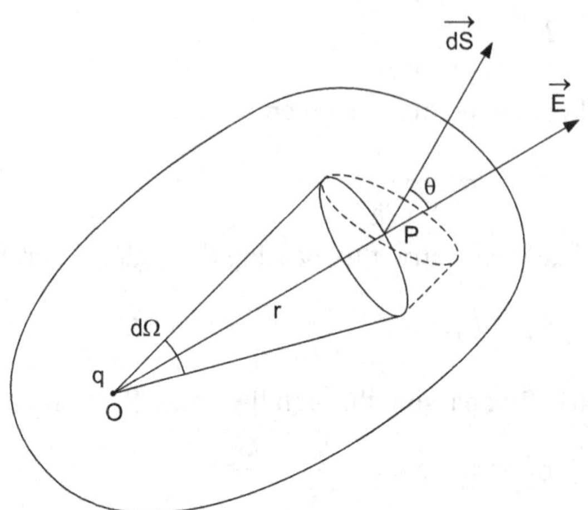

Fig. 3.9. Illustration of the flux, from a charge q, passing through solid angle in irregular body.

Now, the electric filed at point P is given by

$$E = \frac{q}{4\pi\epsilon_0}\frac{1}{r^2} \qquad \ldots(3.26)$$

The flux through the element dS which makes an angle θ from the electric field $\vec{E}$, is $d\phi_e = \vec{E}\cdot\vec{dS}$ and total flux is then,

$$\phi_e = \int_S \vec{E}\cdot\vec{dS}$$

$$= \int_S E\, dS \cos\theta \qquad \ldots(2.27)$$

Substituting the value of E from the Eq. (3.26) in the Eq. (2.27), we get

$$\phi_e = \frac{q}{4\pi\epsilon_0}\int_S \frac{dS\cos\theta}{r^2}$$

$$= \frac{q}{4\pi\epsilon_0}\int_S d\Omega = \frac{q}{4\pi\,\epsilon_0}\cdot 4\pi \qquad \ldots(2.28)$$

Hence,

$$\boxed{\phi_e = \frac{q}{\epsilon_0}} \qquad \ldots(2.29)$$

It terms of charge density, Gauss's law many be given as

$$\phi_e = \oint_S \vec{E} \cdot \vec{dS} = \frac{1}{\epsilon_0} \int_V \rho\, dV \quad ...(2.30)$$

where $q = \int_V \rho\, dV$ is the total charge enclosed by the volume and ρ is called volume charge density.

3.4. GAUSSIAN SYMMETRICAL SURFACES

Gauss's law is applicable for any charge distribution in a closed surface. By knowing the charge distribution, we can evaluate the surface integration $\oint \vec{E} \cdot \vec{dS}$ for the flux passing through the gaussian surface, a symmetrical surface. We have following gaussian surfaces for corresponding symmetries.

System	Gaussian Surface
Sphere	Sphere
Infinite Rod	Coarcial cylinder
Plane Sheet	Pill box

3.5. ELECTRIC FIELD OF A SPHERICAL CHARGE DISTRIBUTION

Gauss's law can be used to find the electric field of a sphere with a uniform charge distribution. Now, consider a sphere of radius a with a charge q distributed uniformly over the sphere. Thus, the electric field is directed away from the surface of the sphere. It is noted that the electric field must be constant at all the points on the surface of the sphere. Now, we have two cases for evaluating the electric field, viz,

Case I: We evaluate the electric field at a point outside the sphere, in the region $r > a$ as shown in fig. 3.10.

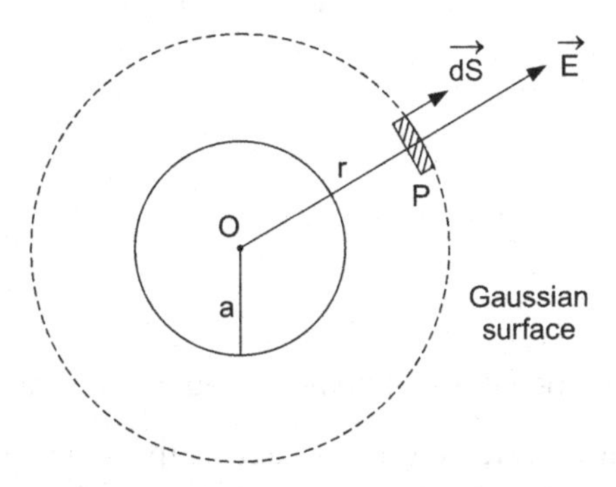

Fig. 3.10. Electric field at a point $P(r > a)$.

Now, we draw a gaussian surface of the radius r, such that, it encloses all the charges present on the surface of the sphere. The electric field at a point P is evaluated as

$$\oint \vec{E} \cdot \vec{dS} = \frac{q}{\epsilon_0} \quad ...(3.31)$$

$$E \oint dS = \frac{q}{\epsilon_0}$$

or

$$E \cdot 4\pi r^2 = \frac{q}{\epsilon_0}$$

or

$$E = \frac{1}{4\pi \epsilon_0} \frac{q}{r^2} \quad ...(3.32)$$

Since the charge is distributed uniformly over the surface on the sphere, we define charge density

$$\sigma = \frac{\text{Charge}}{\text{Surface area}} \quad ...(3.33)$$

or

$$\sigma = \frac{q}{4\pi a^2} \quad ...(3.34)$$

The electric field at the surface may be obtained by substituting $r = a$ in the Eq. (3.32), we get

$$E = \frac{1}{4\pi \epsilon_0} \frac{q}{a^2} = \frac{\sigma}{\epsilon_0} \quad ...(3.35)$$

Case II: We shall, now evaluate the electric field at a point P lying inside the sphere ($r < a$), and we draw the gaussian surface which passes through the point P as shown in fig. 3.11.

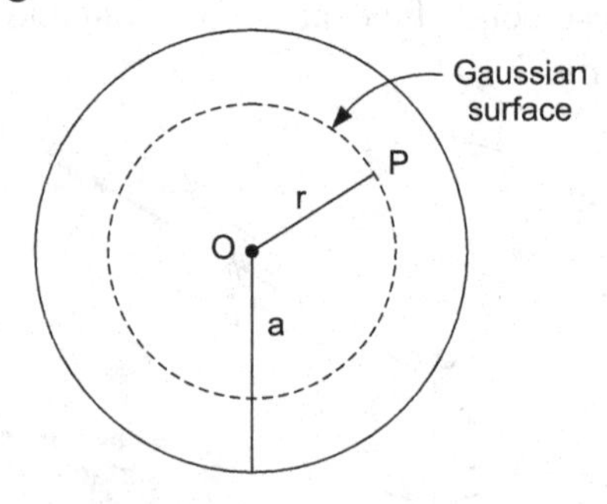

Fig. 3.11. Electric field at a point p lying in the region $r < a$.

Since all the charge q reside on the surface of the sphere, and the gaussian surface does not enclose any charge, hence $q = 0$. The electric field is then, $$\int \vec{E}.\vec{dS} = \frac{q}{\epsilon_0} = 0$$

or $$E = 0 \quad \text{for} \quad r < a \qquad ...(3.36)$$

The plot of electric field E with the distance is shown in fig. 3.12.

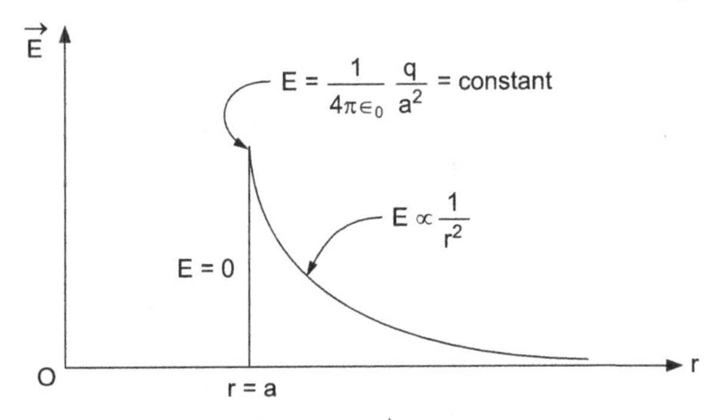

Fig. 3.12. Plot of $\vec{E}$ Vs distance r.

In the Fig. 3.12., It is shown that E varies as $\frac{1}{r^2}$ for the region $r > a$ and $E = 0$ for $r < a$. It is clear that E is constant at $r = a$.

3.6. ELECTRIC FIELD OF AN INFINITE LONG WIRE (LINE CHARGE)

Let there be an infinite long wire of negligible radius having a uniform line charge density λ, such that

$$\lambda = \frac{q}{L} \qquad ...(3.37)$$

The long wire consists of cylindrical symmetry as shown in fig. 3.13 and the magnitude of the electric field is constant over the cylindrical surface.

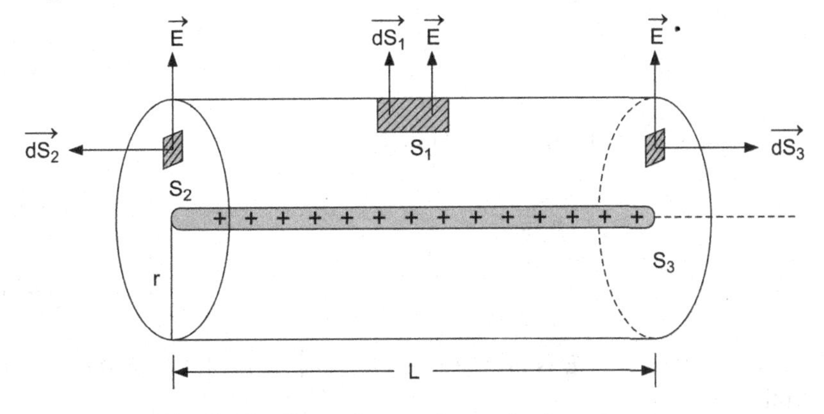

Fig. 3.13. Gaussian surface of a line charge.

It can be seen that the electric field points radially away from the wire. As illustrated in the fig. 3.13, the gaussian surface consists of three parts for computing the electric field, viz,

(1) curved surface S_1,

(2) left surface S_2 and

(3) right surface S_3.

Now, applying Guass's Law on the cylindrical surface, then, the electric flux through the Gaussian surface is

$$\phi_e = \oint_S \vec{E} \cdot \vec{dS} \qquad ...(3.38)$$

$$= \oint_{S_1} \vec{E} \cdot \vec{dS_1} + \oint_{S_2} \vec{E} \cdot \vec{dS_2} + \oint_{S_3} \vec{E} \cdot \vec{dS_3}$$

$$= \oint_{S_1} E\, dS_1 \cos 0 + \oint_{S_2} E\, dS \cos 90 + \oint_{S_3} E\, dS \cos 90$$

$$= E \oint_{S_1} dS_1 + 0 + 0$$

$$\phi_e = E \cdot 2\pi r\, L \qquad ...(3.39)$$

The second and third integral for the surfaces S_1 and S_2 become zero, since the electric field E is perpendicular to the surface elements $\vec{dS_1}$ and $\vec{dS_2}$. Moreover,

$$\phi_e = \frac{q}{\epsilon_0} \qquad ...(3.40)$$

$$\therefore \qquad E \cdot 2\pi r L = \frac{q}{\epsilon_0}$$

or

$$E = \frac{q}{2\pi \epsilon_0 L} \cdot \frac{1}{r} \qquad ...(3.41)$$

From the Eq. (3.37), the Eq. (3.41) takes form,

$$E = \frac{\lambda}{2\pi \epsilon_0 r} \qquad ...(3.42)$$

or

$$E \propto \frac{1}{r} \qquad ...(3.43)$$

Thus, the electric field E is proportional to $\left(\frac{1}{r}\right)$. The plot of E versus r is shown in fig. 3.14.

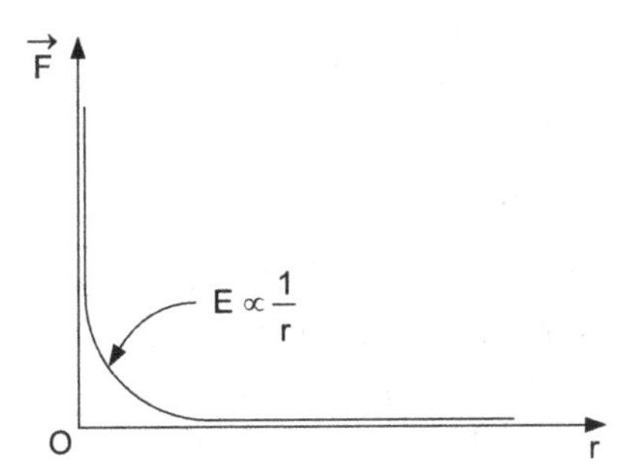

Fig. 3.14. *E* vs distance *r*.

Example 3.1. Obtain Coulomb's law using Gauss's law of electrostatics.

Solution: We may obtain coulomb's law from the Gauss's law of electrostatics. For this, suppose that an isolated positive charge q is situated at a point o as shown in fig. 3.15.

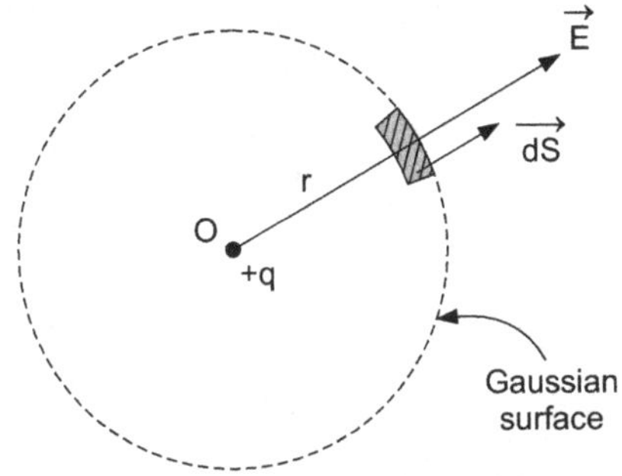

Fig. 3.15. Gaussian surface for isolated charge

We have a spherical gaussian surface for a positive isolated charge. The electric field lines point outward from the charge. The Guass's law is given by

$$\oint_S \vec{E} \cdot \vec{dS} = \frac{q}{\epsilon_0}$$

since, the gaussian surface encloses the net charge q, we have

$$E \oint_S dS = \frac{q}{\epsilon_0}$$

or $$E \cdot 4\pi r^2 = \frac{q}{\epsilon_0}$$

or $$E = \frac{1}{4\pi\epsilon_0} \frac{q}{r^2}$$

Now, if we put a point charge Q on the gaussian surface, then, the force on the point charge Q will be

$$\vec{F} = Q\vec{E}$$

or $$\vec{F} = \frac{1}{4\pi\epsilon_0}\frac{qQ}{r^2}\cdot\hat{r}$$

this is known as coulomb's law.

3.7. ELECTRIC FIELD NEAR AN INFINITE PLANE SHEET

Gauss's law can be used to compute the electric field near a large plane sheet with a uniform surface charge density. Consider a large plane sheet with a uniform charge density σ in *yz* plane as shown in fig 3.16, where charge density σ is given by

$$\sigma = \frac{q}{S} \quad ...(3.44)$$

Here, S is the area of the plane sheet.

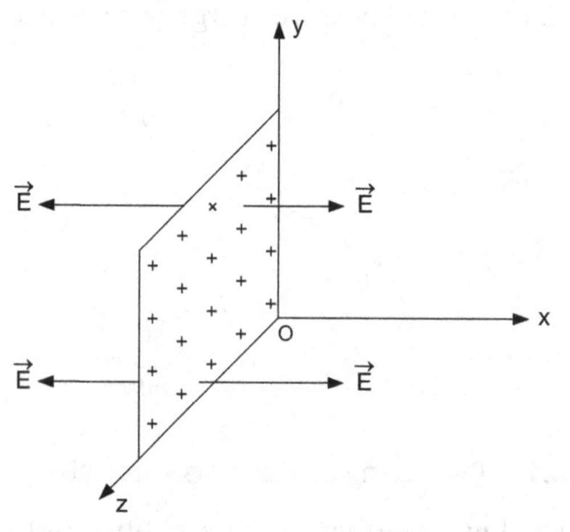

Fig. 3.16. Infinite plane sheet

It has a planner symmetry, and the gaussian surface is a pill box as shown in fig. 3.17.

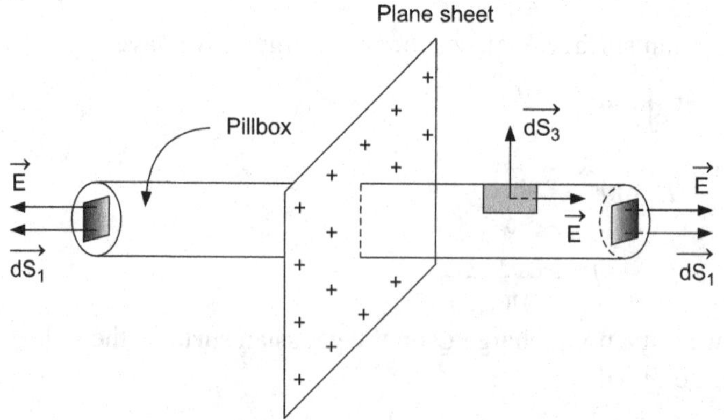

Fig. 3.17. Gaussian pill box

We know that the electric field must be perpendicular and uniform and points outward from the charged plane sheet. Thus, we write

$$\vec{E} = E_{\underline{0}} \quad \text{...(3.45)}$$

The gaussian pill box contains three parts of its surface, viz.

(1) Surface S_1

(2) Surface S_2

(3) a curved surface S_3.

According to Gauss's law, we have

$$\oint_S \vec{E} \cdot \vec{dS} = \frac{q}{\epsilon_0} \quad \text{...(3.46)}$$

or
$$\oint_{S_1} \vec{E} \cdot \vec{dS_1} + \oint_{S_2} \vec{E} \cdot \vec{dS_2} + \oint_{S_3} \vec{E} \cdot \vec{dS_3} = \frac{q}{\epsilon_0}$$

$$ES + ES + 0 = \frac{q}{\epsilon_0}$$

or
$$E = \frac{q}{2S\,\epsilon_0} \quad \text{...(3.47)}$$

Using Eq. (3.44), we have

$$\boxed{E = \frac{\sigma}{2\,\epsilon_0}} \quad \text{...(3.48)}$$

The direction of the electric field is along $\pm x$ axis and E is constant.

Example 3.2. Compute the electric field for two distant parallel infinite non conducting plane sheets using Gauss's law.

Solution: We have three cases for the two plane sheets.

Case I: Consider two positively charged parallel plane sheet, both having a uniform charge density σ as shown in fig. 3.18.

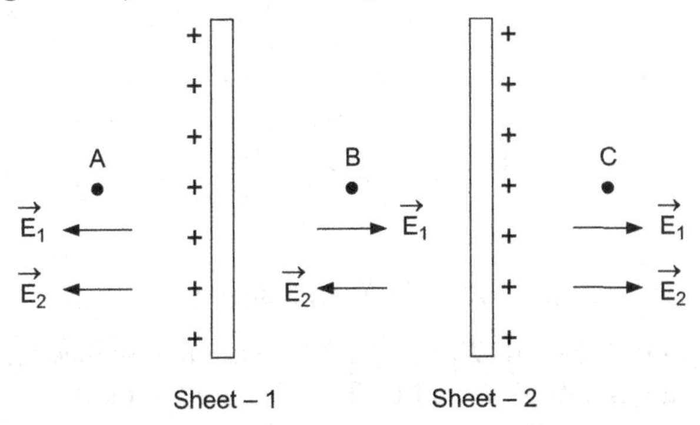

Fig. 3.18. Two positively charged sheets

Now, suppose that $\vec{E}_1$ and $\vec{E}_2$ are the electric fields due to the charged sheets 1 and 2 respectively and $\vec{E}_1$ and $\vec{E}_2$ are given by

$$E_1 = \frac{\sigma}{2\epsilon_0}$$

$$E_2 = \frac{\sigma}{2\epsilon_0}$$

Therefore, we shall compute the electric field at the points A, B and C. The electric field at the point A is,

$$E_A = E_1 + E_2$$

or $$E_A = \frac{\sigma}{2\epsilon_0} + \frac{\sigma}{2\epsilon_0} = \frac{\sigma}{\epsilon_0}$$

The electric field at the point B is given by

$$E_B = E_1 + (-E_2)$$

$$= \frac{\sigma}{2\epsilon_0} - \frac{\sigma}{2\epsilon_0} = 0$$

and the electric field at the point C is

$$E_C = E_1 + E_2$$

or $$E_C = \frac{\sigma}{2\epsilon_0} + \frac{\sigma}{2\epsilon_0} = \frac{\sigma}{\epsilon_0}$$

Case II: Now, let the sheet-1 be positively charged and the sheet-2 negatively charged as shown in fig. 3.19.

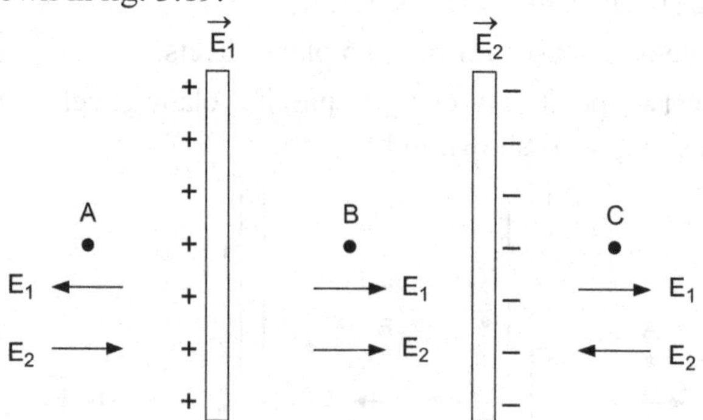

Fig. 3.19. Two charged sheets.

The directions of the fields E_1 and E_2 are computed by placing a positive test charge q_0 at the points A, B and C. To calculate the electric fields at the point A, B and C, applying the superposition principle,

Thus,

$$E_A = E_1 + (-E_2)$$
$$= \frac{\sigma}{2\epsilon_0} - \frac{\sigma}{2\epsilon_0} = 0$$

and

$$E_B = E_1 + E_2$$
$$= \frac{\sigma}{2\epsilon_0} + \frac{\sigma}{2\epsilon_0} = \frac{\sigma}{\epsilon_0}$$

The electric field at point C is

$$E_C = E_1 + (-E_2)$$

or

$$E_C = \frac{\sigma}{2\epsilon_0} - \frac{\sigma}{2\epsilon_0} = 0$$

Case III: If the both plane sheets are negatively charged, as shown in fig. 3.20.

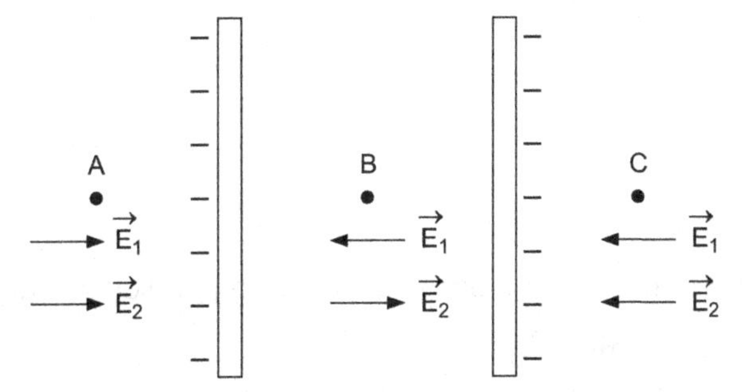

Fig. 3.20. Same charged sheets.

The electric field at the points A, B and C is given by

at point A,

$$E_A = E_1 + E_2$$
$$= \frac{\sigma}{\epsilon_0}$$

at point B,

$$E_A = E_1 - E_2$$
$$= 0$$

at point C,

$$E_C = E_1 + E_2$$
$$= \frac{\sigma}{\epsilon_0}$$

3.8. GAUSS'S LAW IN DIFFERENTIAL FORM

The Gauss's law is applied for the computation of the symmetrical fields. There are many problems in electrostatics which can be solved using Gauss's law. But there is a restriction that the field should have a symmetry. The Gauss's law in integral form is very useful for solving many problems, however, the differential form of Gauss's law is very useful for calculating the fields in rectangular coordinates. Now, consider a parallopiped of sides *dx*, *dy* and *dz* along the axes as shown in fig. 3.21.

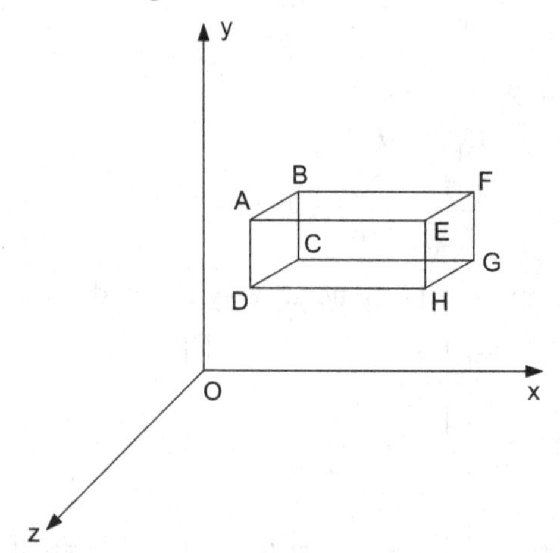

Fig. 3.21. Electric flux passing through a parallelopiped.

The electric field is given by

$$\vec{E} = E_x \hat{i} + E_y \hat{j} + E_z \hat{k} \qquad ...(3.49)$$

The electric flux at the surface *ABCD* is

$$E_x - \frac{1}{2} \cdot \frac{\partial E_x}{\partial x} dx$$

The electric flux entering per second through the face *ABCD* is

$$\left(E_x - \frac{1}{2} \frac{\partial E_x}{\partial x} dx \right) dydz \qquad ...(3.50)$$

and the flux leaving per second through the face *EFGH* face will be

$$\left(E_x + \frac{1}{2} \frac{\partial E_x}{\partial x} dx \right) dydz \qquad ...(3.51)$$

Thus, the net flux in the x-direction is obtained by subtracting the Eq (3.50) from the Eq. (3.51), we have

$$\frac{\partial E_x}{\partial x} dx\, dy\, dx \qquad ...(3.52)$$

Similarly, net flux in y and z directions are given by

$$\frac{\partial E_y}{\partial y} dx\,dy\,dx \qquad ...(3.53)$$

and

$$\frac{\partial E_z}{\partial z} dx\,dy\,dx \qquad ...(3.54)$$

Thus, the electric flux passing through the parallelopiped is give by

$$\left(\frac{\partial E_x}{\partial x} + \frac{\partial E_y}{\partial y} + \frac{\partial E_z}{\partial z}\right) dx\,dy\,dz \qquad ...(3.55)$$

or

$$\nabla \cdot \vec{E}\, dV \qquad ...(3.56)$$

where ∇ is called del operator and is given by

$$\nabla \equiv \frac{\partial}{\partial x}\hat{i} + \frac{\partial}{\partial y}\hat{j} + \frac{\partial}{\partial z}\hat{k} \qquad ...(3.57)$$

But,

$$\phi = \frac{q}{\epsilon_0}$$

$$= \frac{\rho\, dV}{\epsilon_0} \qquad ...(3.58)$$

Combining the Eq. (3.56) and (3.58), we get

$$\boxed{\nabla \cdot \vec{E} = \frac{\rho}{\epsilon_0}} \qquad ...(3.59)$$

The Eq. (3.59) can be proved using a concept of continuous charge distribution. In integral form, Gauss's low is given by

$$\oint_S \vec{E} \cdot \vec{dS} = \frac{q}{\epsilon_0} \qquad ...(3.60)$$

the net charge q, in terms of the charge density is given by

$$q = \int_V \rho\, dV \qquad ...(3.61)$$

Thus, the Eq. (3.60) takes the form as

$$\oint_S \vec{E} \cdot \vec{dS} = \frac{1}{\epsilon_0}\int_V \rho\, dV \qquad ...(3.62)$$

Using Gauss's divergence theorem to change the surface integral into volume integral, as,

$$\int_S \vec{E} \cdot \vec{dS} = \int_S \nabla \cdot \vec{E}\, dV \qquad ...(3.63)$$

Thus, in the light of the Eq. (3.63), the Eq. (3.62) becomes

$$\int_S \nabla \cdot \vec{E}\, dV = \frac{1}{\epsilon_0}\int_V \rho\, dV \qquad ...(3.64)$$

Now, for any arbitrary volume, we may write the Eq. (3.64) as

$$\boxed{\nabla \cdot E = \frac{\rho}{\epsilon_0}} \qquad ...(3.65)$$

Here, ρ is called the charge per unit volume at a point where the electric field is $\vec{E}$.

The $\nabla \cdot \vec{E}$ represents the electric flux per unit volume, and the $\nabla \cdot E$ is proportional to the volume charge density ρ.

Moreover,
$$E = \frac{J}{\sigma} \qquad ...(3.66)$$

Thus, the Eq. (3.65) reduces to

$$\nabla \cdot \vec{J} = \frac{\rho\sigma}{\epsilon_0} \qquad ...(3.67)$$

We know that the charge continuity equation is given by

$$\nabla \cdot \vec{J} = -\frac{\partial \rho}{\partial t} \qquad ...(3.68)$$

thus, we get

$$-\frac{\partial \rho}{\partial t} = \frac{\rho\sigma}{\epsilon_0}$$

or

$$\frac{\partial \rho}{\rho} = -\frac{\sigma}{\epsilon_0}\partial t \qquad ...(3.69)$$

integrating

$$\int_{\rho_0}^{\rho} \frac{\partial \rho}{\rho} = -\frac{\sigma}{\epsilon_0}\int_0^t dt$$

we get,

$$\boxed{\rho = \rho_0\, e^{-t\sigma/\epsilon_0}} \qquad ...(3.70)$$

3.9. ELECTRIC FIELD DUE TO NON-CONDUCTING SPHERE

Consider a non-conducting sphere of the radius a having uniform charge density ρ. The symmetry of the charge distribution is spherical and the electric field points radially outward. Now, we have two cases as,

Case I: In this case, we shall calculate the electric field for the region $r < a$. We define the volume charge density as,

$$\text{Volume charge density } \rho = \frac{\text{net charge}}{\text{volume}}$$

or
$$\rho = \frac{q}{\left(\frac{4}{3}\pi a^3\right)} \qquad ...(3.71)$$

where q is the total charge contained by the sphere of radius a.

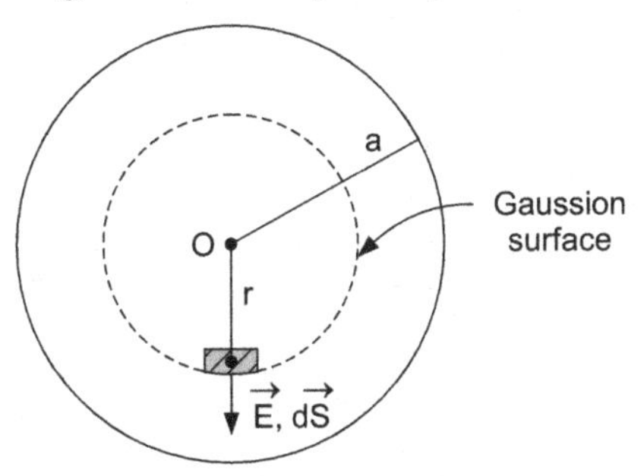

Fig. 3.22. Sphere of radius *a*.

For the region $r < a$, let q' be the charge enclosed by the gaussian surface of radius r as shown in fig. 3.22.

Thus,

$$q' = \int_V \rho \, dV \qquad ...(3.72)$$

$$= \rho \int_V \rho \, dV = \frac{q}{\left(\frac{4}{3}\pi a^3\right)} \cdot \frac{4}{3}\pi r^3$$

$$q' = q\left(\frac{r}{a}\right)^3 \qquad ...(3.73)$$

According to Gauss's law,

$$\oint \vec{E} \cdot \vec{dS} = \frac{q'}{\epsilon_0} \qquad ...(3.74)$$

or
$$E \cdot 4\pi r^2 = \frac{1}{\epsilon_0} q\left(\frac{r}{a}\right)^3$$

or
$$E = \frac{1}{4\pi \epsilon_0} \frac{r}{a^3} \qquad ...(3.75)$$

The Eq. (3.75) shows that the electric field varies linearly with the distance r and at $r = a$ it becomes constant.

Case II: Now, we shall compute the electric field intensity for the region $r > a$. Consider a point P, where E is to be calculated and we draw the gaussian surface passing through the point P as shown in fig. 3.23. In this case, all the charge q is enclosed by the gaussian surface.

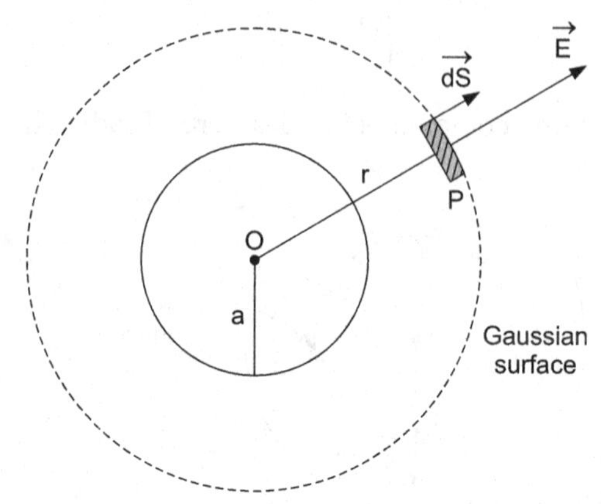

Fig. 3.23. Gaussian surface for $r > a$.

Now, Gauss's law is

$$\oint \vec{E} \cdot \vec{dS} = \frac{q}{\epsilon_0} \quad ...(3.76)$$

$$E \cdot 4\pi r^2 = \frac{q}{\epsilon_0}$$

or

$$E = \frac{1}{4\pi \epsilon_0} \frac{q}{r^2} \quad ...(3.77)$$

The plot of E versus r is shown in fig. 3.24. we can see that there is a discontinuity at $r = a$.

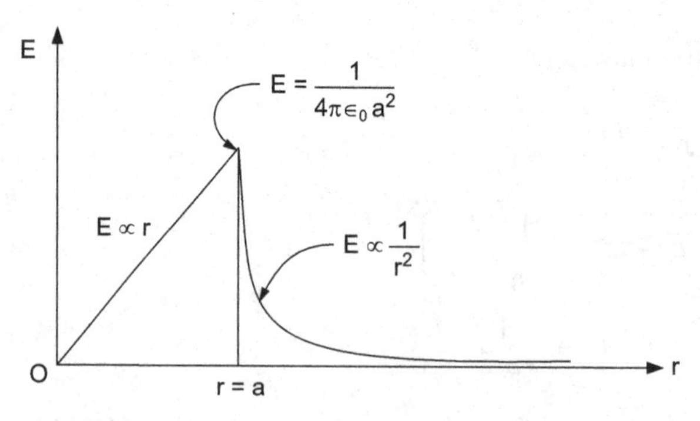

Fig. 3.24. Variation of E with the distance r.

Example 3.3. Suppose that a point charge q is located at the centre of the flat surface of a hemi-sphere of the radius a, Fig. 3.25. Find the electric flux

(a) through the curved surface and

(b) through the flat surface.

Solution: The electric field at the curved surface of the hemi-sphere is given by

$$E = \frac{1}{4\pi \epsilon_0} \frac{q}{a^2}$$

This electric field points radially outward direction.

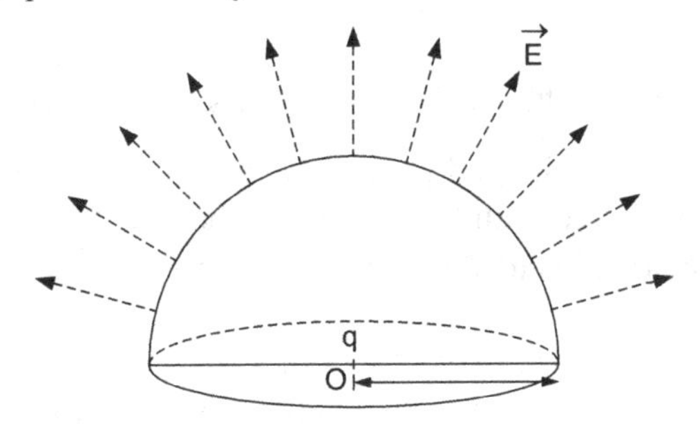

Fig. 3.25. Hemi-sphere.

(a) The flux through the curved surface will be

$$\begin{aligned} \phi_1 &= \oint \vec{E} \cdot \vec{dS} \\ &= E \cdot S \\ &= \frac{1}{4\pi \epsilon_0} \cdot \frac{r}{a^2} \cdot 2\pi a^2 \\ \phi_1 &= \frac{q}{2 \epsilon_0} \end{aligned}$$

(b) The flux through the flat surface is then given by

$$\begin{aligned} \phi_2 &= -\phi_1 \\ &= -\frac{q}{2 \epsilon_0} \end{aligned}$$

since, $\phi_1 + \phi_2 = 0$

3.10. CHARGE ON CONDUCTORS

We are already familiar with the fact that the electrons are free to move inside the conductor. It is remarkable fact that the electric field is zero inside an isolated conductor just as the gravitational attraction inside a sphere is zero. It is clear that the charge resides on the surface of a conductor. To describe it in a pictorial way, consider a charged conductor as shown in fig. 3.26.

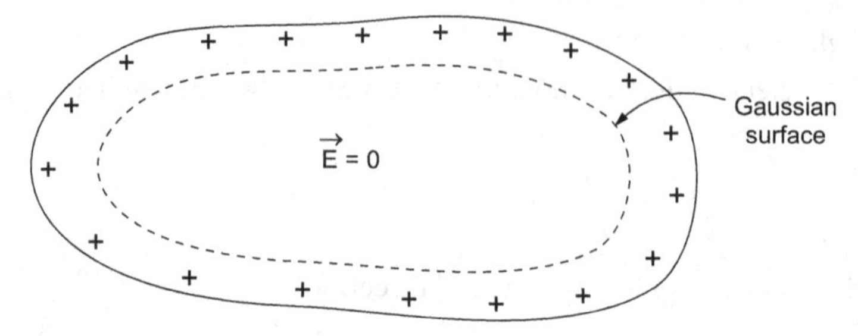

Fig. 3.26. The electric field $\vec{E}$ is zero inside the conductor.

Since the conductor is charged positively, the mutual repulsion between the charges derives the excess charge to the surface of the conductor and as a result, there is no charge inside the gaussian surface, thus the electric field

$$\vec{E} = 0 \qquad ...(3.78)$$

inside the conductor, shown by fig. 3.26. On the other hand, suppose that there is a charge present in-side a non-conductor, we have a gaussian surface which encloses the net charge and therefore, according to Gauss's law,

$$\oint \vec{E} \cdot \vec{dS} = \sum_i \frac{q_i}{\epsilon_0} \qquad ...(3.79)$$

Thus, the electric field will not be zero. In case of good conductor, the charge equilibrium state is achieved very rapidly, because the excess charge would flow to the surface of the conductor quickly. Moreover, consider a conducting sphere in a uniform electric field E_C as shown in fig. 3.27.

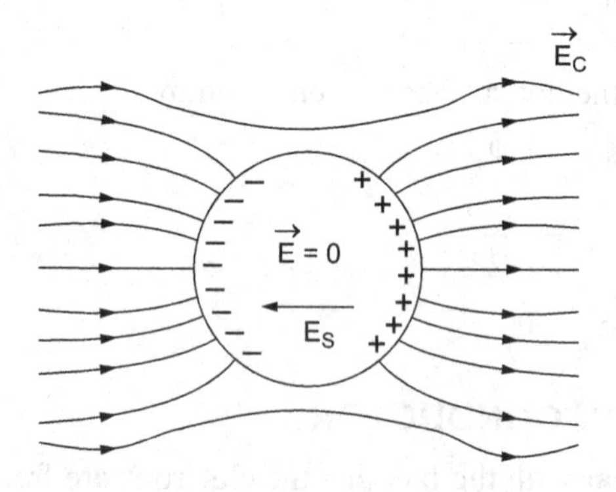

Fig. 3.27. The pattern of field lines in a conductor.

As a result, the positive and negative charges move towards the corresponding polar regions of the sphere and creating an electric field E_S in opposite direction. This is due to the fact that the electric field lines entering the conducting sphere produce negative charges and leaving from the right produce positive charges at the surface of the sphere. The motion of the charges is such that the induced electric field E_S will cancel the external field E_C inside the conductor. Hence, $\vec{E} = 0$ inside the conductor. However, there is no penetration of the electric field lines into the conductor.

Furthermore, if there is an empty cavity inside a conductor, from Gauss's law

$$\oint \vec{E} \cdot \vec{dS} = \frac{q}{\epsilon_0} = 0 \qquad ...(3.80)$$

Thus,

$$E = 0 \qquad ...(3.81)$$

Because, there is no charge in the cavity, $q = 0$, $\phi_e = 0$ and $E = 0$.

Again, suppose that a charge cavity is present inside the conductor as shown in fig. 3.28.

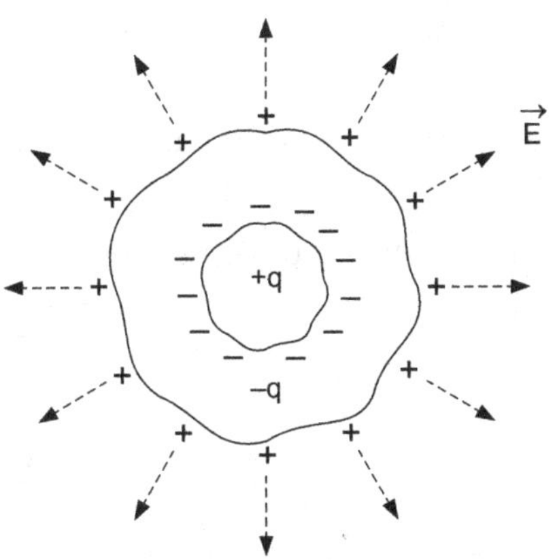

Fig. 3.28. A charge cavity in the conductor.

Due to the presence of $+q$ charge inside the cavity, the $-q$ charge is induced inside the surface of the cavity. As a result, $+q$ charge appears on the surface of the conductor as shown in fig. 3.28. Now, consider a case, where a charge $+q'$ is present on the surface on the conductor. There should be no confusion between cavity and the conductor. The total charge on the surface will be equal to $q + q'$. Moreover, since charge is present in the cavity, the gaussian surface encloses the net charge $+q - q = 0$. Again, the electric field

$\vec{E}$ is just perpendicular to the surface of the conductor and pointing outward, as shown in fig. 3.29.

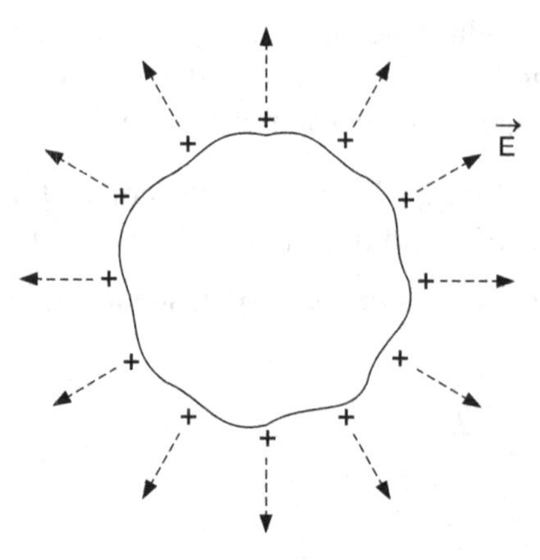

Fig. 3.29. Direction of $\vec{E}$ of a charged conductor.

The tangential component E_t of the electric field $\vec{E}$ is zero at the surface of the conductor, if it were non-zero initially, the charge on the surface will move in such a way that it becomes zero quickly and only the normal component of the electric field exists.

Example 3.4. For a conductor having uniform charge density σ, calculate the electric field.

Solution: We can consider a conductor of any shape, Fig. 3.30.

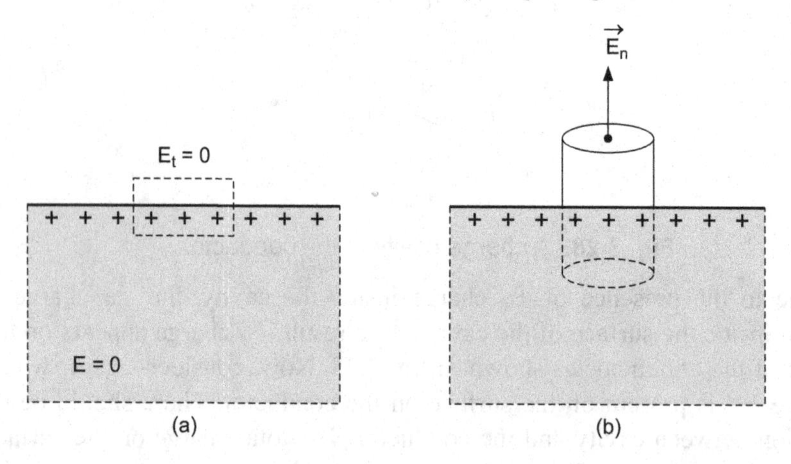

Fig. 3.30. $E_t = 0$ and normal component of electric field E_n exists.

Let q be the charge on the conductor, then

$$\sigma = \frac{q}{S}$$

where S is the surface area of the conductor. To compute the electric field, draw a pill-box, and $E_t = 0$. According to Gauss's law,

$$\oint \vec{E} \cdot \vec{dS} = \frac{q}{\epsilon_0}$$

or

$$E_n S = \frac{q}{\epsilon_0}$$

or

$$E_n = \frac{\sigma}{\epsilon_0}.$$

Example 3.5. The electric field for a given region is proportional to the square root of the distance, That is,

$$E \propto x^{1/2}$$

or

$$E = k\,x^{1/2}$$

then, compute

(a) The electric flux through the face of the cube of side L, Fig. 3.31.

(b) Net charge confined to the cube.

Solution: by Gauss's law, the flux through the small surface dS is given by

$$d\phi = \vec{E} . \vec{dS}$$

or

$$\phi_e = \int \vec{E} . \vec{dS}$$

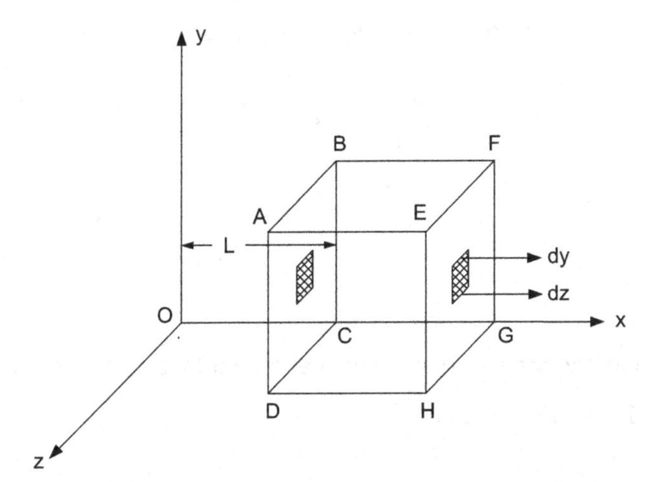

Fig. 3.31. Flux through the surface of the cube.

Net electric flux through the cube would be

$$\phi_{net} = \phi_{EFGH} - \phi_{ABCD}$$

But $$\phi_{ABCD} = \int_{ABCD} \vec{E}.\vec{dS}$$

$$= K L^{1/2} \int_0^L dy \int_0^L dz$$

$$= K L^{5/2}$$

Again,

$$\phi_{EFGH} = \int_{EFGH} \vec{E}.\vec{dS}$$

$$= K (2L)^{1/2} \int_0^L dy \int_0^L dz$$

$$= K\sqrt{2}\ L^{5/2}$$

net flux through the cube is

$$\phi_{net} = K L^{5/2} (\sqrt{2} - 1)$$

But $$\phi_{net} = \frac{\Sigma q}{\epsilon_0}$$

$$\therefore \quad \Sigma q = K\epsilon_0 L^{5/2} (\sqrt{2} - 1)$$

Example 3.6. Show that tangential component of the electric field is zero at the surface of the conductor of any arbitrary shape.

Solution: Consider a conductor of any arbitrary shape as shown in fig. 3.32. By Gauss Law, the excess charge must be distributed over the surface of the conductor, and no electric field may exist within a conductor. i.e.

$$E = 0 \quad \text{(inside the conductor).}$$

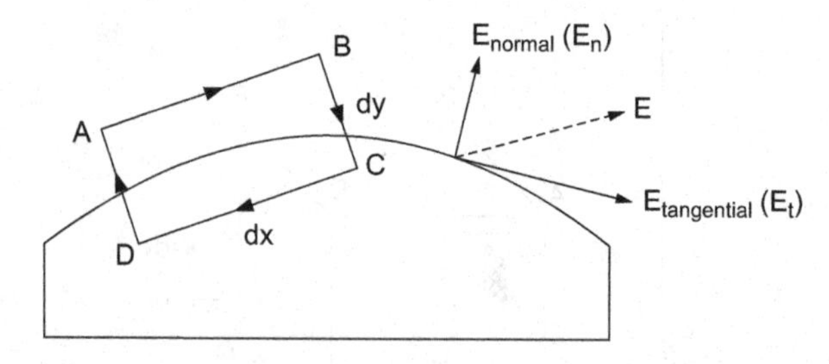

Fig. 3.32. Components of electric field at the surface of the conductor.

Now, by Maxwell's equation,

$$\oint_l \vec{E} \cdot \vec{dl} = 0$$

or $\oint_{ABCDA} \vec{E} \cdot \vec{dl} = 0$

or $E_t\,dx - E_n\,dy + 0.dx + E_n\,dy = 0$

As $dx \to 0$, we have

$$Et = 0$$

Thus, the tangential component of the electric field on the surface of the conductor is zero and the surface of conductor will remain in electrostatic equilibrium.

Example 3.7. A point charge is placed at the centre of the cube. Calculate the electric flux passing through the cube.

Solution: Consider a cube of sides L and a point charge is situated at the centre of the cube as shown in fig. 3.33.

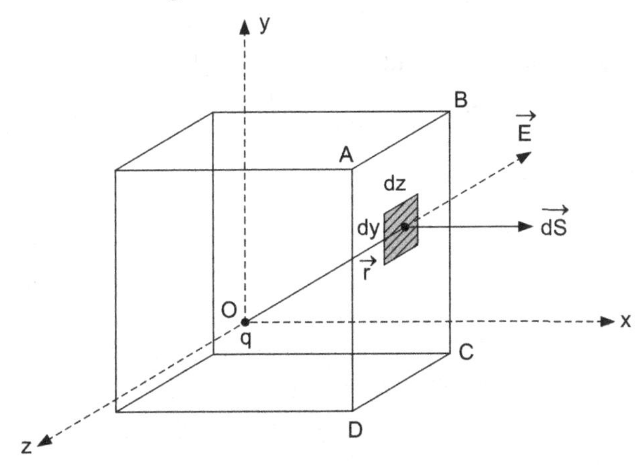

Fig. 3.33. Cube.

Again, consider an area element $\vec{dS}$ whose position vector is $\vec{r}$. Therefore, $\vec{dS}$ is given by

$$\vec{dS} = dy\,dz\,\hat{i}$$

the electric field due to point charge q is given as

$$\vec{E} = \frac{1}{4\pi\epsilon_0}\frac{q}{r^3}\vec{r}$$

$$= \frac{1}{4\pi\epsilon_0}\frac{q}{r^3}(x\hat{i} + y\hat{j} + z\hat{k})$$

where $r^2 = x^2 + y^2 + z^2$

The electric flux through small area $\vec{dS}$ would be

$$d\phi = \vec{E} \cdot \vec{dS}$$

$$= \frac{1}{4\pi\epsilon_0} \frac{q\,(x\hat{i} + y\hat{j} + z\hat{k}) \cdot dy\,dz\,\hat{i}}{(x^2 + y^2 + z^2)^{3/2}}$$

$$d\phi_e = \frac{1}{4\pi\epsilon_0} \frac{q\,x\,dy\,dz}{(x^2 + y^2 + z^2)^{3/2}}$$

But $x = L$ (side of cube)

$$\therefore \quad d\phi_e = \frac{qL}{4\pi\epsilon_0} \frac{dy\,dz}{(L^2 + y^2 + z^2)^{3/2}}$$

Now, total flux through the surface (face) $ABCD$ is

$$\phi_{ABCD} = \frac{4qL}{4\pi\epsilon_0} \int_0^L \int_0^L \frac{dy\,dz}{(L^2 + y^2 + z^2)^{3/2}}$$

$$= \frac{qL}{\pi\epsilon_0} \int_0^L dy \left[\frac{z}{(L^2 + y^2)(L^2 + y^2 + z^2)^{1/2}} \right]_0^L$$

$$= \frac{qL^2}{\pi\epsilon_0} \int_0^L \frac{dy}{(L^2 + y^2)(2L^2 + y^2)^{1/2}}$$

$$= \frac{qL^2}{\pi\epsilon_0} \cdot \frac{1}{L^2} \left[\tan^{-1} \frac{y}{(2L^2 + y^2)^{1/2}} \right]_0^L$$

$$= \frac{q}{\pi\epsilon_0} \tan^{-1} \frac{1}{\sqrt{3}}$$

$$= \frac{q}{\pi\epsilon_0} \cdot \frac{\pi}{6} = \frac{q}{6\epsilon_0}$$

net flux through the cube is the 6 times of the flux through one face of the cube.

$$\therefore \quad \phi_{\text{cube}} = \frac{q}{6\epsilon_0} \cdot 6$$

or $$\phi_{\text{cube}} = \frac{q}{\epsilon_0}.$$

EXERCISES

3.1. Compute the charge density ρ, if the electric field is given by

$$\vec{E} = E_0 x\hat{i}$$

Hint: $$\nabla \cdot \vec{E} = \frac{\rho}{\epsilon_0}.$$

3.2. If in a certain volume of the space, the ten thousand electric field lines enter and four thousand leaving the volume, compute the net charge within the volume.

Hint: $\phi_e = \frac{q}{\epsilon_0}$, $q = \phi \cdot \epsilon_o = 6000 \times 8.85 \times 10^{-12}$ coulomb.

3.3. A spherical charge distribution has the charge density

$$\rho = \begin{cases} \rho_0 (a-r) & 0 \leq r \leq a \\ 0 & \text{otherwise} \end{cases}$$

Find the expression for flux.

Hint: $\phi_e = \frac{q}{\epsilon_0} = \frac{1}{\epsilon_0}\int_V \rho \, dV = \frac{1}{\epsilon_0}\int_0^a \int_0^\pi \int_0^{2\pi} \rho \, r^2 dr \sin\theta \, d\theta \, d\phi$

3.4. Prove that there can be no existence of the net electric charge inside a hollow conductor

Hint: Gauss's Law.

3.5. Determine the expression for the electric field for a conducting sphere for (*a*) $r < a$ (*b*) $r \geq a$, where *a* is the radius of the sphere.

3.6. Compute the electric field at a point outside a long cylinder of radius a and having a uniform charge density.

3.7. For a given volume, the electric field is given by

$$\vec{E} = 2x\hat{i} + y\hat{j} + \hat{k}$$

Compute the charge density.

3.8. Two non conducting charged plates *X* and *Y* are placed at a distance *d* apart as shown in fig. 3.34. Both plates have uniform charge distribution. Find the electric field at the points *A*, *B* and *C*.

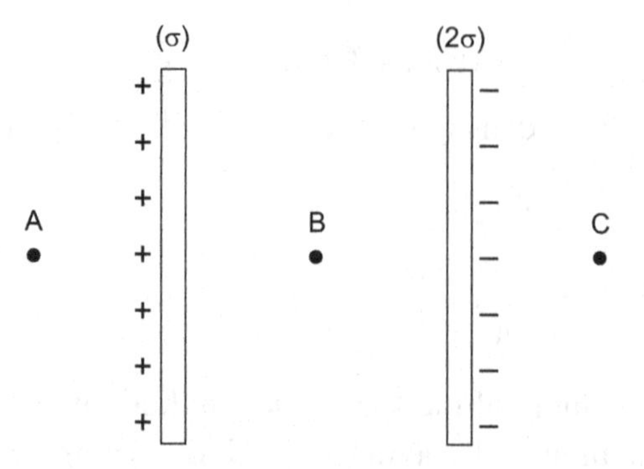

Fig. 3.34. Charged plates.

3.9. Suppose that a square of the edge '*a*' is placed at a distance '*a*' from a positive charge *q* as shown in fig. 3.35. Compute the electric flux passing through the square.

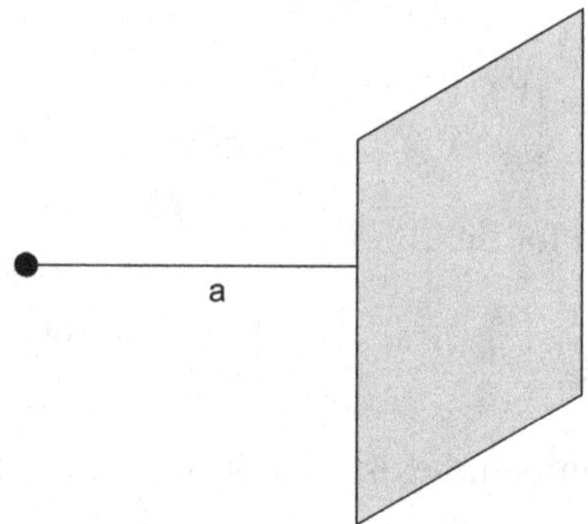

Fig. 3.35. A square plane.

3.10. A sphere of uniform charge density ρ and radius a contains a spherical cavity of radius $\frac{R}{2}$ as shown in fig. 3.36. Compute the electric field at point *A* and *B*.

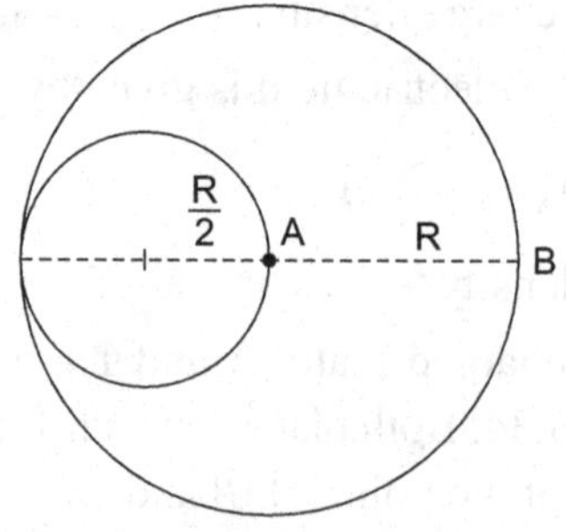

Fig. 3.36. Spherical cavity.

■ ■ ■

4 Electric Potentials

CHAPTER

We are familiar with the electrostatic force that may be presented in terms of the electric field. Since electric field is a vector field, the electrostatic force is conservative force. In the experimental physics, the concept of the electric field may be transformed into the electric potential. The concept of electric potential or potential difference plays an important role in electrical engineering.

4.1. LINE ELEMENT AND LINE INTEGRAL

When an electrostatic force acts between the two point charges, we have associated with the electrostatic potential energy of the system. Thus, the concept of the potential energy is concerned with the work done by the conservative force. To calculate the amount of work done when a conservative force moves a particle along a path in the space, the line integral plays an important role. To understand the concept of the potential, suppose that a particle of mass m moves from the initial point A to the final point B under the influence of the force $\vec{F}$, as shown in fig. 4.1.

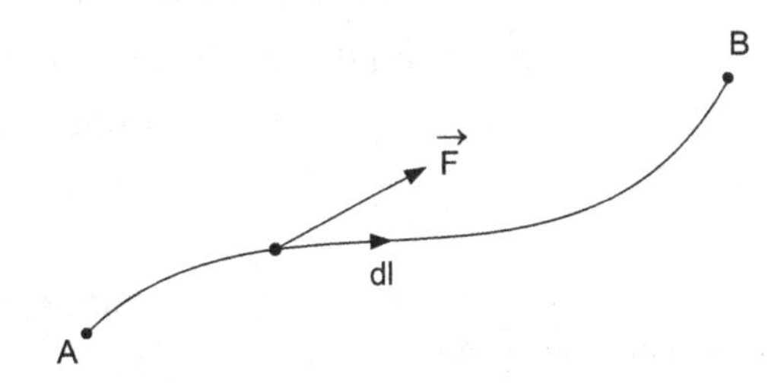

Fig. 4.1. The motion of a particle along the path *AB*.

To calculate the work, the path AB is divided into several small segments and using one such path element dl. dl is also known as line element and is given by

$$\vec{dl} = dx\hat{i} + dy\hat{j} + dz\hat{k} \qquad ...(4.1)$$

where dy, dy and dz are the path elements in rectangular co-ordinates. The line element, in cylindrical coordinates, is

$$\vec{dl} = \hat{r}\,dr + \hat{\phi}\,rd\phi + \hat{k}\,dz \qquad ...(4.2)$$

and in spherical polar coordinates is given by

$$\vec{dl} = \hat{r}\,dr + \hat{\theta}\,rd\theta + \hat{\phi}\,r\sin\theta\,d\phi \qquad ...(4.3)$$

Now, the scalar product of the force and the line element dl is $\vec{F}\cdot\vec{dl}$ which represents the work done for the small displacement $\vec{dl}$. That is,

$$dw = \vec{F}\cdot\vec{dl} \qquad ...(4.4)$$

The total work done in moving a particle from the initial point A to final point B is then, given by

$$w = \int_A^B \vec{F}\cdot\vec{dl} \qquad ...(4.5)$$

or

$$w = \int_A^B F_x\,dx + \int_A^B F_y\,dy + \int_A^B F_z\,dz \qquad ...(4.6)$$

Since work done by the force $\vec{F}$ depends on the initial and final points A and B respectively, and is independent of path, the force $\vec{F}$ is called the conservative force. Moreover, if the work done by force is path dependent, it is known as non-conservative force.

4.2. ELECTRIC POTENTIAL AND POTENTIAL DIFFERENCE

Suppose that a positive point charge q_0 is allowed to move in an electric field $\vec{E}$, an electrostatic force $\vec{F} = q_0 E$ acts on the charge q_0 and this force $\vec{F}$ is governed by the coulomb's law. If the charge q_0 moves from an initial point A to the final point B, the potential difference is, often, written as

$$dV = V_B - V_A \qquad ...(4.7)$$

Let W_{AB} be the work done by the electrostatic force in moving a positive charge q_0 from the point A to B, Thus, we write,

$$\boxed{V_B - V_A = \frac{W_{AB}}{q_0}} \qquad ...(4.8)$$

Therefore, the potential difference between any two points A and B is equal to the work done in moving a positive charge q_0 from the point A to B against the field. Moreover, since field is conservative and $\vec{F} = q_0\vec{E}$, then, work done by the electrostatic field in moving the charge q_0 through a small displacement dl is given by

$$dw = -\vec{F} \cdot \vec{dl}$$

$$= -q_0 \vec{E} \cdot \vec{dl} \qquad \text{...(4.9)}$$

The negative sign denotes that the electric field $\vec{E}$ points in the direction of decreasing the electric potential.

Thus, the total work done in moving the charge q_0 from the point A to B is,

$$W_{AB} = \int_A^B dw \qquad \text{...(4.10)}$$

$$= \int_A^B -q_0 \vec{E} \cdot \vec{dl}$$

$$= -q_0 \int_A^B \vec{E} \cdot \vec{dl} \qquad \text{...(4.11)}$$

The potential difference is

$$V_B - V_A = \frac{W_{AB}}{q_0} \qquad \text{...(4.12)}$$

Hence,

$$\boxed{V_B - V_A = -\int_A^B \vec{E} \cdot \vec{dl}} \qquad \text{...(4.13)}$$

If we assign a point in the space as a reference point where the potential is assumed to be zero, then, this point will be infinite where $V \rightarrow 0$. Consider the point A at infinite, so that $V_A = 0$, from the Eq. (4.13) we write,

$$V_B = V = - -\int_{\infty}^B \vec{E} \cdot \vec{dl}$$

or

$$\boxed{V = -\int_{\infty}^B \vec{E} \cdot \vec{dl}} \qquad \text{...(4.14)}$$

The Eq. (4.14) states the electric potential at any point in the field. It states that the electric potential at any point in the electrostatic field is equal to the work done in bringing a positive charge q_0 from the infinite to the point in the field. If the potential difference is negative, it means that the work is being done by the force $\vec{F} = q_0 E$, and if dV is positive, there is a gain in the potential energy. Since potential is a scalar quantity, it is easy to calculate rather than electric field. The unit of work is joule and that of charge is coulomb. Then, unit of potential or potential difference is volt. Now,

$$1 \text{ volt} = \frac{1 \text{ joule}}{1 \text{ coulomb}}$$

In CGS system, unit is stat volt.

Example 4.1. If the work of 2×10^{-4} J is required in carrying a charge of 2μC from a point *A* to point *B* calculate the potential difference and which point is at higher potential.

Solution:
$$V_B - V_A = \frac{W_{AB}}{q_0} = \frac{2\times10^{-4}\text{J}}{2\times10^{-6}\text{C}} = 100 \text{ volts}$$

Since $V_B - V_A = +\text{ve}$, V_B is at higher potential.

4.3. ELECTRIC POTENTIAL OF A POINT CHARGE

The electric field may be computed directly for the distribution of charges. However, in computing the electric field for a uniform charge distribution, there is a mathematical difficulty in solving the problems. We have, now, a concept of the electric potential in solving the field related problems. Thus, two fundamental laws are there.

(1) For the system of stationary charges, the work done by the conservative force in carrying a positive test charge q_0 around a closed path is zero.

(2) The electric flux density is expressed in terms of Gauss's law.

To obtain an expression for the electric potential due to a point charge *q*, we must compute the potential difference between *A* and *B* as shown in fig. 4.2.

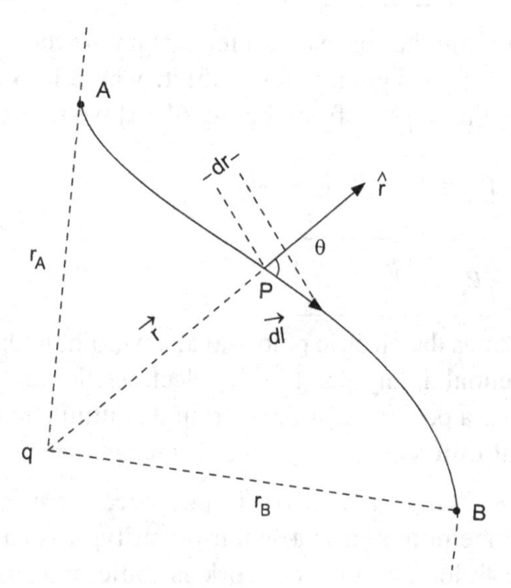

Fig. 4.2. The potential difference between two points *A* and *B*.

For the charge q, the electric field at a point P is given by

$$\vec{E} = \frac{1}{4\pi \epsilon_o} \frac{q}{r^2} \hat{r} \qquad \text{...(4.15)}$$

where $\hat{r}$ is a unit vector along the position vector $\vec{r}$.

Since $$\hat{r} \cdot \vec{dl} = dl \cos\theta = dr \qquad \text{...(4.16)}$$

then, potential difference between the point A and B is given by

$$dV = V_B - V_A = -\int_A^B \vec{E} \cdot \vec{dl} \qquad \text{...(4.17)}$$

$$= -\int_{r_A}^{r_B} \frac{q}{4\pi \epsilon_0} \frac{dr}{r^2}$$

$$= \frac{q}{4\pi \epsilon_0} \left[\frac{1}{r_B} - \frac{1}{r_A} \right] \qquad \text{...(4.18)}$$

Thus, dV is the potential at a point B with respect to A. If it is assumed that the point A is at infinity, $V_A = 0$ because at $r_A \to \infty$, $V_A \to 0$. Therefore, we write

$$V_B - V(\infty) = \frac{q}{4\pi \epsilon_0 r_B} \qquad \text{...(4.19)}$$

we can choose an arbitrary path the evaluate the electric potential, in the reference, of the Eq. (4.19), the potential at any arbitrary point will be

$$V(r) = \frac{1}{4\pi \epsilon_0} \frac{q}{r} \qquad \text{...(4.20)}$$

If we look at the equation (4.17), it is obvious that the line integral $\oint \vec{E} \cdot \vec{dl}$ must be zero for any closed path, that is,

$$\oint \vec{E} \cdot \vec{dl} = 0 \qquad \text{...(4.21)}$$

Thus, the potential difference $dV = V_B - V_A$ does not depend on the path, it is the difference of the potentials at the points A and B. The potential is negative or positive, it depends on the sign of the charge.

4.4. POTENTIAL DUE TO CONTINUOUS CHARGE DISTRIBUTION

In the previous section, we have obtained the expression for the potential difference between two points A and B. If there are several charges q_1, q_2, q_3 ... are located at the points with the position vectors $\vec{r_1}, \vec{r_2}, \vec{r_3}$ ..., the potentials are additive, Thus, we write

$$V = \frac{1}{4\pi\epsilon_0}\frac{q_1}{|\vec{r}-\vec{r_1}|} + \frac{1}{4\pi\epsilon_o}\frac{q_2}{|\vec{r}-\vec{r_2}|} + \ldots \quad \text{...(4.22)}$$

or

$$V = \frac{1}{4\pi\epsilon_0}\sum_i \frac{q_i}{|\vec{r}-\vec{r_i}|} \quad \text{...(4.23)}$$

For the uniform charge distribution, we must have charge elements with the different charge densities. Thus,

$$V = \frac{1}{4\pi\epsilon_0}\int_V \frac{\rho\, dV}{|\vec{r_1}-\vec{r_2}|} \quad \text{...(4.24)}$$

or

$$V = \frac{1}{4\pi\epsilon_0}\int_S \frac{\sigma\, dS}{|\vec{r_1}-\vec{r_2}|} \quad \text{...(4.25)}$$

and

$$V = \frac{1}{4\pi\epsilon_0}\int_l \frac{\lambda\, dl}{|\vec{r_1}-\vec{r_2}|} \quad \text{...(4.26)}$$

Example 4.2. Suppose that the three charges $q_1 = 0.1\ \mu C$, $q_2 = 0.2\ \mu C$ and $q_3 = -0.1\ \mu C$ are situated at the three corners of a square of side 2 m as shown in fig. 4.3.

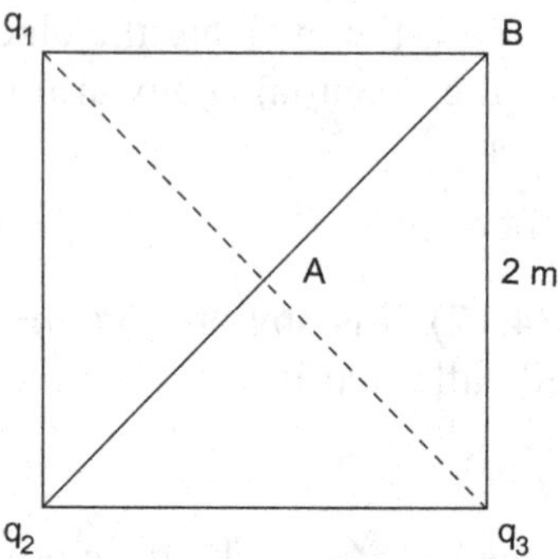

Fig. 4.3. A square.

Find the potential at the points A and B and also $V_B - V_A$.

Solution: The potential at the point A will be

$$V_A = \frac{1}{4\pi\epsilon_o}\left(\frac{q_1}{r_1} + \frac{q_2}{r_2} + \frac{q_3}{r_3}\right)$$

$$= 9\times10^9\left(\frac{0.1}{2} + \frac{0.2}{2\sqrt{2}} - \frac{0.1}{\sqrt{2}}\right)\times10^{-6}$$

$$= 9 \times 10^9 \times \sqrt{2} \times 10^{-7}$$
$$= 12.72 \times 10^2 = 1272 \text{ volts}$$
$$V_B = \frac{1}{4\pi\epsilon_o}\left(\frac{q_1}{r_1} + \frac{q_2}{r_2} + \frac{q_3}{r_3}\right)$$
$$= 9 \times 10^9 \left(\frac{0.1}{2} + \frac{0.2}{2\sqrt{2}} - \frac{0.1}{\sqrt{2}}\right) \times 10^{-6}$$
$$= 9 \times 10^2 \times \frac{1}{\sqrt{2}} = 636 \text{ volts}$$
$$\therefore \qquad V_B - V_A = 636 - 1272 = -636$$

The V_B is at lower potential.

4.5. NEGATIVE GRADIENT OF THE POTENTIAL

Let V be the function of the rectangular coordinates as

$$V = V(x, y, z) \qquad \text{...(4.27)}$$

Now, we use the concept of partial differentiation and we take

$$dV = \frac{\partial V}{\partial x} dx + \frac{\partial V}{\partial y} dy + \frac{\partial V}{\partial z} dz \qquad \text{...(4.28)}$$

The Eq. (4.28) may be given in cylindrical and spherical polar coordinates,

$$\left.\begin{aligned} dV &= \frac{\partial V}{\partial r} dr + \frac{\partial V}{\partial \phi} d\phi + \frac{\partial V}{\partial z} dz \\ \text{and} \qquad dV &= \frac{\partial V}{\partial r} dr + \frac{\partial V}{\partial \theta} d\theta + \frac{\partial V}{\partial \phi} d\phi \end{aligned}\right\} \qquad \text{...(4.29)}$$

The potential difference between two points A and B is given by

$$dV = -\vec{E} \cdot \vec{dl} \qquad \text{...(4.30)}$$

Here, the electric field $\vec{E}$ and the line element $\vec{dl}$ are expressed (in the rectangular coordinate system) as,

$$\vec{E} = E_x\hat{i} + E_y\hat{j} + E_z\hat{k} \qquad \text{...(4.31)}$$

and
$$\vec{dl} = dx\,\hat{i} + dy\,\hat{j} + dz\,\hat{k}$$

The components of $\vec{E}$ are shown in fig. 4.4.

Using the Eq. (4.31), The Eq (4.30) takes the form

$$dV = -(E_x dx + E_y dy + E_z dz) \qquad \text{...(4.32)}$$

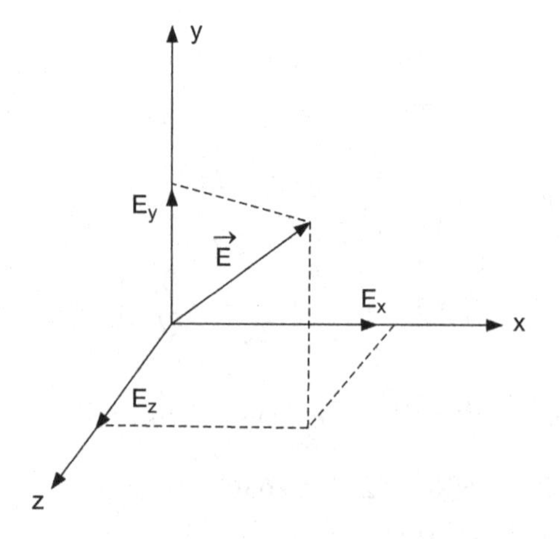

Fig. 4.4. The components of the electric field $\vec{E}$ in cartesian coordinates system.

Now, from the Eq. (4.32), the components of $\vec{E}$ are given by

$$E_x = -\frac{dV}{dx},\ E_y = \frac{-dV}{dy}, \text{ and } E_z = \frac{-dV}{dz} \quad ...(4.33)$$

Moreover, $\vec{E}$ is then,

$$\vec{E} = E_x\hat{i} + E_y\hat{j} + E_z\hat{k} \quad ...(4.34)$$

$$= -\left[\frac{dV}{dx}\hat{i} + \frac{dV}{dy}\hat{j} + \frac{dV}{dz}\hat{k}\right]$$

$$= -\left(\frac{\partial}{\partial x}\hat{i} + \frac{\partial}{\partial y}\hat{j} + \frac{\partial}{\partial z}\hat{k}\right)V$$

or

$$\boxed{\vec{E} = -\nabla V} \quad ...(4.35)$$

Again, $\vec{E} = -\text{grad } V$ ...(4.36)

Thus, the rate of change of potential $V(x, y, z)$ in any direction at a given point is called the gradient of the potential at that point.

The Eq. (4.35) can be represented in cylindrical and spherical polar coordinates.

$$\left.\begin{aligned} \vec{E} &= -\left(\hat{r}\frac{\partial}{\partial r} + \hat{\phi}\frac{1}{r}\frac{\partial}{\partial \phi} + \hat{k}\frac{\partial}{\partial z}\right)V \\ \text{and} \quad \vec{E} &= -\left(\hat{r}\frac{\partial}{\partial r} + \hat{\theta}\frac{1}{r}\frac{\partial}{\partial \theta} + \hat{\phi}\frac{1}{r\sin\theta}\frac{\partial}{\partial \phi}\right)V \end{aligned}\right\} \quad ...(4.37)$$

Physical Interpretation of $\vec{E} = -\nabla V$

The grad V is an operation of the ∇ on the potential at a point. However, it provides the information in the region near the point P(say) where it is evaluated. We have following properties of the ∇V,

(1) The grad V is pointed in the direction of increasing potential.

(2) For the surface where V is constant, ∇V is assumed to be perpendicular to this surface. It is known as the concept of equipotential surface where $\vec{E}$ is always perpendicular to the surface.

(3) The magnitude of the grad V is equal to the maximum of the rate of change of V. For example

$$|\nabla V| = \left|\frac{dV}{dx}\right| = \text{maximum for } E \text{ in the direction of } x.$$

(4) If E is perpendicular to the dx,

$$\frac{dE}{dx} = 0 \qquad ...(4.38)$$

or $$E = \text{constant}$$

4.6. ELECTRIC POTENTIAL DUE TO A DIPOLE

Consider a pair of equal and opposite charges separated by a distance of $2a$ as shown in fig. 4.5.

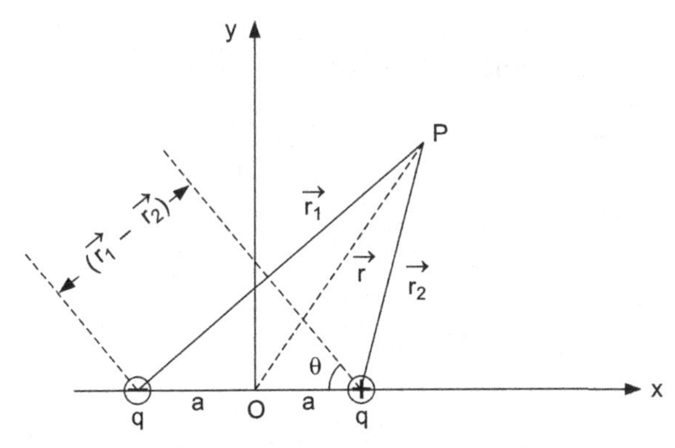

Fig. 4.5. An electric dipole

The potential at a point P due to $-q$ charge is

$$V_1 = \frac{1}{4\pi\epsilon_0}\frac{(-q)}{r_1} \qquad ...(4.39)$$

and the potential at point P due to the $+q$ charge is given by

$$V_2 = \frac{1}{4\pi \epsilon_0} \frac{q}{r_2} \quad ...(4.40)$$

Thus, the net potential at the point P will be

$$V = V_1 + V_2$$

$$= \frac{1}{4\pi \epsilon_o}\left(\frac{1}{r_2} - \frac{1}{r_1}\right)$$

$$= \frac{1}{4\pi \epsilon_o} \frac{|\vec{r_1} - \vec{r_2}|}{r_1 r_2} \quad ...(4.41)$$

If the distance between two charges $2a$, is very small as compared to the distance r of the point P, we may write, $r_1 r_2 = r^2$ and $|\vec{r_1} - \vec{r_2}| = 2a \cos \theta$.

Thus, we get

$$V = \frac{1}{4\pi \epsilon_0} \frac{2aq \cos\theta}{r^2} \quad ...(4.42)$$

Since dipole moment of the system is given by

$$p = 2aq, \quad ...(4.43)$$

using the Eq. (4.43), the Eq. (4.42) takes the form

$$\boxed{V = \frac{1}{4\pi \epsilon_0} \frac{p \cos\theta}{r^2}} \quad ...(4.44)$$

Moreover, the electric field is then calculated as

$$\vec{E} = -\nabla V = -\left(\hat{r}\frac{\partial V}{\partial r} + \hat{\theta}\frac{1}{r}\frac{\partial V}{\partial \theta}\right) \quad ...(4.45)$$

Here, $\frac{\partial V}{\partial r} = -\frac{2p \cos\theta}{4\pi \epsilon_0 r^3}$

and $\frac{\partial V}{\partial \theta} = -\frac{p \sin\theta}{4\pi \epsilon_0 r^2}$ ...(4.46)

Substituting the Eq. (4.46) in the Eq. (4.45) we get,

$$\vec{E} = \frac{p}{4\pi \epsilon_0 r^3}(2 \cos\theta \hat{r} + \sin\theta \hat{\theta}) \quad ...(4.47)$$

$$\therefore \quad |\vec{E}| = \frac{p}{4\pi \epsilon_0 r^3}(3 \cos^2\theta + 1) \quad ...(4.48)$$

which is the expression for the electric field.

Example 4.3. Consider an electric dipole as shown in fig. 4.6. Then, find

(a) the potential at any point on the axis of the dipole.

(b) the electric potential at the centre of the dipole and,

(c) if the negative charge is replaced by a positive charge of same magnitude, the potential at the centre of the dipole.

Solution: The dipole moment is given by

$$p = 2aq$$

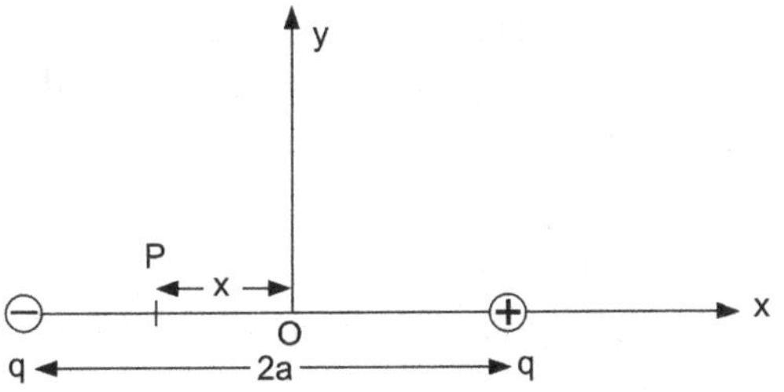

Fig. 4.6. Electric diple.

(a) To find the electric potential at a point P, we have to find out the potential at the point P due to both charges, the potential at P due to $-q$ charge is

$$V_1 = \frac{1}{4\pi\epsilon_0}\frac{(+q)}{(a+x)}$$

and the potential due to $+q$ is

$$V_1 = \frac{1}{4\pi\epsilon_0}\frac{(+q)}{(a+x)}$$

Thus, the net potential at P is then,

$$V = V_1 + V_2 = \frac{q}{4\pi\epsilon_0}\left[\frac{1}{(a+x)} - \frac{1}{(a-x)}\right]$$

$$V = \frac{2qx}{4\pi\epsilon_0 (a^2 - x^2)}$$

(b) The potential at point '0', $x = 0$

$$V = 0$$

Moreover, $$V(q) = \frac{1}{4\pi\epsilon_0}\frac{q}{a}$$

and $$V(-q) = \frac{1}{4\pi\epsilon_0}\frac{(-q)}{a}$$

Thus, $$V = V(q) + V(-q) = 0$$

(c) If the charge $-q$ is replaced by $+q$, then, the potential at O due to the both charges are given by

$$V_1 = V_2 = \frac{1}{4\pi\epsilon_0}\frac{q}{a}$$

$$\therefore \qquad V = V_1 + V_2 = \frac{2q}{4\pi\epsilon_0 a}$$

4.7. THE EQUIPOTENTIAL SURFACES

The equipotential surface is a surface that has same electrical potential at the every point lying on the surface. We know that the potential difference at the two points A and B is given by

$$V_B - V_A = -\int_A^B \vec{E} \cdot \vec{dl} \qquad ...(4.49)$$

Now, according to the definition of the equipotential surface,

$$V_A = V_B$$

Thus, we write the Eq. (4.49) as

$$\int_A^B \vec{E} \cdot \vec{dl} = 0 \qquad ...(4.50)$$

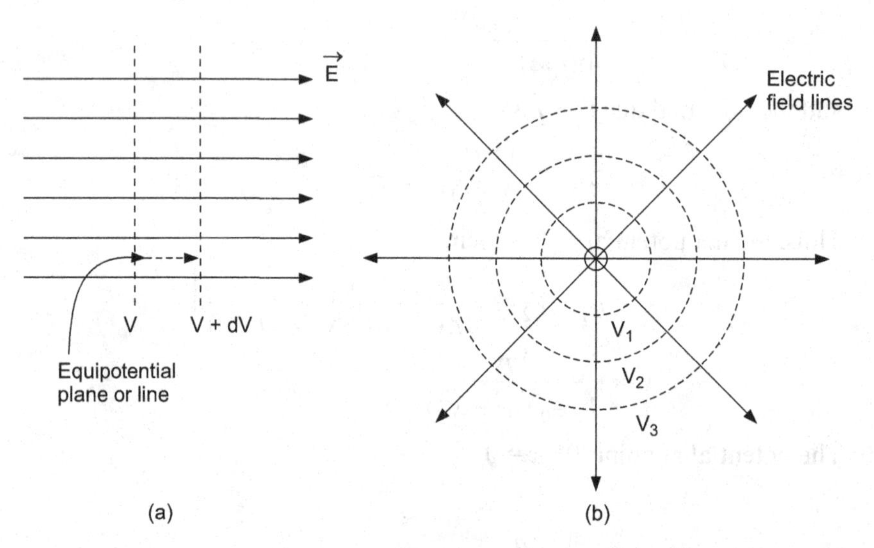

Fig. 4.7. Equipotential surface (a) Equipotential plane or line (b) for positive charge, V_1, V_2 and V_3.

The Eq. (4.50) shows that the electric field E is perpendicular to the displacement. It means that the electric lines of force are perpendicular to the equipotential surface as shown in fig. 4.7. Moreover, by the Eq. (4.50), we have

$$\int_A^B \vec{F} \cdot \vec{dl} = 0 \qquad ...(4.51)$$

The Eq. (4.51) predicts that no work is done in moving a charge from a point A to B along the equipotential line (a line on the equipotential surface).

From the fig. 4.7, it is clear that the electric field $\vec{E}$ is always perpendicular to the surface. In fig. 4.7 (a) we have two equipotential surfaces or lines that have potentials V and $V + dV$. However, in fig 4.7 (b), there are three spherical equipotential surfaces at the potentials V_1, V_2 and V_3. We have mentioned earlier that the potential is always same at every point on the equipotential surface but it is not true in case of the electric field. The magnitude of the electric field may be different at the different points on the equipotential surface.

Furthermore, we may prove that the electric field $\vec{E}$ is perpendicular to an arbitrary equipotential surface. For this, consider the equipotential curves as shown in fig. 4.8. These curves are characterised by the constant potential* $V(x, y, z)$.

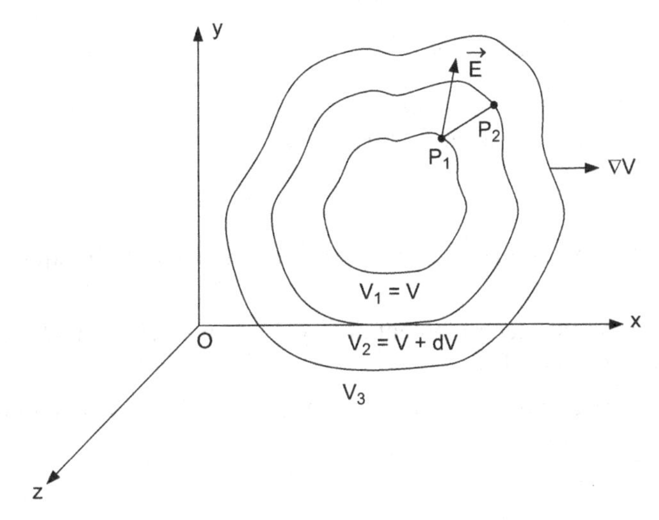

Fig. 4.8 Arbitrary equipotential surfaces.

Suppose that there are three surfaces at the constant potentials $V_1 = V$, $V_2 = V + dV$ and V_3. The electric field $\vec{E}$ can be written as a negative gradient of the electric potential, that is,

$$\vec{E} = -\nabla V \qquad ...(4.52)$$

* Classical Electricity and magnetism by Panofsky and Phillips, Addison-Wesley

Let the potentials at the points P_1 and P_2 be $V(x, y, z)$ and $V(x + dx, y + dy, z + dz)$ respectively. Thus, the potential difference between the points P_1 and P_2 is given by

$$dV = V(x + dx, y + dy, z + dz) - V(x, y, z) \quad ...(4.53)$$

The first term is expanded using Taylor series and neglecting the higher order terms, we get

$$dV = \left[V(x, y, z) + \frac{\partial V}{\partial x}dx + \frac{\partial V}{\partial y}dy + \frac{\partial V}{\partial z}dz\right] - V(x, y, z)$$

or
$$dV = \frac{\partial V}{\partial x}dx + \frac{\partial V}{\partial y}dy + \frac{\partial V}{\partial z}dz \quad ...(4.54)$$

$$= \left(\frac{\partial V}{\partial x}\hat{i} + \frac{\partial V}{\partial y}\hat{j} + \frac{\partial V}{\partial z}\hat{k}\right)\cdot(dx\hat{i} + dy\hat{j} + dz\hat{k})$$

or
$$dV = \nabla V \cdot \vec{dl} \quad ...(4.55)$$

on using the Eq. (4.52), the Eq. (4.55) takes form as

$$dV = -\vec{E}\cdot\vec{dl} \quad ...(4.56)$$

Since the surface is equipotential and the displacement $\vec{dl}$ is assumed to be along the tangent to the surface, thus, we have $dV = 0$, hence

$$\vec{E}\cdot\vec{dl} = 0 \quad ...(4.57)$$

The Eq. (4.57) shows that the $\vec{E}$ is always perpendicular to the equipotential surface.

Example 4.4. Find the electric potential at a point on a perpendicular bisector of a uniformly charged rod.

Solution: Consider a uniformly charged rod of length $2L$ as shown in fig. 4.9.

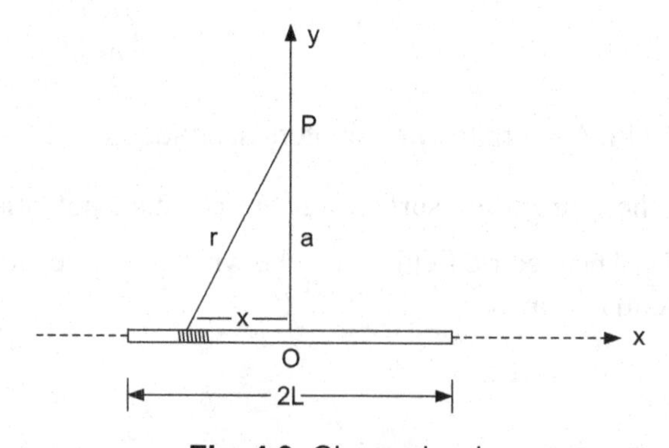

Fig. 4.9. Charged rod.

Let dx be the length element at a distance x from the origin. The length element has the charge

$$dq = \lambda\, dx$$

Thus, the potential at a point due to the charge element is given by

$$dV = \frac{1}{4\pi \in_0} \frac{dq}{r}$$

since $$r^2 = x^2 + a^2$$

$\therefore$ $$dV = \frac{1}{4\pi \in_0} \frac{\lambda dx}{(a^2 + x^2)^{1/2}}$$

Total potential is then,

$$V = \int dV$$

$$= \int_{-L}^{L} \frac{\lambda}{4\pi \in_0} \frac{dx}{(a^2 + x^2)^{1/2}}$$

$$= \frac{2\lambda}{4\pi \in_0} \int_0^L \frac{dx}{(a^2 + x^2)^{1/2}}$$

Using integration

$$\int \frac{dx}{(a^2 + x^2)^{1/2}} = \log_e[x + \sqrt{a^2 + x^2}]$$

Thus,

$$V = \frac{\lambda}{2\pi \in_0} \left[\log(x^2 + \sqrt{a^2 + x^2})\right]_0^L$$

or $$V = \frac{\lambda}{2\pi \in_0} \log_e\left(\frac{L + \sqrt{a^2 + L^2}}{a}\right)$$

4.8. PROPERTIES OF EQUIPOTENTIAL SURFACES

The properties of the equipotential surface are as follows;

(1) The equipotential surfaces are closer in the region of the strong electric field and farther in the region of weak electric field.

(2) The electric field lines are perpendicular to the equipotential surfaces.

(3) The field lines are pointed from the higher potential to the lower potential.

(4) No work is done in moving a charge particle along the equipotential surface.

(5) Two equipotential surfaces cannot intersect each other otherwise two values of the potential exist at the point of intersection, which is impossible.

(6) The equipotential surfaces for a constant electric field are the planes perpendicular to the electric field.

(7) For a positive point charge, the equipotential surfaces are concentric spheres as shown in fig. 4.7 (b).

(8) The tangential component E_t of the electric field is zero, if E_t is non zero, the work done in moving a test charge along the surface would not be zero.

4.9. ELECTRIC POTENTIAL ENERGY

Since the electric field is a conservative field, the external work done in moving a charge from a point A to B, does not depend on the path between A and B. Thus, the potential energy is a function of the position only. The electric potential energy at a point in the electric field is defined as the amount of work done in carrying a unit positive charge from the infinitely to that point. Mathematically, we may write,

$$U = -\int_{\infty}^{P} q\vec{E}\cdot\vec{dl} \quad ...(4.58)$$

Moreover, suppose that a charge q_o moves from the point A to the point B, then, the change in potential energy of the system is given by

$$dU = q_0(V_B - V_A) \quad ...(4.59)$$

or

$$U = -q_0\int_{A}^{B} \vec{E}\cdot\vec{dl} \quad ...(4.60)$$

If q_0 is negative, the potential energy U is positive. Thus, the potential energy of the system increases. From the Eq. (4.58), it is clear that $U = +W_{ext}$, that is, on moving the charge from infinity to the point P, work W is positive. That is, the work done by the external agent is positive, when assembling a system of charges. Now, we want to assemble a system of two charges q_1 and q_2 as shown in fig. 4.10.

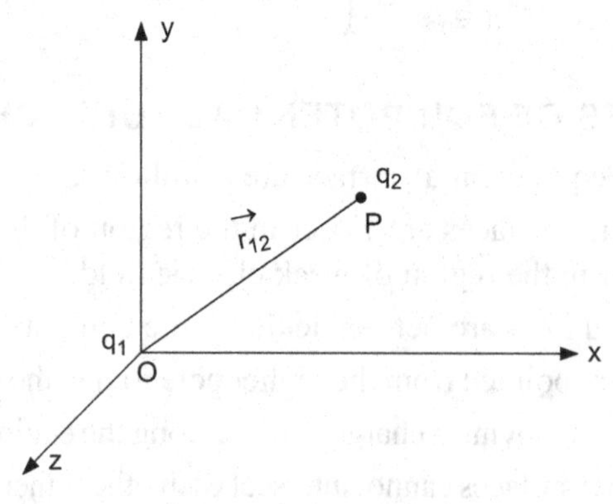

Fig. 4.10. Assembly of two charges q_1 and q_2.

The potential at the point P due to the charge q_1 is

$$V_1 = \frac{1}{4\pi\epsilon_0}\frac{q_1}{r_{12}} \quad ...(4.61)$$

Now, work done by the external agent in carrying the charge q_2 from the infinite to the point P is given by

$$W_1 = V_1 q_2 \quad ...(4.62)$$

or

$$U_{12} = W_1 = \frac{1}{4\pi\epsilon_0}\frac{q_1 q_2}{r_{12}} \quad ...(4.63)$$

If the sign of the charges q_1 and q_2 are same, there is an electrostatic repulsion between the charges and this is overcome by doing a positive work against the repulsion. In this way U_{12} positive. Further-more the third charge can be added to this system of two charges and the work is done against the interaction exerted by the charge q_1 and q_2, Fig. 4.11. Thus.

W_2 = Potential energy at the point Q due to the charges q_1 and q_2.

or

$$W_2 = \frac{1}{4\pi\epsilon_0}\frac{q_1 q_2}{r_{13}} + \frac{1}{4\pi\epsilon_0}\frac{q_2 q_3}{r_{23}} \quad ...(4.64)$$

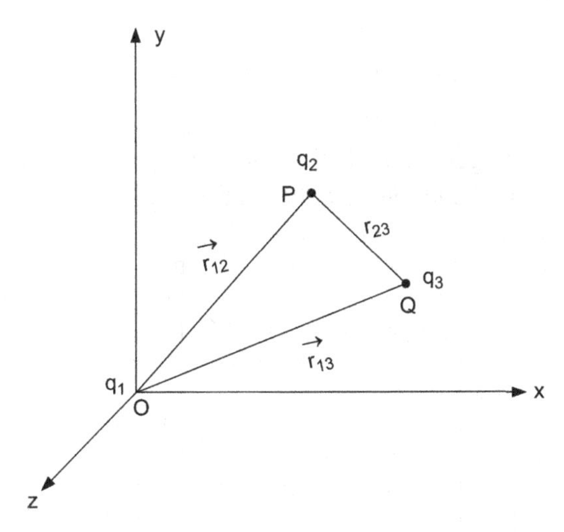

Fig. 4.11. Assembly of three charges q_1, q_2 and q_3.

The potential energy for the assembly of three charges will be

$$U = W_1 + W_2$$

$$= \frac{1}{4\pi\epsilon_0}\left(\frac{q_1 q_2}{r_{12}} + \frac{q_2 q_3}{r_{23}} + \frac{q_1 q_3}{r_{13}}\right) \quad ...(4.65)$$

In this way, we may extend the above expression, Eq. (4.65) for the N point charges

$$\boxed{U = \frac{1}{2} \cdot \frac{1}{4\pi\epsilon_0} \sum_{\substack{i, j=1 \\ i \neq j}}^{N} \frac{q_i q_j}{r_{ij}}} \quad ...(4.66)$$

The expression for U contains $\frac{1}{2}$ because summation counted each pair twice. For example.

$$\frac{q_1 q_2}{r_{12}} = \frac{q_2 q_1}{r_{21}}$$

To avoid this ambiguity, dividing the expression by 2. Moreover,

$$U = \frac{1}{2} \sum_i q_i \left(\frac{1}{4\pi\epsilon_0} \sum_j \frac{q_j}{r_{ij}} \right) \quad ...(4.67)$$

or

$$U = \frac{1}{2} \sum_i q_i V \quad ...(4.68)$$

where V_i is the potential at a point i due to all charges. The unit of the electrostatic potential energy is electronvolt (eV).

$$1 \text{ eV} = e \times 1 \text{ volt}$$
$$= 1.6 \times 10^{-19} \text{ C} \times 1 \text{ volt}$$

or

$$1 \text{ eV} = 1.6 \times 10^{-19} \text{ joules} \quad ...(4.69)$$

Example 4.5. Four charges are placed at the corners of a square of side a as shown in fig. 4.12. Find the potential energy of the system.

Solution:

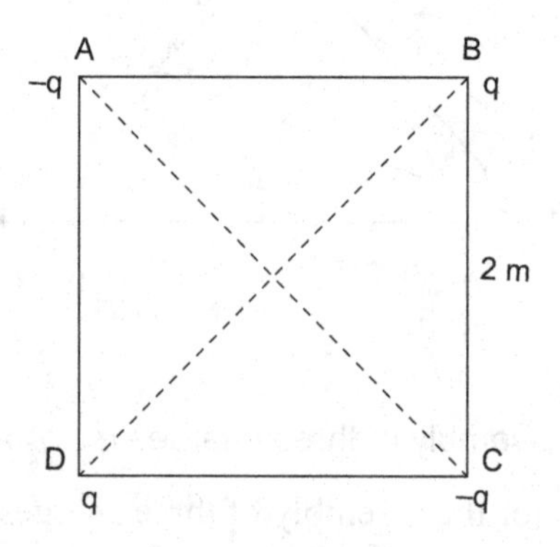

Fig. 4.12. A system of charges.

Let $ABCD$ be a square of side a, we have

$$AB = BC = CD = DA = a$$

and $$AC = BD = \sqrt{2}\,a$$

The potential energy of the system is

$$U = \frac{1}{4\pi\epsilon_0}\left[\frac{-q^2}{AB} - \frac{q^2}{BC} - \frac{q^2}{CD} - \frac{q^2}{DA} + \frac{q^2}{AC} + \frac{q^2}{BD}\right]$$

$$= \frac{q^2}{4\pi\epsilon_0}\left[-\frac{1}{a} - \frac{1}{a} - \frac{1}{a} - \frac{1}{a} + \frac{1}{\sqrt{2}a} + \frac{1}{\sqrt{2}a}\right]$$

$$= \frac{q^2}{4\pi\epsilon_0}\left[-\frac{4}{a} + \frac{1}{\sqrt{2}a}\right]$$

$$U = \frac{q^2}{4\pi\epsilon_0}\frac{(\sqrt{2}-4)}{a}$$

Example 4.6. A charge q_0 is placed at a point A in a uniform electric field as shown in Fig. 4.13. This charge is allowed to move from a point A to the point B and covers a distance d. Find the

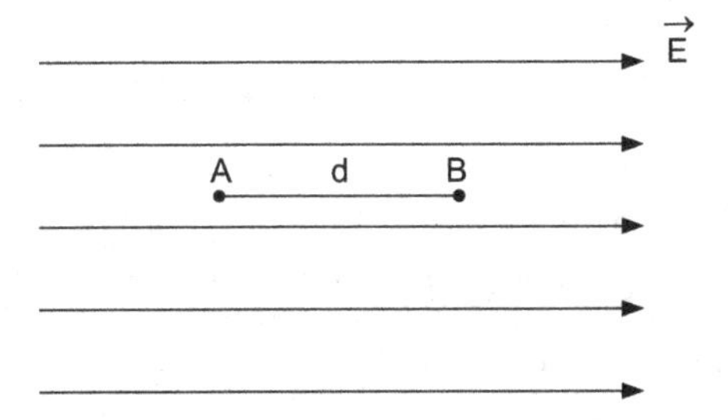

Fig. 4.13. Charge in the electric field.

(1) potential difference, $V_B - V_A$.

(2) which point is at higher potential.

(3) workdone is moving a charge q_0 from A to B.

(4) change in potential energy of the charge.

(5) change in kinetic energy of the charge

(6) velocity of the charge.

Solution: Suppose that an electric force $\vec{F} = q_0\vec{E}$ carries a positive charge q_0 from a point A to B, in the direction parallel to the field. The distance between points A and B is d. In carrying the charge q_0 from A to B the work done is

(1) $$W = q_0(V_B - V_A)$$

or $$V_B - V_A = \frac{W}{q_0} = -\int_A^B \vec{E} \cdot \vec{dl}$$

$$= -Ed$$

Thus,

$$V_B - V_A = -Ed$$

(2) Since $V_B - V_A = -Ed$, a negative quantity, thus the point A is at higher potential.

(3) The workdone in moving a charge q_0 from A to B is

$$W = q_0(V_B - V_A)$$

or $$W = -q_0 Ed$$

(4) When charge q_o moves in the direction of the electric field, the kinetic energy of the charge increases and the electric field lines always point from a higher potential to the lower potential, the change in potential energy of the system is

$$dU = q_0(V_B - V_A)$$

$$= -q_0 Ed$$

If q_o is positive, dU is negative, Thus the potential energy of the charge decreases.

(5) The change in K.E. is

$$K_B - K_A = q_0 dV = q_0(V_B - V_A)$$

or $$\Delta K = -q_0 Ed.$$

(6) Since $K_A = 0$,

Thus, $$K_B = \frac{1}{2}mv^2$$

where m is the mass of the charge particle,

or $$\frac{1}{2}mv^2 = q_0 Ed$$

or $$v = \frac{2q_0 Ed}{m}$$

Example 4.7. A charge q is situated at the vertix A of an equilateral triangle of side a as shown in fig. 4.14. compute the work done in carrying a charge q_0 from the point B to C.

Solution:

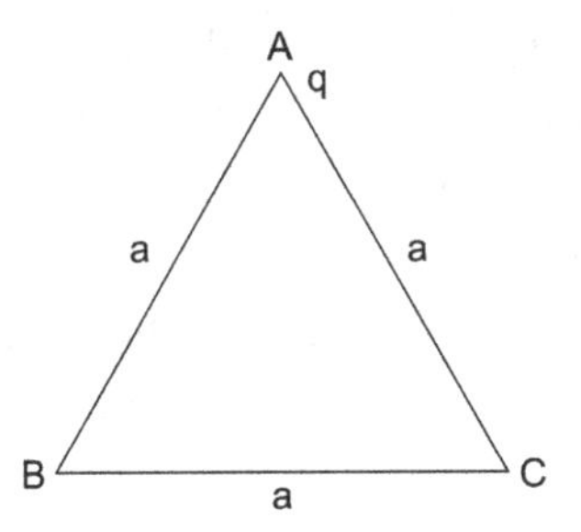

Fig. 4.14. Equilateral triangle.

The potentials at B and C are given by

$$V_B = \frac{1}{4\pi\epsilon_0}\frac{q}{a}$$

and
$$V_C = \frac{1}{4\pi\epsilon_0}\frac{q}{a}$$

The work done in moving a charge q_0 from the point B to C is given by

$$W_{BC} = q_0(V_C - V_B) = 0.$$

since,
$$V_C = V_B.$$

Example 4.8. Suppose that the eight identical water drops are charged to a same potential V. If these eight drops coalesce into one drop, Find its potential.

Solution: Let r and R be the radius of small and large drops respectively. Then,

$$\frac{4}{3}\pi R^3 = 8\cdot\frac{4}{3}\pi r^3$$

or
$$R = 2r$$

The potential of each small drop is

$$V = \frac{1}{4\pi\epsilon_0}\frac{q}{r}$$

and the potential of the large drop is then

$$V' = \frac{1}{4\pi\epsilon_0}\frac{q}{R}$$

where $Q = 8q$ and $R = 2r$,

Thus,
$$V' = \frac{1}{4\pi\epsilon_0}\cdot\frac{8q}{2r}$$

$$= 4\cdot\frac{1}{4\pi\epsilon_0}\frac{q}{r}$$

or
$$V' = 4V.$$

Example 4.9. A non conducting sphere of the radius a has a uniform charge density ρ. Find the potential

(a) at a point outside the sphere.

(b) at a point inside the sphere.

Solution: Consider a sphere of the radius a as shown in fig. 4.15.

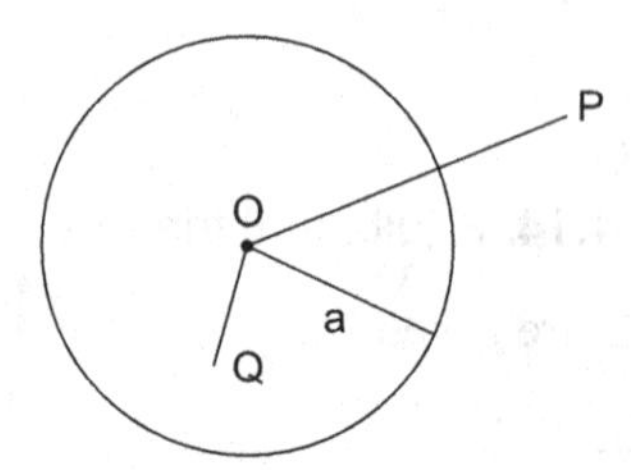

Fig. 4.15. None-conducting sphere.

The electric fields for the point P and Q are given by

$$E = \begin{cases} \dfrac{q}{4\pi\epsilon_0 r^2}\hat{r} \text{ for } r > a \\ \dfrac{q}{4\pi\epsilon_0}\dfrac{\vec{r}}{a^3}\hat{r} \text{ for } r > a \end{cases}$$

where $$\rho = \frac{q}{\left(\frac{4}{3}\right)\pi a^3}$$

(a) The electric potential at the point P is given by

$$V_P(r) - V(\infty) = -\int_\infty^r \vec{E}\,\vec{dr}$$

$$= -\int_\infty^r \frac{q}{4\pi\epsilon_0}\frac{dr}{r^2}$$

$$= \frac{1}{4\pi\epsilon_0}\frac{q}{r}$$

Since $V(\infty) = 0$, again, the potential at the point Q is then,

$$V_Q - V(\infty) = -\int_\infty^r \vec{E}\,\vec{dr}$$

or $$V_Q - V(\infty) = -\int_\infty^a E\,(r > a)\,dr - \int_a^r E\,(r < a)\,dr$$

$$= -\int_\infty^a \frac{q}{4\pi\epsilon_0}\frac{dr}{r^2} - \int_a^r \frac{q}{4\pi\epsilon_0}\frac{rdr}{a^3}$$

$$= \frac{1}{4\pi\epsilon_0} \cdot \frac{q}{a} - \frac{1}{4\pi\epsilon_0} \frac{q}{2a^3}(r^2 - a^2)$$

or $$V_Q = \frac{1}{8\pi\epsilon_0} \frac{q}{a}\left(3 - \frac{r^2}{a^2}\right)$$

Example 4.10. If a charge q is distributed uniformly over the surface of a conducting sphere of the radius a. Find the potential inside and outside the sphere. Sketch the potential also.

Solution: The electric field for a conducting sphere is given by

$$E = \begin{cases} \frac{1}{4\pi\epsilon_0} \cdot \frac{q}{r^2}\hat{r}, & r > a \\ 0, & r < a \end{cases}$$

Now, the electric potential for the region $r > a$ is

$$V(r) - V(\infty) = \int_\infty^r \frac{q}{4\pi\epsilon_0} \frac{dr}{r^2}$$

$$V(r) = \frac{1}{4\pi\epsilon_0} \frac{q}{r}$$

The electric potential for the region $r < a$ is given by

$$V(r) - V(\infty) = \int_\infty^a E(r > a)\, dr - \int_a^r E(r < a)\, dr$$

$$= -\int_\infty^a \frac{q}{4\pi\epsilon_0} \frac{dr}{r^2}$$

$$V(r) = \frac{1}{4\pi\epsilon_0} \frac{q}{a}$$

A plot of $V(r)$ versus r is shown in fig. 4.16.

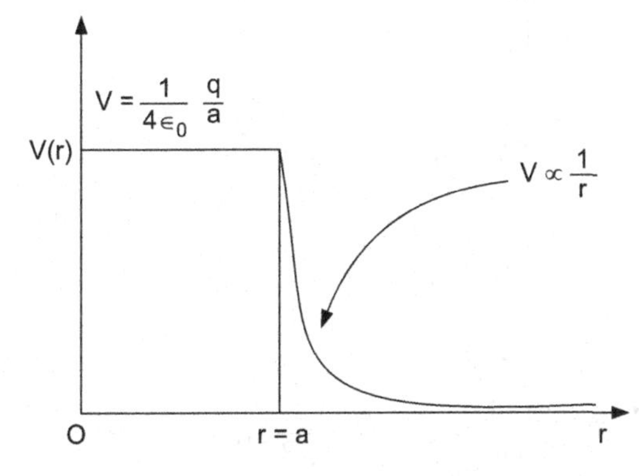

Fig. 4.16. Plot of V with r.

Example 4.11. Find the potential at a point lying on the axis of a uniformly charged ring of the radius *a*.

Solution: Consider a uniformly charged ring of radius a as shown in fig. 4.17. Let *dl* be a charge element on the circumference of the ring. The charge of the element is

$$dq = \lambda dl$$

where λ is a line charge density.

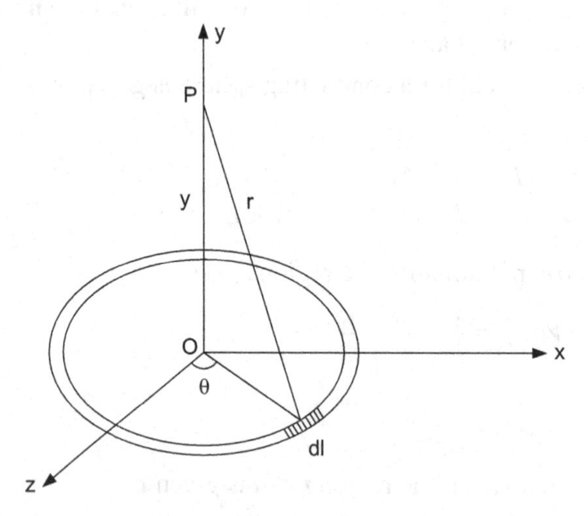

Fig. 4.17. Charged ring.

Thus, the potential at a point on the *y*-axis due to charge element *dl*

$$dV = \frac{1}{4\pi\epsilon_0}\frac{dq}{r}$$

$$= \frac{1}{4\pi\epsilon_0}\frac{\lambda dl}{(a^2+y^2)^{1/2}},$$

net potential at *P* is then obtained as

$$V = \int dV$$

$$= \frac{1}{4\pi\epsilon_0}\frac{\lambda}{(a^2+y^2)^{1/2}}\int dl$$

$$= \frac{1}{4\pi\epsilon_0}\frac{\lambda . 2\pi a}{(a^2+y^2)^{1/2}}$$

since $\lambda = \frac{a}{2\pi a},$

$\therefore$ $$V = \frac{1}{4\pi\epsilon_0}\frac{q}{(a^2+y^2)^{1/2}}, a < 1$$

Example 4.12. Consider two spherical conductors with the radii r_1 and r_2. A charge q is uniformly distributed on these two conductors and both are connected by a wire. Find the

(1) charge on the sphere

(2) ration of electric field (E_1/E_2).

Solution: Let q_1 and q_2 be the charges on the spheres of the radii r_1 and r_2 respectively. Since charge flows continuously until an equilibrium is reached, both conductors are at same potential, Fig. 4.18.

$$\therefore \qquad q = q_1 + q_2$$

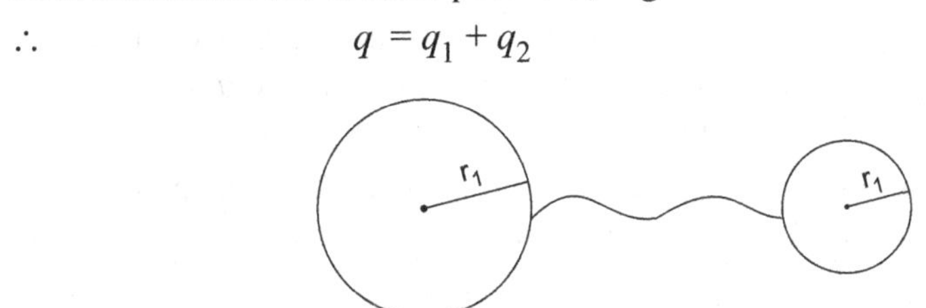

Fig. 4.18. Two spherical conductors.

we have,

$$V_1 = V_2$$

$$\frac{1}{4\pi\epsilon_0}\frac{q_1}{r_1} = \frac{1}{4\pi\epsilon_0}\frac{q_2}{r_2}$$

or

$$\frac{q_1}{q_2} = \frac{r_1}{r_2}$$

and

$$q = q_1 + q_2$$

Thus, we get

$$q_1 = \frac{r_1}{r_1 + r_2}q$$

and

$$q_2 = \frac{r_2}{r_1 + r_2}q$$

(2) The electric fields at the surface of both spheres are

$$E_1 = \frac{1}{4\pi\epsilon_0}\frac{q_1}{r_1^2}$$

and

$$E_2 = \frac{1}{4\pi\epsilon_0}\frac{q_2}{r_2^2}$$

on dividing, we get

$$\frac{E_1}{E_2} = \left(\frac{r_2}{r_1}\right)^2 \frac{q_1}{q_2}$$

$$= \left(\frac{r_2}{r_1}\right)^2 \frac{r_1}{r_2}$$

or
$$\frac{E_1}{E_2} = \frac{r_2}{r_1}$$

Example 4.13. Consider two concentric spherical conducting shells with the radii r_1 and r_2, where $r_1 > r_2$. The outer shell consists of a charge q. Compute the charge of the inner shell if it is grounded.

Solution: Suppose that Q is the charge on the inner shell. Then, the potential of the inner shell is the sum of the potentials due to charges q and Q. Thus,

$$V = \frac{1}{4\pi\epsilon_0}\cdot\frac{q}{r_1} + \frac{1}{4\pi\epsilon_0}\cdot\frac{Q}{r_2}$$

Since inner shell is grounded, $V = 0$

$$\frac{q}{r_1} + \frac{Q}{r_2} = 0$$

or
$$Q = -\left(\frac{r_1}{r_2}\right)q$$

Example 4.14. In spherical polar coordinates, the potential at a point is given by

$$V = \frac{V_0 \cos\theta \sin\phi}{r^2}$$

where V_o is a constant. Find the components of the electric field.

Solution: In spherical polar coordinates, the components of the electric field intensity are given by

$$E_r = -\frac{dV}{dr},$$

$$E_\theta = -\frac{1}{r}\frac{dV}{d\theta},$$

and

$$E_\phi = -\frac{1}{r\sin\theta}\frac{dV}{d\phi}$$

$$\therefore \quad E_r = -\frac{d}{dr}\left(\frac{V_0\cos\theta\sin\phi}{r^2}\right)$$

$$= \frac{2V_0 \cos\theta \sin\phi}{r^3}$$

$$E_\theta = -\frac{1}{r}\frac{d}{d\theta}\left(\frac{V_0 \cos\theta \sin\phi}{r^2}\right)$$

$$= \frac{V_0 \sin\theta \sin\phi}{r^3}$$

and

$$E_\phi = -\frac{1}{r\sin\theta}\frac{d}{d\phi}\left(\frac{V_0 \cos\theta \sin\phi}{r^2}\right)$$

$$= \frac{-V_0 \cot\theta \cos\phi}{r^3}$$

Example 4.15. Suppose that the two rings of some radius are placed coaxially at a distance a. If q_1 and q_2 are the charges of the rings, compute the work done in bringing a charge q_0 from the centre of one ring to the another.

Solution: Let V_1 and V_2 be the potentials at the centres of rings and a be the radius of each ring, fig. 4.19, then

$$V_1 = \frac{1}{4\pi\epsilon_0}\left(\frac{q_1}{a} + \frac{q_2}{\sqrt{2}\,a}\right)$$

and

$$V_2 = \frac{1}{4\pi\epsilon_0}\left(\frac{q_2}{a} + \frac{q_1}{\sqrt{2}\,a}\right)$$

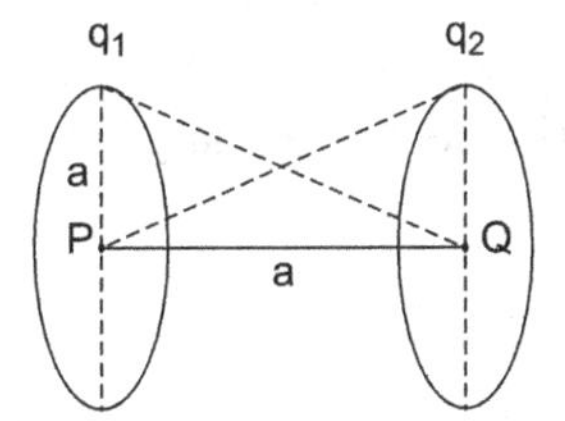

Fig. 4.19. Coaxial rings.

The potential difference

$$dV = V_1 - V_2$$

$$= \frac{(q_1 - q_2)}{4\pi\epsilon_o\, a}\left[\frac{\sqrt{2}-1}{\sqrt{2}}\right]$$

Now, work done in bringing a charge q_0 from P to Q will be

$$W = q_0\, dV$$

$$= \frac{q_0(q_1 - q_2)}{4\pi \in_0 a}\left(\frac{\sqrt{2}-1}{\sqrt{2}}\right)$$

EXERCISES

4.1. Describe the potential and potential difference and show that

$$V(r) - V(\infty) = -\int_{\infty}^{r} \vec{E} \cdot \vec{dr}$$

4.2. Derive an expression for the potential at any point at a distance r from the centre of the electric dipole.

4.3. What is the difference between electric potential and the potential energy and obtain an expression for the electric potential energy of a system consisting of N point charges.

4.4. The electric field for a non-conducting sphere having uniform charge density is given by

$$\vec{E} = \begin{cases} \frac{Kq}{r^2}\hat{r} & \text{for } r > a \\ \frac{Kq\,\vec{r}}{a^3} & \text{for } r > a \end{cases}$$

where a is the radius of the sphere. Find the electric potential and sketch the potential also.

4.5. Show that the total work needed to charge a sphere of radius a to q_0 is given by

$$W = \frac{1}{2} q_0 V$$

Hint:

$$V = \frac{q}{4\pi\in_0 a}$$

$$W = \int_0^{q_0} V\, dq = \int_0^{q_0} \frac{1}{4\pi\in_0} \frac{q}{a}\, dq = \frac{1}{2} q_0 V$$

4.6. For a conservative field, show that

$$E = -\nabla V$$

4.7. The electric potential at a point in a plane is given by

(a) $$V = \frac{\alpha y}{(y^2 + z^2)^{3/12}} + \frac{\beta}{(y^2 + z^2)^{1/2}}$$

where α, and β are constants. Find the components of the electric field intensity.

4.8. A particle of charge $2C$ is placed in a field as given by

$$\vec{E} = a\,(yz\hat{i} + zx\hat{j} + xy\hat{k})$$

where $a = 4$ N/m^2. If the corresponding potential at the origin is 15V, Compute the potential at the point (3, 2, 4).

Hint: $$V = -\int E \cdot dr = xyz$$

4.9. The four charges are placed at the four corners of a square of side a in the xy plane as shown in fig. 4.20.

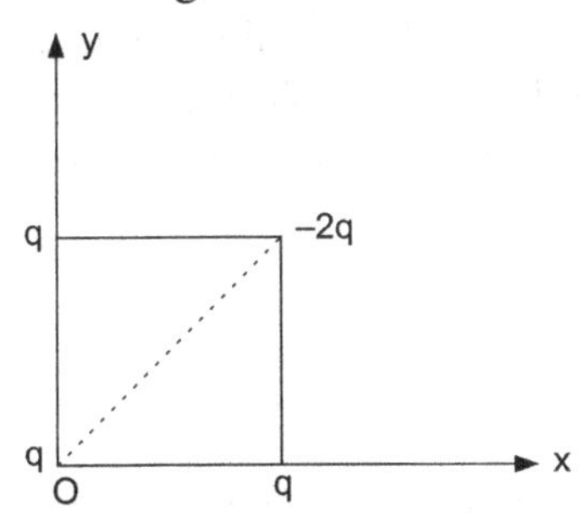

Fig. 4.20. A square.

Hint: $\vec{p}$ acts from $-q$ to $+q$, $\vec{p}_1 = -aq\hat{i}$, $\vec{p}_2 = -aq\hat{j}$, $\vec{p} = \vec{p}_1 + \vec{p}_2$

4.10. Consider twenty seven identical water drops at the same potential V and all the drops coalesce to form a big drop. Find the potential of the big drop. If $V = 100$ Volts estimate the potential of the big drop. **Ans.** 900V.

4.11. Two charges 2μC and –6μC are separated by a distance of 4 m. Find the position of the null point from the charge 2μC. **Ans.** 1.0 m.

4.12. Three point charges 2μC, –4μC and 6μC are placed at the three vertices of an equilateral triangle of side 4 m. Compute the

(a) Work needed to assemble the system

(b) electrostatic potential energy.

4.13. Four charges 2μC, –1μC, 3μC and 6μC are placed at the corners of a square of side 1 m. Find the electrostatic energy.

4.14. A disc has a uniform charge density σ. Compute the potential at a point lying on the axis of the disc of the radius a, sketch the potential also.

4.15. Find the potential at the centre of a uniformly charged ring of the radius a.

4.16. How much work is done to assemble eight identical point charges, each of magnitude q, at the corners of a cube of side a.

Hint: $W = q$V, find electrostatic energy.

4.17. The electric field is given as

$$\vec{E} = 2\hat{i} + 3\hat{j} + 4\hat{k}$$

Compute the potential difference between the points $A(1, 2, 0)$ and $B(2, 3, 3)$.

Hint: $V_B - V_A = -\int \vec{E}\, \vec{dr}$

$$= -\int E_x\, dx - \int E_y\, dy - \int E_z\, dz = 17 \text{ volts.}$$

4.18. Consider two hollow concentric conducting spheres of radii r_1 and r_2 as shown in fig. 4.21. If the inner sphere is charged with q and the outer sphere is grounded. Compute the potential at any point P in the region $r > r_2$.

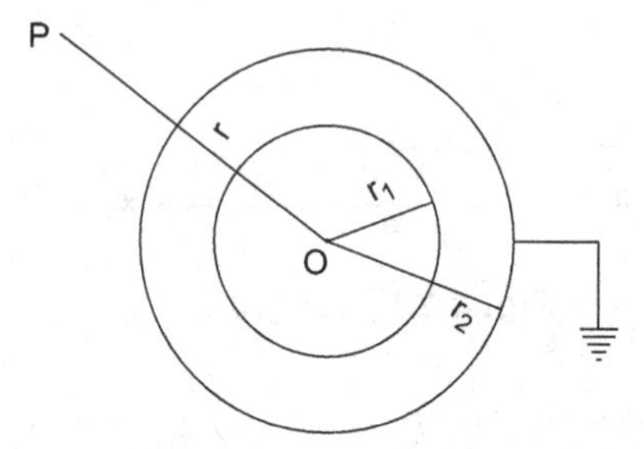

Fig. 4.21. Hollow conducting spheres.

Hint: The outer sphere gets charge $-q$, potential at P is $V = \frac{Kq}{r} - \frac{Kq}{r} = 0$

4.19. Compute the electric field for yukawa potential,

$$V = q_0 V_0\, e^{-ar}$$

4.20. Compute the electric field at a point (1, 1, 1) from the origin for the following potential

(a) $V = x^2\hat{i} + y^2\hat{j} + z^2\hat{k}$

(b) $V = x^2\hat{i} + zy^2\hat{j} + \hat{k}$

4.21. Suppose that there is a spherical cavity of the radius a in a conductor as shown in fig. 4.22. If the conductor has a uniform surface charge density

σ, compute the electric potentials and fields at the point P, Q and R. Where the potential at the point R is V_0.

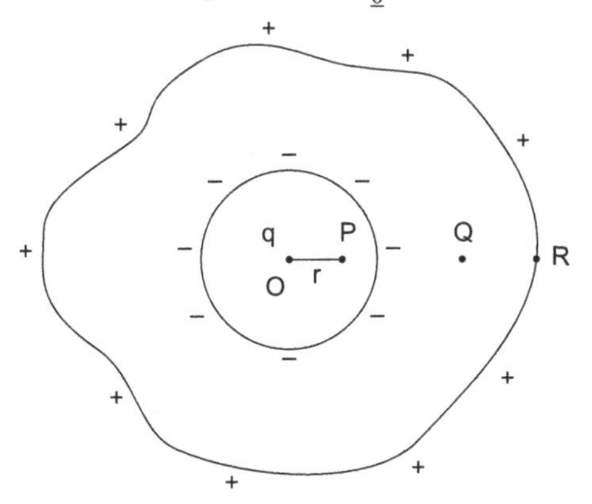

Fig. 4.22. A cavity.

Solution: Due to induction, the cavity surface has $-q$ charge.

The potential at R is V_0

The electric field at R is

$$E_R = \frac{\sigma}{\epsilon_0}$$

The electric field inside the conductor (at point Q) is

$$E_a = 0$$

and the potential $V_Q = V_0$

Moreover, potential at the point P will be

$$V_P = \frac{q}{4\pi\epsilon_0 r} - \frac{q}{4\pi\epsilon_0 a} + V_0$$

and

$$E_P = -\frac{dV_P}{dr}$$

$$= \frac{q}{4\pi\epsilon_0} \frac{1}{r^2} \hat{r}$$

■ ■ ■

CHAPTER 5 Methods for the Solution of Electrostatic Problems

We have already solved the various potential or field problems with known charge distribution. If the potential is known, charge q, electrostatic force, electric field etc may be computed easily. In solving such problems, an integration method is employed, that is, in other words, we take an integration over a definite charge distribution. We known that the electric field is given by

$$E = -\nabla V \qquad ...(5.1)$$

that is, E can be taken as a negative gradient of the potential, this is because, the electric field is a conservative field. In this way, we can assume that the potential function V is associated with a given region of space. If the region of the space is not bounded, it has some difficulty in obtaining the solution to the potential problems. That is, there is a no unique solution to the problem. The uniqueness theorem helps in obtaining the solution to the electrostatic problems. This solution can be obtained using certain boundary conditions. In this way, a complex problem may be solved to a high degree of accuracy. Thus, Laplace and Poisson's equations provide a good solution of the potential problems with known boundary conditions.

5.1. UNIQUENESS THEOREM

The uniqueness theorem states that, within a given boundary, if these exists a solution to the Laplace equation is only the unique and exact solution. In other words, a solution to the potential equation satisfying the boundary conditions is only an exact solution. However, there can be the number of solutions of the Laplace equation, but there will be a solution which satisfies the boundary conditions of the particular problem. The uniqueness theorem may also be applied to the Poisson's equation with known boundary conditions. We can prove this theorem by contradiction.

Suppose that there are two solutions V_1 and V_2 for the Laplace equation and both solutions satisfy the boundary conditions. Now, we have

$$\nabla^2 V = 0 \qquad ...(5.2)$$

Since V_1 and V_2 both are solutions to the Eq. (5.2), we write

$$\left.\begin{aligned} \nabla^2 V_1 &= 0 \\ \text{and} \qquad \nabla^2 V_2 &= 0 \end{aligned}\right\} \qquad ...(5.3)$$

At the boundary, the solutions V_1 and V_2, and their normal derivatives are equal, that is,

$$\left.\begin{aligned} V_1 &= V_2 \\ \text{and} \qquad \nabla V_1 &= \nabla V_2 \end{aligned}\right\} \qquad ...(5.4)$$

Now, according to Gauss's theorem,

$$\iint_S \vec{F}\cdot\vec{dS} = \iiint_V \nabla\cdot\vec{F}\, dv \qquad ...(5.5)$$

Taking $\vec{F} = V\nabla V$, and substituting in the Eq. (5.5.) we get

$$\iint_S V\nabla V\cdot\vec{dS} = \iiint_V \nabla\cdot(V\nabla V)\, dV$$

$$= \iiint_V (\nabla V\cdot\nabla V + V\nabla^2 V)\, dV$$

$$= \iiint_V [(\nabla V)^2 - V\nabla^2 V]\, dV \qquad ...(5.6)$$

Now, we substitute $V = V_1 - V_2$ in the Eq. (5.6) we get,

$$\iint_S (V_1 - V_2)\,\nabla\,(V_1 - V_2)\cdot\vec{dS} = \iiint_V [\{\nabla\cdot(V_1 - V_2)\}^2 + (V_1 - V_2)\,\nabla^2\,(V_1 - V_2)]\, dV \qquad ...(5.7)$$

Using boundary condition on the surface

$$V_1 = V_2$$

then the Eq. (5.7) reduces to

$$\iiint_V [\nabla(V_1 - V_2)]^2\, dV = 0 \qquad ...(5.8)$$

or
$$\nabla(V_1 - V_2) = 0 \qquad ...(5.9)$$

or
$$\nabla V_1 = \nabla V_2 \qquad ...(5.10)$$

or
$$V_1 = V_2 + C \text{ (a constant)} \qquad ...(5.11)$$

The Eq. (5.11) shows that the two potential solutions V_1 and V_2 are differing with a positive constant C only and it does not make any contribution to the gradient of the potential, since

$$\nabla C = 0 \qquad ...(5.12)$$

Thus, we conclude that the potentials V_1 and V_2 provide the same electric field $\vec{E}$. Hence, there exists a unique solution of the Laplace equation. The uniqueness theorem may be proved using Poisson equation with the similar steps. When we solve any potential problem, we should have following in our mind.

(1) A potential equation which is to be solved.

(2) The region of space, that is, for example, a sphere, a cone, an infinite plate etc.

(3) Boundary conditions associated with the problem.

5.2. POISSON'S AND LAPLACE'S EQUATIONS

In electrostatics, Gauss's law in differential form is given by

$$\nabla \cdot \vec{E} = \frac{\rho}{\epsilon_0} \quad \text{...(5.13)}$$

since electric field is a conservative field, we can write

$$\vec{E} = -\nabla V \quad \text{...(5.14)}$$

Substituting the value of $\vec{E}$ from the eq. (5.14) in the Eq. (5.13), we get

$$\nabla \cdot (\nabla V) = -\frac{\rho}{\epsilon_0}$$

or

$$\boxed{\nabla^2 V = -\frac{\rho}{\epsilon_0}} \quad \text{...(5.15)}$$

The Eq. (5.15) is known as Poisson's equation, and ρ is volume charge density for free charges.

In Cartesian Coordinates System

Since,

$$\nabla^2 = \frac{\partial^2}{\partial x^2} + \frac{\partial^2}{\partial y^2} + \frac{\partial^2}{\partial z^2}$$

thus,

$$\frac{\partial^2}{\partial x^2} + \frac{\partial^2}{\partial y^2} + \frac{\partial^2}{\partial z^2} = -\frac{\rho}{\epsilon_0} \quad \text{...(5.16)}$$

In Cylindrical Coordinates

The transformation equations are given by

$$x = r\cos\phi \qquad r^2 = x^2 + y^2$$

$$y = r\sin\phi \qquad \tan\phi = \frac{y}{x} \quad \text{...(5.17)}$$

Therefore,

$$\nabla^2 V = \frac{1}{r}\frac{\partial}{\partial r}\left(r\frac{\partial V}{\partial r}\right) + \frac{1}{r^2}\frac{\partial^2 V}{\partial \phi^2} + \frac{\partial^2 V}{\partial z^2} = -\frac{\rho}{\epsilon_0} \quad ...(5.18)$$

In Spherical Polar Coordinates

The transformation equations are

$$\begin{aligned} x &= r \sin\theta \cos\phi \\ y &= r \sin\theta \sin\phi \\ z &= r\cos\theta \end{aligned} \quad ...(5.19)$$

and

$$\nabla^2 V = \frac{1}{r}\frac{\partial}{\partial r}\left(r^2\frac{\partial V}{\partial r}\right) + \frac{1}{r^2 \sin\theta}\frac{\partial}{\partial \theta}\left(\sin\theta\frac{\partial V}{\partial \theta}\right) + \frac{1}{r^2\sin^2\theta}\frac{\partial^2 V}{\partial \phi^2} = \frac{-\rho}{\epsilon_0} \quad ...(5.20)$$

In a region where there is no free charge, $\rho = 0$ thus the Eq. (5.15) reduces to

$$\boxed{\nabla^2 V = 0} \quad ...(5.21)$$

which is Laplace's equation. Thus, the following steps are taken to solve the potential equation.

(a) Since V is a function of variables $V = V(x, y, z)$ or $V = V(r\ \phi\ z)$ or $V = V(r, \theta, \phi)$, solve it by integrating directly.

(b) If V has only one variable it can be solved using step (a). Otherwise solve it using he method of separation of variables.

(c) Apply the suitable boundary conditions to get a unique solution of the potential equation.

(d) On obtaining the expression for the potential V, we can obtain other physical parameters given as,

(1) **Electric Field:**

$$\vec{E} = -\nabla V$$

$$E_r = -\frac{\partial V}{\partial r}$$

$$E_\theta = -\frac{1}{r}\frac{\partial V}{\partial \theta} \quad ...(5.22)$$

and $$E_\phi = -\frac{1}{r\sin\theta}\frac{\partial V}{\partial \phi}$$

(2) **Surface Charge Density.**

$$\sigma = -\epsilon_0 \frac{dV}{dr} \quad ...(5.23)$$

(3) **Electric Displacement vector:**

$$\vec{D} = -\epsilon_0 \vec{E} \quad \text{...(5.24)}$$

(4) **Current Density:**

$$\vec{J} = \sigma \vec{E} \quad \text{...(5.25)}$$

Example 5.1. Prove that $V = \frac{k}{r}$ is a solution of the Laplace equation.

Solution: Laplace equation is given by

$$\nabla^2 V = 0$$

The radial part of the equation is

$$\frac{1}{r^2}\frac{d}{dr}\left(r^2 \frac{dV}{dr}\right) = 0$$

since, $$V = \frac{k}{r},$$

$$\frac{d}{dr}\left(r^2 \frac{d}{dr}\cdot\frac{k}{r}\right) = 0$$

or $$-\frac{d}{dr}\left(r^2 \cdot \frac{k}{r^2}\right) = 0$$

or $$-\frac{dk}{dr} = 0$$

since k is a constant, so differentiation of a constant is zero. Thus, $V = \frac{k}{r}$ satisfies the Laplace equation.

Example 5.2. In spherical polar coordinates, the potential is given by $V = V_0 \ln r$, compute the volume charge density.

Solution: The Poisson equation is

$$\nabla^2 V = -\frac{\rho}{\epsilon_0}$$

in spherical polar coordinates, it is given by

$$\frac{1}{r^2}\frac{\partial}{\partial r}\left(r^2 \frac{\partial V}{\partial r}\right) = -\frac{\rho}{\epsilon_0}$$

or $$\frac{1}{r^2}\frac{\partial}{\partial r}\left(r^2 \cdot \frac{V_0}{r}\right) = -\frac{\rho}{\epsilon_0}$$

or $$\frac{1}{r^2}\cdot V_0 = -\frac{\rho}{\epsilon_0}$$

or $$\rho = -\frac{\epsilon_0 V_0}{r^2}$$

Example 5.3. The plates of a parallel plate capacitor are located at $x = 0$ and $x = L$. The plate at $x = 0$ is grounded and the plate at $x = L$ has a constant potential V_0. Compute the

(a) potential

(b) electric field

(c) surface charge density σ.

Solution:

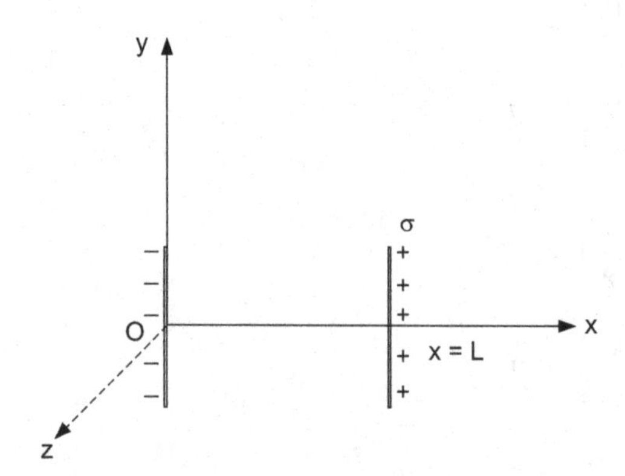

Fig. 5.1. Capacitor.

The Laplace equation is

$$\frac{d^2V}{dx^2} + \frac{d^2V}{dy^2} + \frac{d^2V}{dz^2} = 0$$

since there is no variation of the potential V in y and z directions, we write,

$$\frac{d^2V}{dx^2} = 0$$

on integration, we have,

$$\frac{dV}{dx} = A$$

Where A is the constant of integration, on integrating it again, we get

$$V = Ax + B$$

To determine the constants A and B we require boundary conditions which are

$$V = 0 \quad \text{at } x = 0$$
$$V = V_0 \quad \text{at } x = L$$

Applying boundary conditions, we have

$$V = 0 \quad \text{at } x = 0$$

$\Rightarrow$ $$B = 0$$

Thus,

$$V = Ax$$

Again, $$V = V_0 \quad \text{at} \quad x = L$$

$\Rightarrow$ $$A = \frac{V_0}{L}$$

Thus we get potential as

$$V = \frac{V_0 x}{L}$$

(b) The electric field

$$E_x = -\frac{dV}{dx}\hat{i}$$
$$= \frac{-V_0}{L}\hat{i}$$

(c) The surface charge density is given by

$$\sigma = -\epsilon_0 \frac{dV}{dx}$$

or $$\sigma = -\frac{\epsilon_0 V_0}{L}$$

and $$\sigma = \frac{\epsilon_0 V_0}{L}$$

5.3. SOLUTION OF LAPLACE'S EQUATION IN RECTANGULAR COORDINATES

The Laplace equation, in rectangular coordinates is given by

$$\nabla^2 V = \frac{\partial^2 V}{dx^2} + \frac{d^2 V}{dy^2} + \frac{d^2 V}{dz^2} \qquad \text{...(5.26)}$$

The solution of Laplace equation can be obtained in the rectangular coordinates satisfying the required boundary conditions by the method of separation of the variables.

We assume that the Eq. (5.26) has the solution of the form

$$V(x, y, z) = V_1(x)\, V_2(y)\, V_3(z) \qquad \text{...(5.27)}$$

where $V_1(x)$, $V_2(y)$ and $V_3(z)$ are the functions of x, y, and z respectively. On substitution $V(x, y, z)$ in the Eq. (5.26) we get,

$$V_2V_3\frac{d^2V_1}{dx^2}+V_1V_3\frac{d^2V_2}{dy^2}+V_1V_2\frac{d^2V_3}{dz^2}=0 \quad ...(5.28)$$

Dividing the Eq. (5.28) by $V_1V_2V_3$ we get

$$\frac{1}{V_1}\frac{d^2V_1}{dx^2}+\frac{1}{V_2}\frac{d^2V_2}{dy^2}+\frac{1}{V_3}\frac{d^2V_3}{dz^2}=0 \quad ...(5.29)$$

Since all the three terms are independent of each other, each term must be equal to a constant, thus, we write

$$\frac{1}{V_1(x)}\frac{d^2V_1}{dx^2}=K_1^2 \quad ...(5.30.a)$$

$$\frac{1}{V_1(x)}\frac{d^2V_1}{dx^2}=K_2^2 \quad ...(5.30.b)$$

$$\frac{1}{V_1(x)}\frac{d^2V_1}{dx^2}=K_3^2 \quad ...(5.30.c)$$

with the condition,

$$K_1^2+K_2^2+K_3^2=0 \quad ...(5.31)$$

The differential equations given by the Eq. (5.30) are second order differential equations and the unique solutionis of these equation must depend on the K_1^2, K_2^2 and K_3^2.

Example 5.4. Two parallel conducting disks of radius 30 cm each, are separated by a distance of 10 mm as shown in fig. 5.2. The disk at $x = 0$ is at the potential of 100 V and disk at $x = 10$ mm is at the potential of 200V. Compute the electric field and charge densities on the disks.

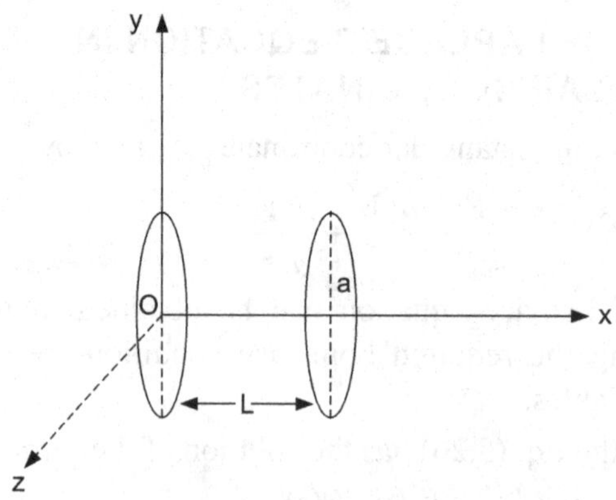

Fig. 5.2. Conducting disks.

Solution: Consider two parallel disks each of radius a separated by a distance L as shown in fig. 5.2.

The Laplace equation for the system is

$$\frac{d^2V}{dx^2} = 0$$

On integration, we get

$$\frac{dV}{dx} = A$$

Integrating again w.r.to x, we get

$$V = Ax + B$$

where A and B are constants and can be determined with the boundary conditions.

$$V = V_1 \quad \text{at} \quad x = 0$$

and

$$V = V_2 \quad \text{at} \quad x = L$$

we get

$$B = V_1, \quad A = \frac{V_2 - V_1}{L}$$

Thus,

$$V = \frac{(V_2 - V_1)x}{L} + V_1$$

The electric field is

$$E = -\frac{dV}{dx} = \frac{(V_2 - V_1)}{L}$$

$$= \frac{(200 - 100)}{10 \times 10^{-3}}$$

$$= 1.0 \times 10^4 \text{ V/m}$$

E points in $-x$ direction.

The surface charge densities

$$\sigma_S = \pm D_n$$

$$= \pm \epsilon_0 E$$

$$= \pm 8.85 \times 10^{-8} \text{ c/m}^2$$

5.4. SOLUTION OF LAPLACE'S EQUATION IN CYLINDRICAL COORDINATES

The Laplace's equation in cylindrical coordinates is given by

$$\frac{1}{r}\frac{d}{dr}\left(r\frac{dV}{dr}\right)+\frac{1}{r^2}\frac{d^2V}{d\phi^2}+\frac{d^2V}{dz^2}=0 \qquad ...(5.32)$$

To solve the Eq. (5.32), we use the method of separation of variables. If we write

$$V(r, \phi, z) = R(r)\,\Phi(\phi)\,Z(z) \qquad ...(5.33)$$

On substituting $V(r, \phi, z)$ and multiplying by $r^2/V(r, \phi, z)$ in the Eq. (5.32) we get

$$\frac{1}{r}\frac{d}{dr}\left(r\frac{dR}{dr}\right)+\frac{r^2}{Z}\frac{d^2Z}{dz^2}=-\frac{1}{\Phi}\frac{d^2\Phi}{d\phi^2} \qquad ...(5.34)$$

The left side of the Eq. (5.34) contains the functions of r and z while the right side has the function of ϕ only which is possible if each side is equal to a constant n^2, say. Thus, we have

$$\frac{d^2\Phi}{d\phi^2}+n^2\Phi=0 \qquad ...(5.35)$$

and

$$\frac{r}{R}\frac{d}{dr}\left(r\frac{dR}{dr}\right)+\frac{r^2}{Z}\frac{d^2Z}{dz^2}=n^2$$

or

$$\frac{1}{rR}\frac{d}{dr}\left(r\frac{dR}{dr}\right)-\frac{n^2}{r^2}=-\frac{1}{Z}\frac{d^2Z}{dz^2} \qquad ...(5.36)$$

The left side of this equation has the function of r only and right side has function of z which is possible if it is equal to a constant $-\lambda^2$, Thus we have

$$\frac{d^2Z}{dz^2}-\lambda^2 Z=0 \qquad ...(5.37)$$

and

$$r\frac{d}{dr}\left(r\frac{dR}{dr}\right)+(\lambda^2r^2-n^2)\,R=0 \qquad ...(5.38)$$

(a) Solution of Φ equation: The solution of $\Phi(\phi)$ equation is given by

$$\Phi(\phi) = A\,e^{\pm in\phi} \qquad ...(5.39)$$

where A is a constant, and n is an integer having values

$$n = 0, \pm1, \pm2, \pm3 \ldots$$

(b) Solution of *Z* equation: The solution of the Eq. (5.37) is given by

$$Z(z) = A\,e^{\lambda z} + Be^{-\lambda z} \qquad ...(5.40)$$

where A and B are constants and can be determined by the boundary conditions of the given problem.

(c) Solution of *R*(*r*) equation: To obtain the solution of the Eq. (5.38) we make a substitution as

$$\rho = \lambda r$$

and

$$\frac{d}{dr} = \frac{\lambda d}{d\rho} \qquad ...(5.41)$$

Thus, we obtain the Eq. (5.37), on substitution the Eq. (5.41) into Eq. (5.37),

$$\rho \frac{d}{d\rho}\left(\rho \frac{dR}{d\rho}\right) + (\rho^2 - n^2)R = 0 \qquad ...(5.42)$$

or

$$\rho^2 \frac{d^2R}{d\rho^2} + \rho \frac{dR}{d\rho} + (\rho^2 - n^2)R = 0 \qquad ..(5.43)$$

Now, we look for a series solution of the Frobenius type, and we assume that the solution of the equation (5.43) is of the form,

$$R(\rho) = \sum_{v=0}^{\infty} a_v \rho^{v+s} \qquad ...(5.44)$$

$$\frac{dR}{d\rho} = \sum_{v=0}^{\infty} a_v (v+s)\, \rho^{v+s-1} \qquad ...(5.45)$$

and

$$\frac{d^2R}{d\rho^2} = \sum_{v=0}^{\infty} a_v (v+s)(v+s-1)\, \rho^{v+s-2} \qquad ...(5.46)$$

on substituting in the Eq. (5.43), we get

$$\sum_{v=0}^{\infty} a_v (v+s)(v+s-1)\, \rho^{v+s} + \sum_{v=0}^{\infty} a_v (v+s)\, \rho^{v+s}$$

$$\sum_{v=0}^{\infty} a_v \rho^{v+s+2} - n^2 \sum_{v=0}^{\infty} a_v \rho^{v+s} = 0 \qquad ...(5.47)$$

To obtain recursion relation, equating the coefficients of ρ^{v+s}, we get

$$[(v+s)(v+s-1) + (v+s) - n^2]\, a_v + a_{v-2} = 0 \qquad ...(5.48)$$

or

$$a_v = \frac{a_{v-2}}{(v+s)^2 - n^2}$$

or $$a_\nu = -\frac{a_{\nu-2}}{(\nu+s+n)(\nu+s-n)} \qquad ...(5.49)$$

The indicial equation is

$$s(s-1)+s-n^2 = 0$$

or $$s = \pm n \qquad ...(5.50)$$

First we look for $s = n$, and the solution is a Bessel function as given by, $\nu = 2$ m, that is series starts with

$$J_n(\rho) = \sum_{n=0}^{\infty} \frac{(-1)^m}{m!\Gamma_{(m+n+1)}} \left(\frac{\rho}{2}\right)^{n+2m} \qquad ...(5.51)$$

a_o and ν is always even. Again, we have

$$J_{-n}(\rho) = (-1)^n J_n(\rho) \qquad ...(5.52)$$

$$J_o(\rho) = \frac{\sin\rho}{\rho}$$

$$J_1(\rho) = \frac{\sin\rho}{\rho} - \frac{\cos\rho}{\rho}$$

and $$J_2(\rho) = \left(\frac{3}{\rho^3} - \frac{1}{\rho}\right)\sin\rho - \frac{3}{\rho^3}\cos\rho$$

However, we have a simple problem of electrostatics.

Example 5.5. Two semi-infinite conducting planes are at an angle $\phi = \pi/3$ and are joined along the z-axis. The plane at the angle $\phi = 0$ is grounded while other plane which is at the angle $\phi = \pi/3$ has potential of 200 V. Compute the potential and electric field.

Solution: The Laplace equation is,

$$\frac{1}{r}\frac{d^2V}{d\phi^2} = 0$$

on integrating twice, we get

$$V = A\phi + B$$

where A and B are constants. The boundary conditions are

$$V = 0, \quad \text{when } \phi = 0$$

and $$V = 200 \text{ V at } \phi = \pi/3$$

Thus, we have,

$$V = \frac{3V_o}{\pi}\phi$$

since $V_0 = 200$ V

$$V = \frac{600}{3.14}\phi$$

or
$$V = 191.08\,\phi$$

The electric field is

$$E = -\frac{1}{r}\frac{dV}{d\phi}\hat{\phi}$$

$$= -\frac{600}{r\pi}\hat{\phi}$$

or
$$E = \frac{-191.08}{r}\hat{\phi}$$

5.5. SOLUTION OF LAPLACE'S EQUATION IN SPHERICAL POLAR COORDINATES

The Laplace's equation in spherical polar coordinates is given by

$$\nabla^2 V = \frac{1}{r^2}\frac{\partial}{\partial r}\left(r^2\frac{\partial V}{\partial r}\right) + \frac{1}{r^2\sin\theta}\frac{\partial}{\partial r}\left(\sin\theta\frac{\partial V}{\partial\theta}\right) + \frac{1}{r^2\sin^2\theta}\frac{\partial^2 V}{\partial\phi^2} = 0 \quad ...(5.53)$$

The Eq. (5.53) can be solved by the method of separation of variables and we assume that

$$V(r, \theta, \phi) = R(r)\,\mathbf{H}(\theta)\,\phi(\phi) \quad ...(5.54)$$

On substituting for V from the Eq. (5.54) in the Eq. (5.53) and dividing by $V(r, \theta, \phi)$, we get,

$$\frac{1}{R}\frac{1}{r^2}\frac{d}{dr}\left(r^2\frac{dR}{dr}\right) + \frac{1}{\mathbf{H}\,r^2\sin\theta}\frac{d}{d\theta}\left(\sin\theta\frac{d\mathbf{H}}{d\theta}\right) + \frac{1}{r^2\Phi\sin^2\theta}\frac{d^2\Phi}{d\phi^2} = 0 \quad ...(5.55)$$

To eliminate $r^2\sin^2\theta$ in Φ term, we multiply by $r^2\sin^2\theta$ through the Eq. (5.55) we get

$$\frac{\sin^2\theta}{R(r)}\frac{d}{dr}\left(r^2\frac{dR}{dr}\right) + \frac{\sin\theta}{\mathbf{H}}\frac{d}{d\theta}\left(\sin\theta\frac{d\mathbf{H}}{d\theta}\right) = -\frac{1}{\Phi}\frac{d^2\Phi}{d\phi^2} \quad ...(5.56)$$

The left hand side contains the functions of r and θ only while right handside has a function of ϕ only and it is possible when each side is equal to a constant m^2, say, we have

$$\frac{d^2\Phi}{d\phi^2} + m^2\Phi = 0 \quad ...(5.57)$$

Now, from the Eq. (5.56), we have

$$\frac{1}{R}\frac{d}{dr}\left(r^2\frac{dR}{dr}\right) = -\frac{1}{\mathbf{H}\sin\theta}\frac{d}{d\theta}\left(\sin\theta\frac{d\mathbf{H}}{d\theta}\right) = -\frac{m^2}{\sin^2\theta} \quad ...(5.58)$$

The left hand side of the Eq (5.58) has function of r only and right hand side is θ only which is possible if it is equal to a constant $n(n + 1)$, Thus

$$\frac{1}{R}\frac{d}{dr}\left(r^2\frac{dR}{dr}\right) = n(n+1)$$

or
$$\frac{d}{dr}\left(r^2\frac{dR}{dr}\right) - n(n+1)R = 0 \quad ...(5.59)$$

and
$$\frac{1}{\mathbf{H}\sin\theta}\frac{d}{d\theta}\left(\sin\theta\frac{d\mathbf{H}}{d\theta}\right) = -n(n+1)$$

or
$$\frac{1}{\sin\theta}\frac{d}{d\theta}\left(\sin\theta\frac{d\mathbf{H}}{d\theta}\right) + \left[n(n+1) - \frac{m^2}{\sin^2\theta}\right]\mathbf{H} = 0 \quad ...(5.60)$$

(a) Solution of Φ equation: The solution of the Eq. (5.57) is given by

$$\Phi = A\,e^{\pm im\phi} \quad ...(5.61)$$

where A is a constant.

(b) Solution of radial equation: Since potential varies as power in r, we have solution of the equation (5.59), as

$$R_n(r) = A\,r^n + Br^{-(n+1)} \quad ...(5.62)$$

where A and B are any arbitrary constants.

(c) Solution of H (θ) equation: Now substituting $x = \cos\theta$ in the Eq. (5.60), we have

$$\therefore \quad \left.\begin{aligned} x &= \cos\theta \\ dx &= -\sin\theta\, d\theta \\ \text{and} \quad \frac{d}{d\theta} &= -\sin\theta\frac{d}{dx} \end{aligned}\right\} \quad ...(5.63)$$

On substituting the Eq. (5.63) in the Eq. (5.60), we get

$$\frac{d}{dx}\left[(1-x^2)\frac{d\mathbf{H}}{dx}\right] + \left[n(n+1) - \frac{m^2}{(1-x^2)}\right]\mathbf{H}(\theta) = 0 \quad ...(5.64)$$

or
$$(1-x)^2\frac{d^2\mathbf{H}}{dx^2} - 2x\frac{d\mathbf{H}}{dx} + n(n+1)\mathbf{H} = 0 \quad ...(5.65)$$

Here we have assumed that the electric field has a azimuthal symmetry, that is, the potential V is independent of the angle ϕ, and replacing function **H** with Legendre' polynomial $Pn(x)$, thus,

$$(1-x)^2 \frac{d^2 P_n}{dx^2} - 2x\frac{dP_n}{dx} + n(n+1)\, P_n(x) = 0 \qquad ...(5.66)$$

which is called Legendre's equation, which has the solution as

$$P_n(x) = P_n(\cos\theta)$$

or

$$P_n(x) = \frac{1}{2^n n!}\frac{d^n}{dx^n}(x^2-1)^n \qquad ...(5.67)$$

The orthogonality condition is

$$\int_{-1}^{1} P_n(x)\, P_m(x)\, dx = \frac{2}{2n+1}\delta_{mn} \qquad ...(5.68)$$

where δ_{mn} is known as Kronecker delta and is defined as

$$\delta_{mn} = \begin{cases} 1 & \text{if } m = n \\ 0 & \text{if } m \neq n \end{cases} \qquad ...(5.69)$$

The first few terms are

$$\begin{aligned} P_0(x) &= 1 \\ P_1(x) &= x \\ P_2(x) &= \frac{1}{2}(3x^2-1) \\ P_3(x) &= \frac{x}{2}(5x^2-3) \\ P_4(x) &= \frac{1}{8}(35x^4-30x^2+3) \\ P_5(x) &= \frac{x}{8}(63x^4-70x^2+15) \end{aligned} \qquad ...(5.70)$$

The complete solution to the potential equation in spherical polar coordinates, when electric field has azimuthal symmetry is given by

$$V(r,\theta) = [A_n r^n + B_n r^{-(n+1)}]\, P_n(\cos\theta) \qquad ...(5.71)$$

This solution depends on the boundary conditions of the given problem.

5.6. A CONDUCTING SPHERE IN A UNIFORM ELECTRIC FIELD

Consider a conducting sphere of the radius a in a uniform electric field pointing in $+z$ direction. We know that the electric field is quite uniform in between the plates of a capacitor. Thus, a sphere is placed in the capacitor as shown in fig. 5.3.

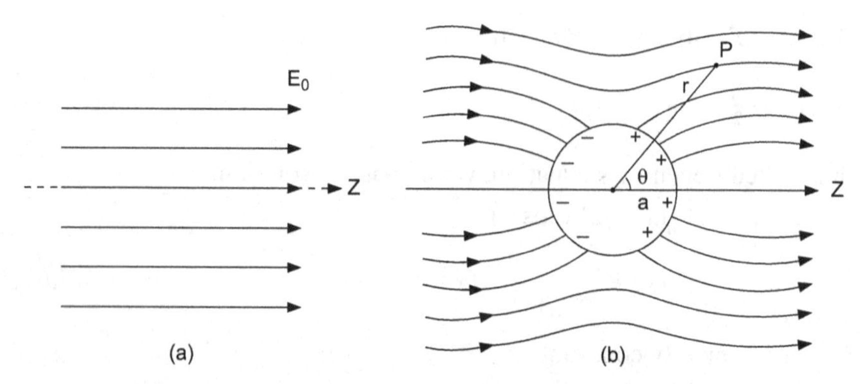

Fig. 5.3. A conducting sphere in a uniform electric field and a uniform electric field points in z-direction.

The conducting sphere distorts the electric field lines. The positive and negative charge in equal amount are induced at the sphere. Thus the conducting sphere has zero potential. Therefore, the boundary conditions are

$$V = \begin{cases} 0 & \text{for } r = a \\ -E_o z = -E_o r\cos\theta & \text{for } r >> a \end{cases} \quad ...(5.72)$$

The Laplace's equation in spherical polar coordinates with axial symmetry is given by

$$\frac{1}{r^2}\frac{\partial}{\partial r}\left(r^2\frac{\partial V}{\partial r}\right)+\frac{1}{r^2\sin\theta}\frac{\partial}{\partial\theta}\left(\sin\theta\frac{\partial V}{\partial\theta}\right)=0 \quad ...(5.73)$$

The general solution of the Eq. (5.73) is given by

$$V(r, \theta) = \sum_{n=0}^{\infty} A_n r^n P_n(\cos\theta) + \sum_{n=0}^{\infty} B_n r^{-(n+1)} P_n(\cos\theta) \quad ...(5.74)$$

At $r = a$, $V = 0$, we have

$$\sum_{n=0}^{\infty} A_n a^n P_n(\cos\theta) + \sum_{n=0}^{\infty} B_n r^{-(n+1)} P_n(\cos\theta) = 0 \quad ...(5.75)$$

Multiplying $P_m(\cos\theta)$ and integrating from $x = \cos\theta = -1$ to $x = \cos\theta = 1$, we get

$$\int_{-1}^{1} A_n a^n P_n(\cos\theta) P_m(\cos\theta)\, d(\cos\theta) + \int_{-1}^{1} B_n a^{-(n+1)} P_n(\cos\theta) P_m(\cos\theta)\, d(\cos\theta) = 0 \quad ...(5.76)$$

Using the orthogenality of the Legendre's polynomial as,

$$\int_{-1}^{1} P_m(\cos\theta)\, P_n(\cos\theta)\, d(\cos\theta) = \begin{cases} \dfrac{2}{2n+1} & \text{if } m = n \\ 0 & \text{if } m \neq n \end{cases} \quad ...(5.77)$$

The Eq. (5.76) reduces to

$$A_n a^n \left(\frac{2}{2n+1}\right) + B_n a^{-(n+1)} \left(\frac{2}{2n+1}\right) = 0 \quad ...(5.78)$$

From the Eq. (5.74), it is clear that the solutions exist only for $n = 0$ and $n = 1$ which satisfy the boundary condition. Thus, we have

$$V = A_0 + \frac{B_0}{r} + A_1\, r\cos\theta + \frac{B_1 \cos\theta}{r^2} \quad ...(5.79)$$

The boundary condition at $r \to \infty$, gives the potential

$$V = -E_0\, r\cos\theta = A_0 + A_1\, r\cos\theta \quad ...(5.80)$$

This is true for all values of θ, $A_0 = 0$ and $A_1 = -E_0$ since for $n = 1$, we have

$$-E_0\, r\, P_1(\cos\theta) = A_1\, r\, P_1(\cos\theta)$$

or

$$A_1 = -E_o \quad ....(5.81)$$

All A_n are zero (for $n > 1$) and similarly all B_n are zero except B_1. At the $r = a$, $V = 0$, from the Eq. (5.79) we have

$$\frac{B_0}{a} - E_0\, a\cos\theta + \frac{B_1}{a^2}\cos\theta = 0 \quad ...(5.82)$$

since $B_0 = 0, \quad B_1 = E_0 a^3$

Thus, the potential is given by

$$V(r, \theta) = -E_0\, r\cos\theta + \frac{E_0 a^3}{r^2}\cos\theta \quad ...(5.83)$$

The components of the electric field are given by

$$E_r = -\frac{dV}{dr} = E_0\cos\theta + \frac{2a^3 E_0 \cos\theta^3}{r^3}$$

or

$$E_r = E_0\left(1 + \frac{2a^3}{r^3}\right)\cos\theta\, \hat{r} \quad ...(5.84)$$

and

$$E_\theta = -\frac{1}{r}\frac{dV}{d\theta} = -E_0 \sin\theta + \frac{E_0\, a^3 \sin\theta}{r^3}$$

or
$$E_\theta = -E_0\left(1-\frac{a^3}{r^3}\right)\sin\theta \quad ...(5.85)$$

Now the radial and tangential components of the electric field on the surface of the sphere are

$$E_r = -\left.\frac{dV}{dr}\right|_{r=a}$$

or
$$E_r = 3E_0 \cos\theta \quad ...(5.86)$$

and
$$E_\theta = 0 \quad ...(5.87)$$

The surface charge density

$$\sigma = -\epsilon_0 \left.\frac{dV}{dr}\right|_{r=a}$$

or
$$\sigma = 3\,\epsilon_0\, E_0 \cos\theta \quad ...(5.88)$$

Furthermore, the dipole moment is given by

$$\left.\begin{aligned} V(\text{dipole}) &= \frac{E_0 a^3\cos\theta}{r^2} \\ V(\text{dipole}) &\propto \frac{p\cos\theta}{r^2} \end{aligned}\right\} \quad ...(5.89)$$

$$P = E_0\, a^3 \quad ...(5.90)$$

Thus, the dipole moment is proportional to E_0 and a^3, that is, volume of sphere.

5.7. METHOD OF ELECTRICAL IMAGES

In 1848, Lord Kelvin introduced the method of electrical images to solve the problems of the electrostatics. The method of electrical images is a powerful tool for calculating the coulomb force, electric potential, electric field intensity and the charge density of an electrical system. In the method of electrical images, we have a charge and a conducting plane along with the image of the charge. Now, suppose that a positive point charge q is situated at a distance r from a grounded conducting plane which acts as a mirror, then, the image of the charge q is $-q$ will also be at a distance r from the conducting plane as shown in fig. 5.4.

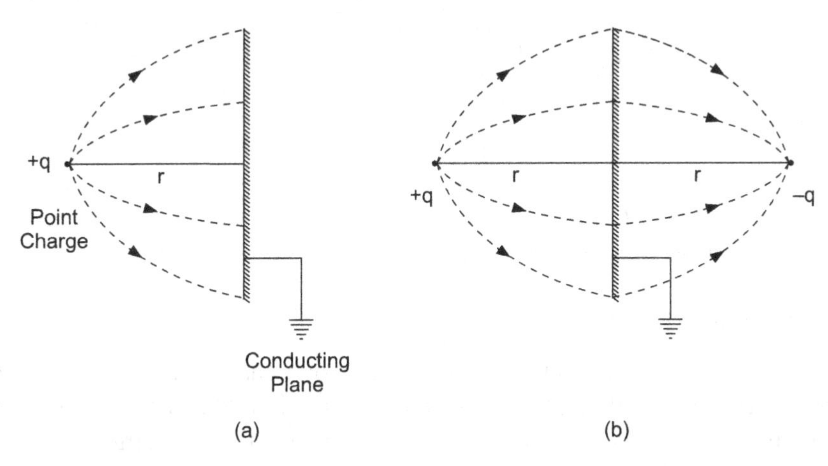

Fig. 5.4. A charge $+q$ near a grounded plane (*a*) and its image (*b*).

The grounded plane has a zero potential and it is observed that there are two electric fields which are equivalent to one another. Furthermore, consider two grounded perpendicular semi-infinite conducting planes as shown in fig. 5.5. If a point charge q is placed near a right angle corner, we obtained three images of the charge $+q$, Fig. 5.5.

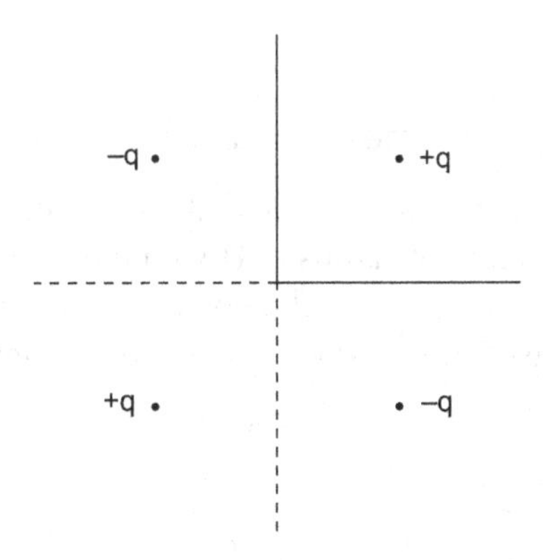

Fig. 5.5. Point charge near a right angle conducting planes and its three images.

In the first quadrant, we have a true electric field due to the image charges. Moreover, when two semi-finite conducting planes are inclined at an angle θ, the number of images of a point charge are given by

$$n = \left(\frac{360}{\theta} - 1\right) \quad ...(5.91)$$

For example, consider a case of two semi-infinite conducting planes are at $\theta = 90°$, then, the number of images of a charge q, Fig. 5.5, will be

$$n = \left(\frac{360}{\theta} - 1\right)$$

or $$n = \left(\frac{360}{90} - 1\right)$$

or $$n = 3$$

There are three images of a charge situated in first-quadrant.

5.8. CONDUCTING SPHERE

The method of electrical images is also useful in demonstrating a procedure of computing the electrical parameters such as potential, field and charge density of many problems. Now, we take a situation in which a point charge is near a grounded conducting sphere, Fig. 5.6.

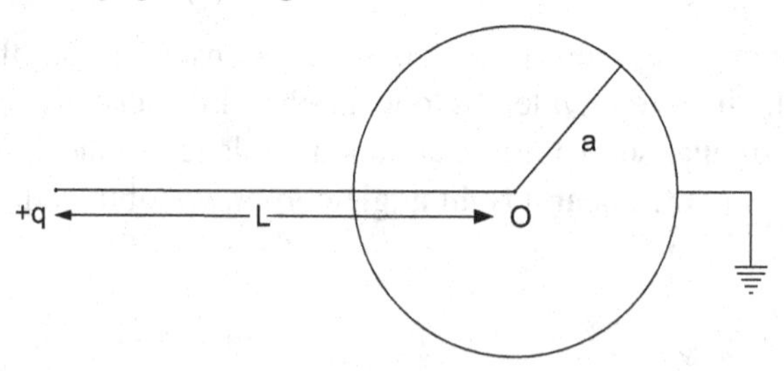

Fig. 5.6. A positive point charge q near a grounded conducting sphere.

Suppose that a point charge q is situated at a distance L from the centre of a grounded conducting sphere of radius a. If we ignore the conducting sphere and try to compute the magnitude and the position of the image charge q' at a distance l from the centre of the conducting sphere, as shown in Fig. 5.7.

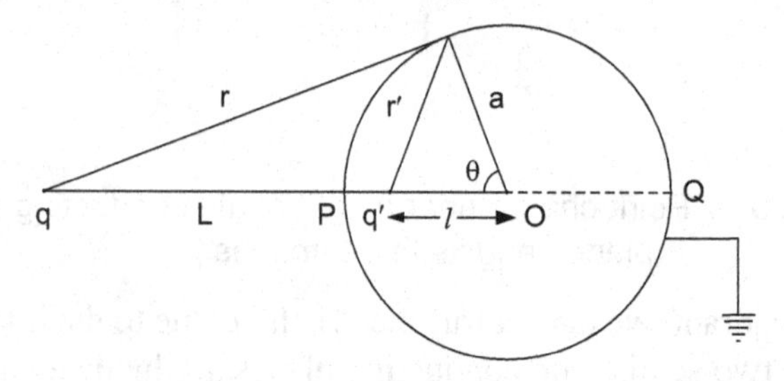

Fig. 5.7. Image charge q' in the sphere.

The symmetry of the sphere suggests that the charge q' must lie on the line joining the charge q and the centre of the conducting sphere. Since the sphere is taken as grounded, the potential at every point on the surface of the sphere is zero. The potential at the point P is

$$\frac{q}{(L-a)}+\frac{q'}{(a-l)}=0 \quad ...(5.92)$$

Again, for the point Q, we write

$$\frac{q}{(L+a)}+\frac{q'}{(a+l)}=0 \quad ...(5.93)$$

Now,

$$\cos\theta = \frac{a^2+l^2-r'^2}{2al} \quad ...(5.94)$$

and

$$\cos\theta = \frac{a^2+(a+L)^2-r^2}{2a(a+L)} \quad ...(5.95)$$

On solving the Eqs. (5.92) and (5.93) we get

$$\left.\begin{aligned} q' &= -\left(\frac{l}{a}\right)q \\ \text{and} \quad q' &= -\left(\frac{a}{L}\right)q \end{aligned}\right\} \quad ...(5.96)$$

Then,

$$l = \frac{a^2}{L} \quad ...(5.97)$$

In the similar triangles having common angle θ, we write

$$\frac{a}{l} = \frac{L}{a}$$

Thus, we have

$$\frac{a}{r} = \frac{l}{r'}$$

Therefore,

$$\frac{q}{r}+\frac{q'}{r'}=0 \quad ...(5.98)$$

or

$$\frac{q}{4\pi\epsilon_0 r}+\frac{q'}{4\pi\epsilon_0 r'}=0 \quad ...(5.99)$$

where the expression for r and r' are

$$\left.\begin{aligned} r^2 &= a^2+(a+L)^2-2a(a+L)\cos\theta \\ \text{and} \quad r'^2 &= a^2+l^2-2al\cos\theta \end{aligned}\right\} \quad ...(5.100)$$

From the Eq. (5.99), it is clear that the total potential is zero on the surface of the sphere. Moreover, the force of attraction is given by

$$F = \frac{qq'}{4\pi\epsilon_0 \, [L+a-l]^2} \qquad ...(5.101)$$

It the sphere were not grounded, there will be a constant potential at the surface of the sphere.

Example 5.6. Two semi-infinite grounded conducting planes are along x and y axes and meet at the origin as shown in fig. 5.8. A positive point charge is situated at the position (a, a), then compute the

(a) The force on the charge q.

(b) Potential at a point $P(x, y)$.

(c) The work done in bringing a charge q from infinity to a point (a, a).

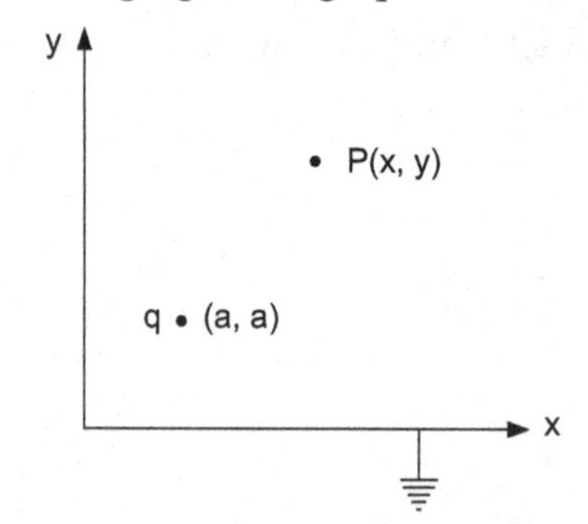

Fig. 5.8. Two conducting planes.

Solution: The images of the charge q is shown in fig. 5.9. and there are three images.

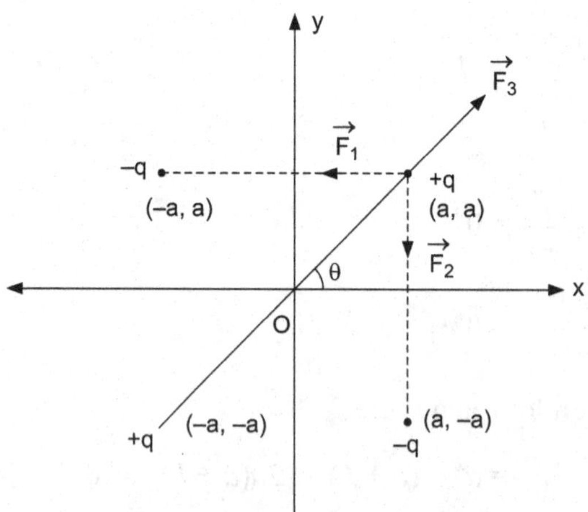

Fig. 5.9. Mirror images of charge q.

(a) The forces $\vec{F_1}, \vec{F_2}$ and $\vec{F_3}$ are

$$\vec{F_1} = \frac{q^2}{4\pi \in_0 (2a)^2}(-\hat{j})$$

$$\vec{F}_2 = \frac{q^2}{4\pi \in_0 (2a)^2}(-j)$$

and

$$\vec{F}_3 = \frac{q^2\, 2a(i+j)}{4\pi \in_0 (8a^2)^{3/2}}$$

The net force on the charge q will be

$$\vec{F} = \vec{F}_1 + \vec{F}_2 + \vec{F}_3$$

$$= \frac{q^2}{16\pi \in_0}\left[\left(\frac{1}{2\sqrt{2}\,a^2} - \frac{1}{a^2}\right)\hat{i} + \left(\frac{1}{2\sqrt{2}\,a^2} - \frac{1}{a^2}\right)\hat{j}\right]$$

$$= \frac{q^2}{16\pi \in_0 a^2}\frac{(1-2\sqrt{2})}{2\sqrt{2}}(\hat{i}+\hat{j})$$

or

$$F = \frac{q^2}{32\sqrt{2}\,\pi \in_0} \cdot \frac{(1-2\sqrt{2})}{a^2}(\hat{i}+\hat{j})$$

(b) To compute the potential at a point $P(x, y)$, suppose that r_1, r_2, r_3 and r_4 are the distances of the point $P(x, y)$ from the charges as shown in fig. 5.10.

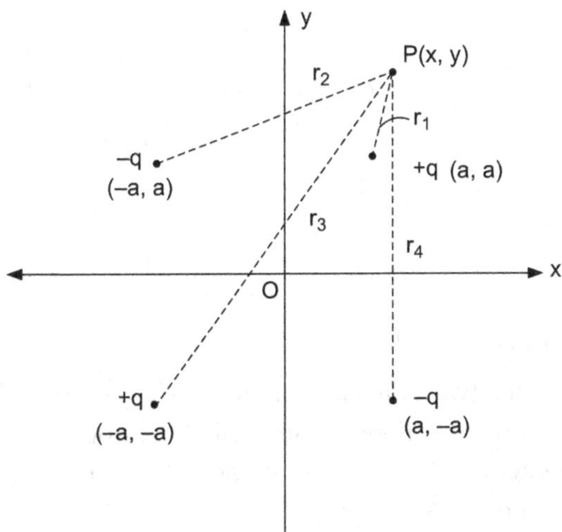

Fig. 5.10. Conducting planes.

The potential at a point P due to the charges is given by

$$V = \frac{q}{4\pi\epsilon_0}\left[\frac{1}{r_1}-\frac{1}{r_2}+\frac{1}{r_3}-\frac{1}{r_4}\right]$$

$$= \frac{q}{4\pi\epsilon_0}\left[\frac{1}{[(x-a)^2+(y-a)^2]^{1/2}}-\frac{1}{[(x+a)^2+(y-a)^2]^{1/2}}\right.$$

$$\left.+\frac{1}{[(x+a)^2+(y+a)^2]^{1/2}}-\frac{1}{[(x-a)^2+(y+a)^2]^{1/2}}\right]$$

(c) The work done in bringing a charge q from infinite to a point $P(a, a)$ will be, Fig. 5.11.

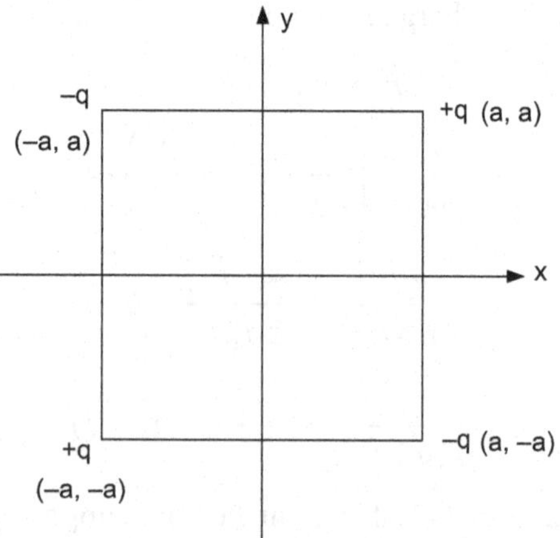

Fig. 5.11. Conducting planes.

$$W = Vq$$

$$= \frac{1}{4\pi\epsilon_0}\left[\frac{-q^2}{2a}-\frac{q^2}{2a}-\frac{q^2}{2a}-\frac{q^2}{2a}+\frac{q^2}{2\sqrt{2}\,a}+\frac{q^2}{2\sqrt{2}\,a}\right]$$

$$= \frac{1}{4\pi\epsilon_0}\left[-\frac{2}{a}+\frac{1}{\sqrt{2}\,a}\right]$$

or $$W = \frac{q^2}{4\pi\epsilon_0\,a}\frac{(1-2\sqrt{2})}{\sqrt{2}}$$

Example 5.7. Consider two concentric metallic spherical shells centered at origin. The inner and outer radii are a and b. In the region $a < r < b$, the charge density $\rho = 0$ and potential $V = 0$ at $r = a$ and $V = V_0$ at $r = b$. Compute the potential in the region $a < r < b$.

Solution: Laplace's equation is

$$\frac{1}{r^2}\frac{d}{dr}\left(r^2\frac{dV}{dr}\right) = 0$$

on integration, we get

$$r^2\frac{dV}{dr} = K$$

or $$\frac{dV}{dr} = \frac{K}{r^2}$$

where K is a constant. Integrating it again, we have

$$V = \frac{-K}{r} + A$$

where $r = a$, $V = 0$,

$\Rightarrow$ $$A = \frac{K}{a}$$

$\therefore$ Potential is

$$V = -K\left(\frac{1}{r} - \frac{1}{a}\right)$$

using boundary condition at $r = b$, $V = V_o$

$$V_0 = -K\left(\frac{1}{b} - \frac{1}{a}\right)$$

or $$V_0 = K\left(\frac{1}{a} - \frac{1}{b}\right)$$

or $$K = \frac{V_0\, ab}{(b-a)}$$

Thus, the potential is

$$V = \frac{V_0 ab}{(b-a)}\left[\frac{1}{a} - \frac{1}{r}\right]$$

Example 5.8. Consider two infinite conducting cones with common vertex having angles $\frac{\pi}{6}$ and $\frac{\pi}{3}$ as shown in fig. 5.12. The inner cone is grounded and $V = V_0$ at $\theta = \frac{\pi}{3}$. Find the expression for potential and electric field.

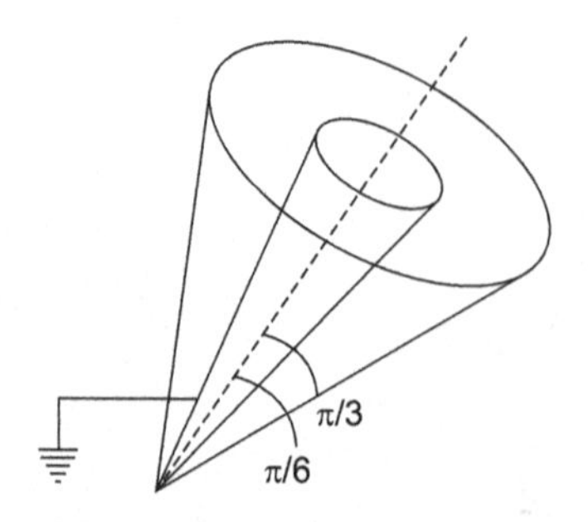

Fig. 5.12. Conducting cone.

Solution: Consider azimuthal symmetry, we have

$$\nabla^2 V = \frac{1}{r^2 \sin\theta}\frac{d}{d\theta}\left(\sin\theta \frac{dV}{d\theta}\right) = 0$$

or $$\frac{d}{d\theta}\left(\sin\theta \frac{dV}{d\theta}\right) = 0$$

on integration we get

$$\sin\theta \frac{dV}{d\theta} = A$$

or $$\frac{dV}{d\theta} = \frac{A}{\sin\theta}$$

where A is a constant. Integrating it again, we get

$$V = A\int \frac{d\theta}{\sin\theta}$$

$$= A\int \frac{d\theta}{2\sin\theta/2\cos\theta/2}$$

Now, dividing numerator and denominator by $\cos^2\theta/2$, we get

$$V = A\int \frac{\sec^2\theta/2\, d\theta}{2\tan\theta/2}$$

or $$V = A\ln\tan\frac{\theta}{2} + B$$

The boundary conditions are

$$V = 0, \quad \theta = \pi/6$$

and $$V = V_0, \quad \theta = \pi/3$$

$$0 = A\ln\tan\pi/12 + B$$

Thus,

$$V = A[\ln \tan \theta/2 - \ln \tan \pi/12]$$

or
$$V = A \ln \left(\frac{\tan \theta/2}{\tan \pi/12} \right)$$

once again,

$$V_o = A \ln \left(\frac{\tan \pi/6}{\tan \pi/12} \right)$$

$$A = \frac{V_0}{\ln \left(\frac{\tan \pi/6}{\tan \pi/12} \right)}$$

$\therefore$
$$V = V_0 \frac{\ln \left(\frac{\tan \theta/2}{\tan \pi/12} \right)}{\ln \left(\frac{\tan \pi/6}{\tan \pi/12} \right)}$$

or
$$V = 1.44\, V_0 \ln \left(\frac{\tan \theta/2}{\tan \pi/12} \right)$$

and
$$E_\theta = -\frac{1.44V_0}{r \sin \theta}$$

Example 5.9. Consider two concentric conducting cylinders of inner radius a and outer radius b as shown in fig. 5.13. If $\rho = 0$ in $a < r < b$ and inner cylinder is grounded, find the expressions for potential and electric field. The potential $V = V_0$ at $r = b$, also compute V and E if $a = 10$ mm, $b = 20$ mm and $V_0 = 100$V.

Solution:

$$\frac{1}{r}\frac{d}{dr}\left(r\frac{dV}{dr} \right) = 0$$

on integration, we get

$$r\frac{dV}{dr} = A$$

or
$$\frac{dV}{dr} = \frac{A}{r}$$

on integration again, we get

$$V = A \ln r + B$$

where A and B are constants and are to be determined with boundary conditions.

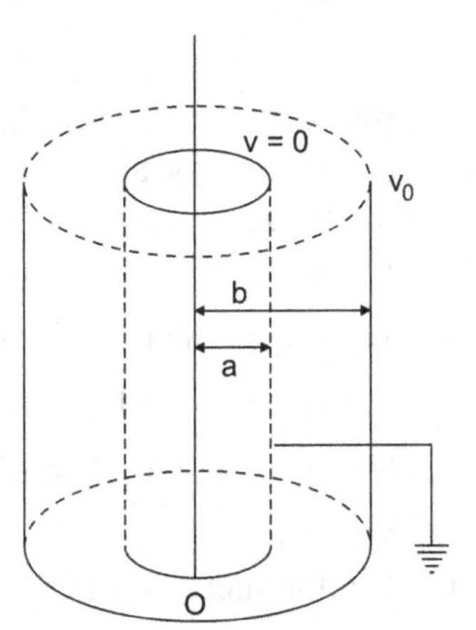

Fig. 5.13. Conducting cylinder.

The boundary conditions are

$$V = 0 \quad r = a$$

$$V = V_0, \; r = b$$

on using boundary conditions

$$B = -A \ln a$$

and

$$A = \frac{V_0}{\ln (b/a)}$$

Thus,

$$V = \frac{V_0}{\ln (b/a)} \ln(r/a)$$

The electric field

$$E_r = -\frac{dV}{dr}$$

or

$$E_r = -\frac{V_0}{a \ln\left(\frac{b}{a}\right)} \cdot \frac{1}{r}$$

Given $a = 10$ mm, $b = 20$ mm, $V_o = 100$ V,

$$V = 144.3 \ln\left(\frac{r}{10}\right) \text{ volts.}$$

and

$$E_r = \frac{1.44 \times 10^4}{r}(-\hat{r})$$

Example 5.10. Consider a thin conducting plane passing through the origin of the coordinates system and perpendicular to the z-axis as shown in fig. 5.14. The two charges q_1 and q_2 are situated at the positions (0, 0, a) and (0, 0, $-a$) respectively. Then compute, using cylindrical coordinates,

(a) The potential at a point $P(x, y, z)$

(b) The Electric field

(c) Net surface charge on the plane as a function of the distance r from the z-axis

(d) Total induced charge

(e) The net force on the charge q_1.

Solution:

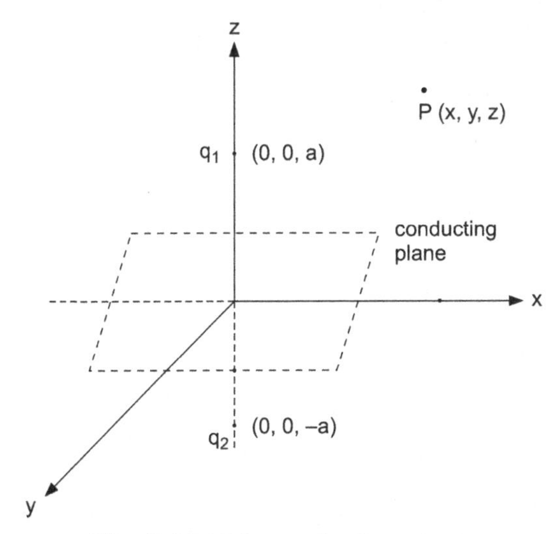

Fig. 5.14. Thin conducting plane.

(a) The potential at a point $P(x, y, z)$ is

$$V(x, y, z) = \frac{1}{4\pi\epsilon_0}\left[\frac{q_1}{[(x^2+y^2+(z-a)^2]^{1/2}} - \frac{q_2}{[(x^2+y^2+(z+a)^2]^{1/2}}\right]$$

Since q_2 is the image of q_1 so it is considered as negative.

(b) The electric field is then given by

$$\vec{E} = -\nabla V(x, y, z)$$

or

$$\vec{E} = -\left[\frac{\partial V}{\partial x}\hat{i} + \frac{\partial V}{\partial y}\hat{j} + \frac{\partial V}{\partial z}\hat{k}\right]$$

$$= \frac{1}{4\pi\epsilon_0}\left[\frac{q_1(x\hat{i}+y\hat{j}+(z-a)\hat{k}}{[x^2+y^2+(z-a)^2]^{3/2}} - \frac{q_2(x\hat{i}+y\hat{j}+(z+a)\hat{k}}{[x^2+y^2+(z+a)^2]^{3/2}}\right]$$

Now,

$$\vec{E}(z=0) = \frac{(q_1+q_2)}{4\pi\epsilon_0}\left[\frac{-2a}{(x^2+y^2+a^2)^{3/2}}\right]\hat{k}$$

$$\vec{E} = \frac{(q_1+q_2)}{4\pi\epsilon_0}\left[\frac{-a}{(r^2+a^2)^{3/2}}\right]\hat{k}$$

Thus, E is perpendicular to the plane.

(c) The surface charge is given by

$$\sigma_S = \epsilon_0 E$$

or $$\sigma_S = \frac{-(q_1+q_2)\,a}{2\pi} \cdot \frac{1}{(r^2+a^2)^{3/2}}$$

where $r^2 = x^2 + y^2$

(d) The total induced charge will be

$$q = \int_{x=-\infty}^{\infty} \int_{y=-\infty}^{\infty} \sigma_S \, dx \, dy$$

$$= \int_{r=0}^{\infty} \int_{\phi=0}^{2\pi} \sigma_S \, r \, dr \, d\phi$$

$$= -(q_1+q_2)\,a \int_0^{\infty} \frac{rdr}{[r^2+a^2]^{3/2}}$$

$$q = -(q_1+q_2) \cdot \frac{a}{a}$$

or $$q = -(q_1+q_2)$$

(e) The force between q_1 and the plane is

$$\vec{F} = \frac{q_1^2}{4\pi\epsilon_0 (2a)^2} \hat{k}$$

or $$\vec{F} = \frac{q_1^2}{16\pi\epsilon_0 a^2} \hat{k}$$

EXERCISES

5.1. Solve the Poisson's equation for the *p-n* junction diode.

5.2. Two semi-infinite conducting planes, which are grounded, intersect each other at an angle 30°. Find the number of images of a charge q placed between them. Draw the diagram also.

5.3. The electric potential of a conducting sphere of radius R in a uniform electric field $E = E_0 \hat{k}$ is given by

$$V(r, \theta) = -E_0 r \cos\theta + E_0 \frac{R^3}{r^2} \cos\theta$$

where r is the distance from the centre of the sphere and θ is the angle that the r makes with z-axis. Then

(a) What is the dipole moment acquired by the sphere.

(b) Compute the radial and tangential components of the electric field, E_r and E_θ, on the surface of the sphere.

(c) Compute the surface charge density.

Hint: $V(\text{dipole}) \propto \dfrac{p\cos\theta}{r^2}, \quad \therefore \quad p = E_0 R^3$

$$E_r = 3\,E_0 \cos\theta$$
$$E_\theta = 0$$
$$\sigma_S = 3\epsilon_0\,E_0 \cos\theta$$

5.4. Consider two metallic spheres each of radius a placed at a distance r from the centres as shown in fig. 5.15. The spheres carry charges q and $-q$, then compute

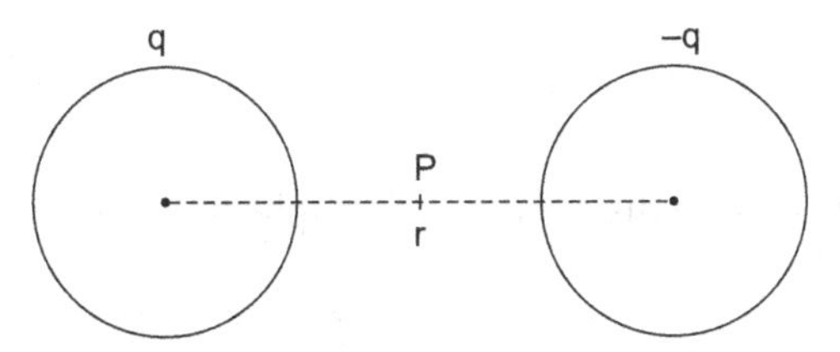

Fig. 5.15. Two metallic spheres.

(a) The potential difference between the spheres

(b) Electric field at the point P.

5.5. Compute the charge distribution in cylindrical coordinates if the potential is given by

$$V = V_0\, r^2\,(3\cos^2\phi - 1)$$

5.6. If potential, in spherical polar coordinates system, is given by

$$V = V_0\, r^2\,(3\cos^2\phi - 1)$$

Compute the charge density.

5.7. Consider a dielectric sphere of radius a and dielectric constant k in the presence of uniform electric field $E = E_0\,\hat{k}$.

The potentials in-side and outside the sphere is given by

$$V_1 = -\frac{3E_0\, r\cos\theta}{k+2}$$

and
$$V_2 = \frac{(k-1)}{(k+2)}\,\frac{E_0\, a^3\cos\theta}{r^2} - E_0\, r\cos\theta$$

(a) Compute the radial and tangential components of the electric field.

(b) Compute the dipole moment acquired by the sphere.

(c) Surface charge density σ_S.

5.8. Compute the electric field and charge distribution if the potential is given by

$$V = \begin{cases} V_0\left(3 - \dfrac{r^2}{a^2}\right) & r < a \\ \dfrac{V_0 a}{r} & r > a \end{cases}$$

where V_0 and a are constants.

Hint: $$E_r = -\frac{dV}{dr},$$

$$\rho = -\in_0 \frac{dV}{dr}$$

5.9. Determine the charge distribution for a spherically symmetric potential

$$V = \frac{V_0}{r} \qquad r > a$$

$$= V_0\left(1 - \frac{r^3}{a^3}\right) \qquad r > a$$

5.10. Compute the electric field and charge density for the potentials

(a) $$V = \frac{\alpha x^2 + \beta y^2 + \gamma z^2}{r^3}$$

(b) $$V = V_0(x^2 + y^2 + z^2)$$

Hint: $$E = -\nabla V$$

$$\sigma = -\in_0 \nabla V$$

5.11. How much work is required to assemble a sphere of the radius R with a volume charge density ρ.

Hint:

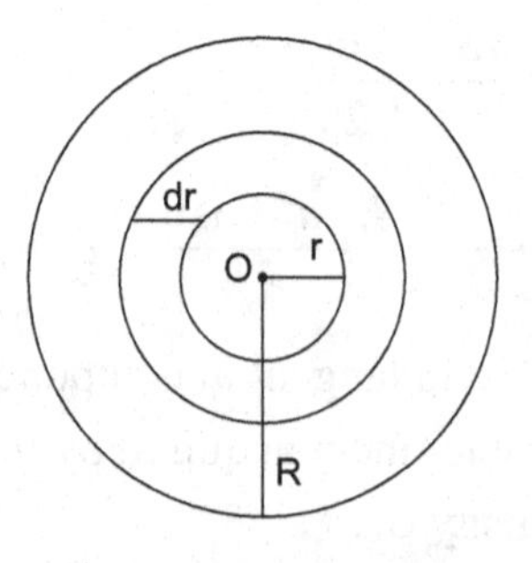

Fig. 5.16. A sphere.

Charge of the layer is

$$dq = \rho . 4\pi r^2 dr$$

$$\therefore \quad V = \frac{q}{4\pi \epsilon_0 r}, \text{ and } \rho = \frac{q}{\frac{4}{3}\pi r^3}$$

$$\therefore \quad V = \frac{1}{3\epsilon_0} r^2 \rho$$

$$\therefore \quad dW = V\,dq = \frac{4\pi}{3\epsilon_0} \rho^2 r^4 dr$$

$$W = \int_0^R dW = \frac{4\pi}{15\epsilon_0} \rho^2 R^5$$

5.12. A wire is connected to the earth from a conducting sphere of radius a as shown in fig. 5.17.

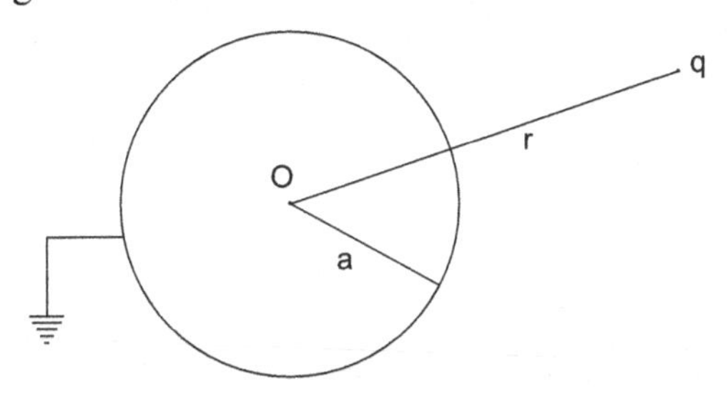

Fig. 5.17. Conducting sphere.

Compute the

(1) potential V

(2) charge density σ

(3) a force on the charge q due to the sphere.

■ ■ ■

6 Capacitors and Dielectrics

CHAPTER

We know that a work is needed to form a group of charges and this system consists of an electrostatic energy. In the early days, Volta used the concept of the electrical capacity in analogy with the heat capacity. The electric potential energy is the result of force between the charges. We can produce a useful device by placing two conductors at a distance that can store an electric energy. Thus, a device that can store the electric energy is called the capacitor. In this chapter we shall study about the capacitor and dielectrics.

6.1. THE CAPACITOR

It is also known as the condensor. The capacitor is an electrical device used for storing the electric charge. A capacitor consists of two conducting plates which are insulated from one another as shown in fig. 6.1. It stores the energy between the plates which are separated by a distance *d*.

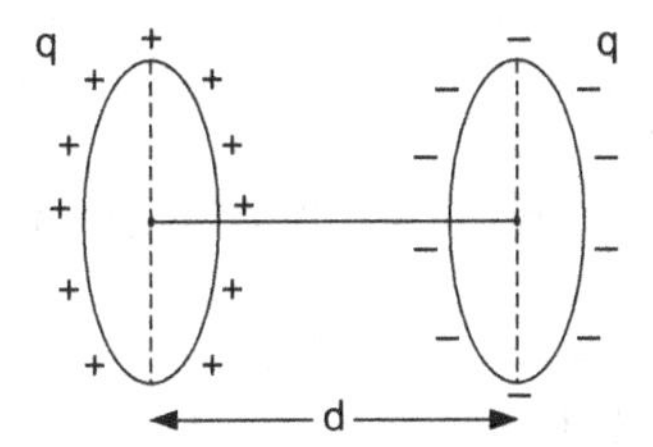

Fig. 6.1. A typical Capacitor

It is a fundamental component of an electric circuit. Our emphasis is on a quantity of importance called capacitance of a capacitor. If a capacitor stores a charge q, it means that one plate of a capacitor is at higher potential ($+q$) and other plate will be at lower potential having a charge $-q$. Thus, there is a potential difference between the plates of the capacitor. Hence, we say that the capacitance is a capacitor is a measure of the capacity of storing the charge for a given potential. However, the net charge on a capacitor is equal to zero $(q - q = 0)$.

(a) **Symbol.** The symbols of the capacitor are shown in fig. 6.2

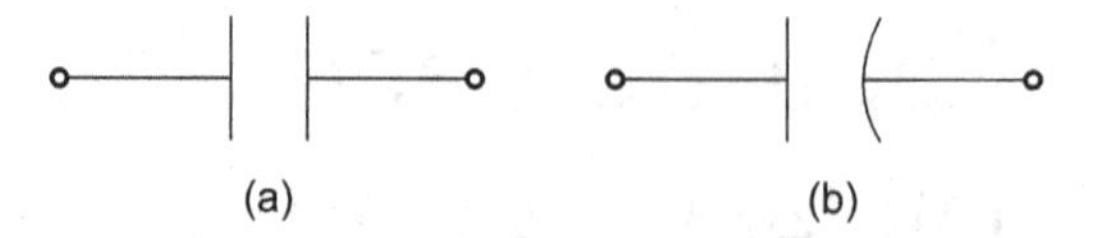

Fig. 6.2. Symbols of a capacitor.

(b) **Types of capacitor:** The capacitors are of different types and these can vary in size and shapes. The types of the capacitors are as follows.

(1) Cylindrical capacitor
(2) Tubular capacitor
(3) Electrolyte capacitor
(4) Paper coated capacitor
(5) Miniature capacitor
(6) Variable capacitor, etc.

The concept of the capacitor is that, it has two conductors carrying equal and opposite charges and these conductors are electrically isolated from each other.

(c) **Capacitance of a Capacitor:** If a charge q is given to one conductor, the charge is moved until it creates a $-q$ charge on the other conductor and a potential difference is produced between the conductors. If there is a potential difference V, then,

$$q \propto V$$

or $$q = CV \quad ...(6.1)$$

where C is a positive proportionality constant and is called capacitance of a capacitor. Moreover,

$$\boxed{C = \frac{q}{V}} \quad ...(6.2)$$

If we increase the charge, the potential difference V between the conductors increases. Thus, the capacitance of a capacitor is defined as the charge required to increase the potential of a capacitor by unity. The capacitance C depends on the following parameters.

(a) shape and size of the capacitor.
(b) material used between the plates (conductors).

However, capacitance C does not depend on the material of the conducting plates.

(d) **Unit of Capacitance:** In SI units, the unit of capacitance is coulombs/volt. The unit of C is farad also, in the honor of Faraday. Keeping always in mind that capacitance C is a positive quantity. Moreover,

$$1 \text{ Farad} = \frac{1 \text{ coulomb}}{1 \text{ volt}}$$

The other small units of C are given as

$$1\ \mu f = 10^{-6} f$$
$$1\ nf = 10^{-9} f$$
$$1\ pf = 10^{-12} f$$

6.2. CAPACITANCE OF A PARALLEL PLATE CAPACITOR

Consider a parallel plate capacitor as shown in fig. 6.3. The configuration contains two parallel plates of equal area and separated by a small distance d.

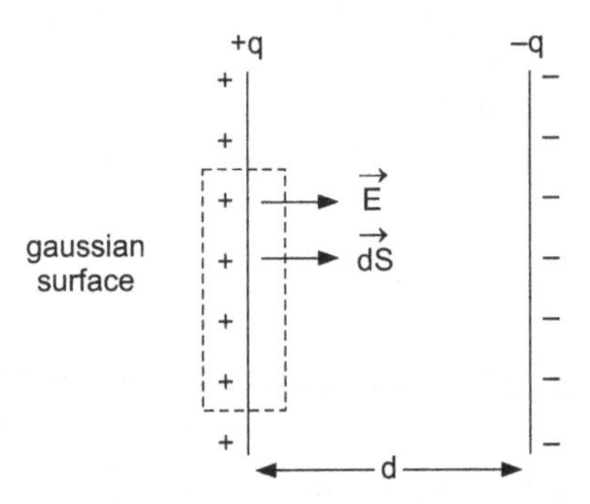

Fig. 6.3. Parallel plate capacitor of plate area A and the separation d.

Initially, the charge on each plate of the capacitor is zero. In the process of charging, the electrons move from one conductor to another and as a result, the conductors have equal and opposite charges. Thus, a uniform electric field is confined in the region between the plates. However, the electric field is not uniform near the edges of the plates as shown in fig. 6.4.

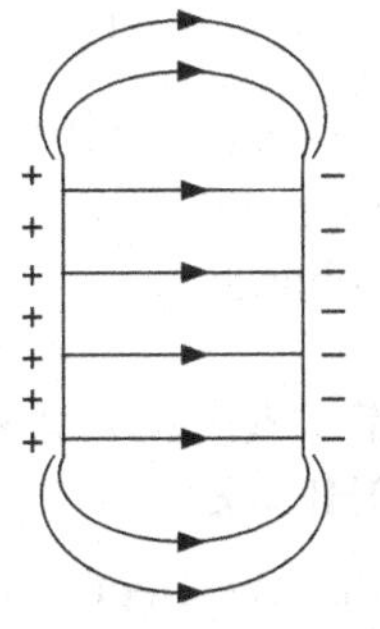

Fig. 6.4. Electric field between the plates of a capacitor

The plate having $+q$ charge is at higher potential and $-q$ charge is at lower potential. To calculate the capacitance C, let A be the area of each plate. According to Gauss's law,

$$\oint \vec{E} \cdot \vec{dS} = \frac{q}{\epsilon_0} \qquad ...(6.3)$$

or

$$EA = \frac{q}{\epsilon_0}$$

Thus,

$$E = \frac{q}{\epsilon_0 A} \qquad ...(6.4)$$

Since electric field is uniform between the plates, the potential across the capacitor is

$$V = \int dV \qquad ...(6.5)$$

$$= -\int_0^d \vec{E} \cdot \vec{dr}$$

$$= \frac{q}{\epsilon_0 A} \cdot d \qquad ...(6.6)$$

The capacitance of a parallel plate capacitor is given by

$$C = \frac{q}{V} = \frac{\epsilon_0 A}{d}$$

or

$$\boxed{C = \frac{\epsilon_0 A}{d}} \qquad ...(6.7)$$

It is experimentally observed that the capacitance C is proportional to the area of the plates A and inversely proportional to the separation between the plates d. That is,

$$C \propto \frac{A}{d} \qquad ...(6.8)$$

It can be seen that the capacitance C is a function of the geometry and the dielectric material used between the plates, and plates having large area can store more charge.

6.3. THE CAPACITANCE OF AN ISOLATED CONDUCTOR (SPHERE)

We know that the electric field lines emerge from the positive charge and end on the negative charge. In case of truly isolated conductor, the electric field lines leave the conductor and extend to infinity. For this, consider an isolated conducting sphere of the radius R. The potential of the conducting sphere of charge q is

$$V = -\int_{\infty}^{R} \vec{E} \cdot \vec{dr} = -\int_{\infty}^{R} \frac{q\, dr}{4\pi \epsilon_0 r^2}$$

$$= \frac{q}{4\pi \epsilon_0 R} \qquad \text{...(6.9)}$$

Thus, the capacitance of an isolated sphere is given by

$$C = \frac{q}{V}$$

or $$\boxed{C = 4\pi \epsilon_0 R} \qquad \text{...(6.10)}$$

The capacitance C depends on the geometry of the sphere and independent of charge q.

Example 6.1. Suppose that there are two concentric spherical shells of the radius a and b. The inner shell of radius a has $+q$ charge and the outer shell of radius b has $-q$ charge. Compute the capacitance of the spherical conductor.

Solution: Suppose that there are two concentric spherical shell, the inner shell of radius a and outer shell of the radius b as shown in fig. 6.5. Now,

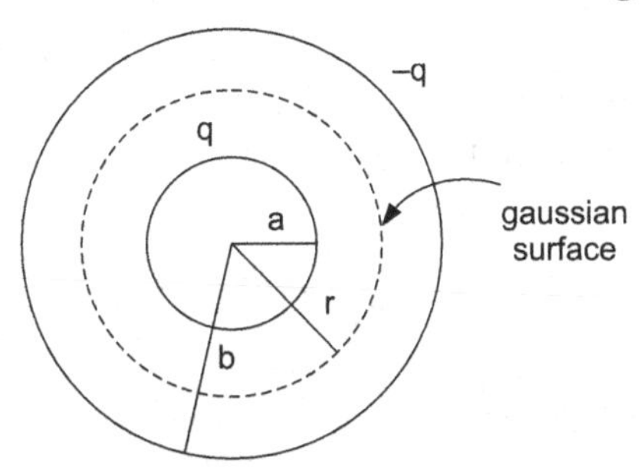

Fig. 6.5. Spherical Capacitor.

applying Gauss's law,

$$\oint E \cdot dS = \frac{q}{\epsilon_0}$$

$$E \cdot 4\pi r^2 = \frac{q}{\epsilon_0}$$

or $$E = \frac{1}{4\pi \epsilon_0} \frac{q}{r^2} \hat{r}$$

The potential difference between two shells is

$$V = \int_a^b dV$$

$$= -\int_a^b \vec{E} \cdot \vec{dr}$$

$$= \frac{q}{4\pi \epsilon_0} \int_a^b -\frac{dr}{r^2}$$

$$V = \frac{q}{4\pi \epsilon_0} \left(\frac{1}{a} - \frac{1}{b} \right)$$

Thus, the capacitance of the spherical capacitor is

$$C = \frac{q}{V}$$

or

$$C = \frac{4\pi \epsilon_0}{\left(\frac{1}{a} - \frac{1}{b} \right)}$$

6.4. CAPACITANCE OF A CYLINDRICAL CAPACITOR

The cylindrical capacitor is a combination of two coaxial cylindrical conductors. Let a and b be the radii of the inner and outer conductors respectively. We assume that the inner conductor has $+q$ charge and outer surface has $-q$ charge as shown in fig. 6.6.

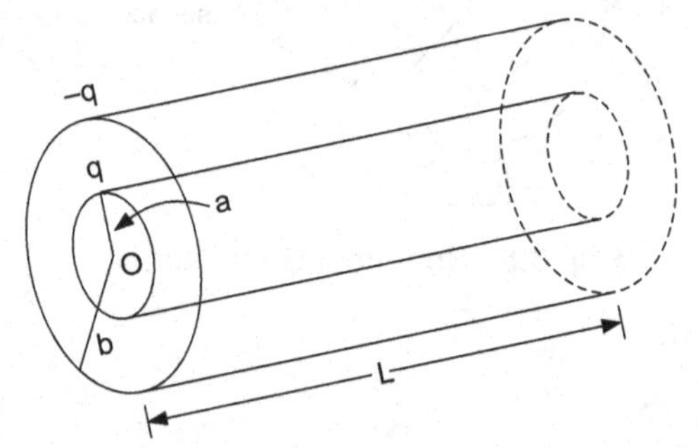

Fig. 6.6. Cylindrical capacitor of length L.

To find the capacitance of a cylindrical capacitor of the length L, first we obtain an expression for the potential V in the region $a < r < b$. According to Gauss's law.

$$\oint E \cdot dS = \frac{q}{\epsilon_0} \qquad ...(6.11)$$

$$E \cdot 2\pi r L = \frac{q}{\epsilon_0}$$

or $$E = \frac{1}{2\pi\epsilon_0 L}\frac{q}{r} \quad ...(6.12)$$

The potential difference between two conductors is

$$V = \int_a^b dV \quad ...(6.13)$$

$$= -\int_a^b \vec{E}\cdot\vec{dl} = +\int_a^b Edr$$

Thus,

$$V = \int_a^b \frac{q}{2\pi\epsilon_0 L}\frac{dr}{r} \quad ...(6.14)$$

$$= \frac{q}{2\pi\epsilon_0 L}\ln(b/a) \quad ...(6.15)$$

∴ Capacitance C is given by

$$C = \frac{q}{V}$$

∴ $$\boxed{C = \frac{2\pi\epsilon_0 L}{\ln b/a}} \quad ...(6.16)$$

Here, we can see that $C \propto L$ and C also depends on the radii of the inner and outer conductors.

6.5. SERIES AND PARALLEL COMBINATIONS OF CAPACITORS

We have many experimental situations where in electrical circuits, we use more than one capacitor. In this way, we have two types of combinations of the capacitors.

(1) Series combination of capacitors

(2) Parallel combination of capacitors.

(1) Series Combination of Capacitors: Fig. 6.7 shows the series combination of the capacitors. Consider three capacitors of capacitances C_1, C_2, and C_3 connected in series. When a potential difference V is applied, the plates of the capacitors acquire $+q$ and $-q$ charges as shown in fig. 6.7. Let V_1, V_2 and V_3 be the potential difference across the capacitors.

Now, the potential V will be the sum of the potential differences across the three individual capacitors. Thus, we have

$$V = V_1 + V_2 + V_3 \quad ...(6.17)$$

where

$$V_1 = \frac{q}{C_1},\ V_2 = \frac{q}{C_2},\ V_3 = \frac{q}{C_3}$$

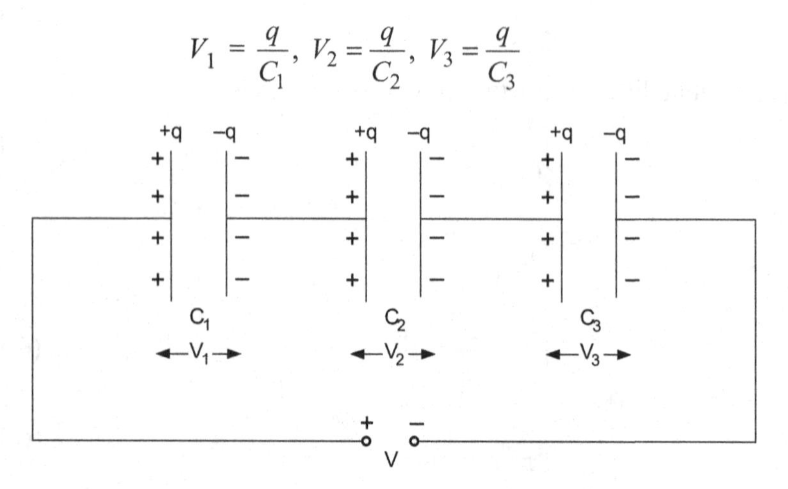

Fig. 6.7. Series combination of capacitors.

Therefore,

$$V = \frac{q}{C_1} + \frac{q}{C_2} + \frac{q}{C_3} \qquad ...(6.18)$$

or

$$\frac{V}{q} = \frac{1}{C_1} + \frac{1}{C_2} + \frac{1}{C_3} \qquad ...(6.19)$$

If the equivalent capacitance of the combination is C, then,

$$C = \frac{q}{V}$$

Substituting it in Eq. (6.19), we get

$$\boxed{\frac{1}{C} = \frac{1}{C_1} + \frac{1}{C_2} + \frac{1}{C_3}} \qquad ...(6.20)$$

It is clear that the equivalent capacitance of a series combination is always less than any individual capacitance in the combination.

(2) Parallel Combination of Capacitors: Consider three capacitors of capacitances C_1, C_2 and C_3 connected in parallel as shown in fig. 6.8.

In parallel combination, we see that the potential difference across the each capacitor will remain same and the charge on each capacitor is different from one another. We can write the total charge as,

$$q = q_1 + q_2 + q_3 \qquad ...(6.21)$$

where $q_1 = C_1V$, $q_2 = C_2V$ and $q_3 = C_3V$. On substituting in the Eq (6.20) we get

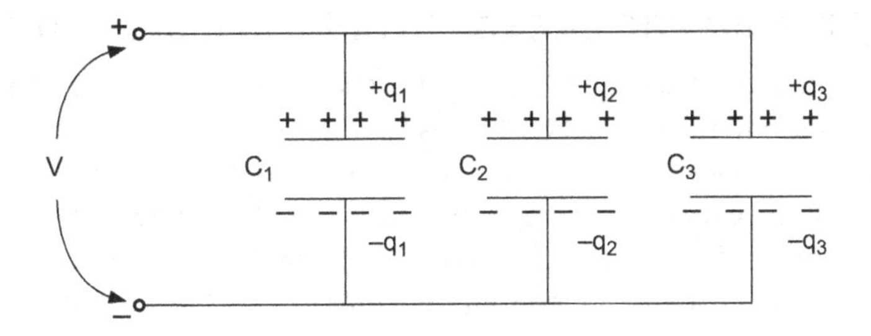

Fig. 6.8. Parallel combination of capacitors

$$q = C_1V + C_2V + C_3V$$

or $$\frac{q}{V} = C_1 + C_2 + C_3 \qquad ...(6.22)$$

If C represents the equivalent capacitance of the parallel combination,

$$C = \frac{q}{V} = C_1 + C_2 + C_3$$

or $$\boxed{C = C_1 + C_2 + C_3} \qquad ...(6.23)$$

It is clear that the equivalent capacitance C is larger than any of the single capacitor.

Example 6.2. Compute the equivalent capacitance of the combination of capacitors between the points A and B as shown in fig. 6.9.

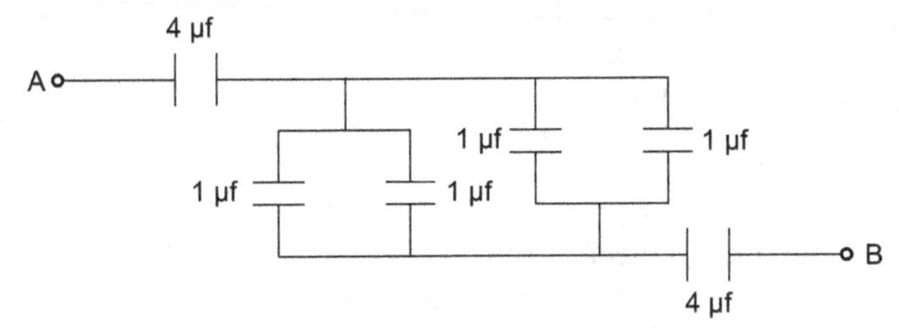

Fig. 6.9. Combination of capacitors.

Here, all the capacitors of 1μf are in parallel, thus

$$C' = C_1 + C_2 + C_3 + C_4$$
$$= 1 + 1 + 1 + 1 = 4\mu f$$

Now, all the capacitors of 4μf are in series, thus, equivalent capacitance is

$$\frac{1}{C} = \frac{1}{4} + \frac{1}{4} + \frac{1}{4} = \frac{3}{4}$$

$$C = \frac{4}{3} = 1.33 \ \mu f.$$

6.6. ELECTROSTATIC ENERGY STORED IN A CAPACITOR

When the process of charging of a capacitor takes place, there is a transfer of the charge from one plate to other. In this way, a capacitor stores the charge. Now, a question arises that, where is the charge stored. The answer of this question is that the charge or energy is stored in the electric field. To calculate the energy stored in the capacitor, suppose that the infinitesimally small charge dq is brought to a capacitor at a constant potential V, so, the work done is

$$dW = V\,dq \qquad \text{...(6.24)}$$

Since V varies during the charging of a capacitor, we have

$$V = \frac{q}{C} \qquad \text{...(625)}$$

Thus, on substituting the V from Eq. (6.25) in Eq. (6.24) we get

$$dW = \frac{q}{C}\,dq \qquad \text{...(6.26)}$$

We assume that the maximum charge on the capacitor is q_0, we can integrate the Eq. (6.26) to obtain the total work done by the electric field to charge the capacitor to q_0. Thus,

$$W = \int_0^{q_0} \frac{q}{C}\,dq$$

or

$$\boxed{W = \frac{1}{2}\frac{q_0^2}{C}} \qquad \text{...(6.27)}$$

Again, $q_0 = CV$

$$\therefore \quad \boxed{W = \frac{1}{2}CV^2} \qquad \text{...(6.28)}$$

This work stored as an electrostatic energy, hence,

$$\boxed{U = \frac{1}{2}\frac{q^2}{C} = \frac{1}{2}CV^2} \qquad \text{...(6.29)}$$

The charge q is in coulomb, C is in farad and V in volts, the energy U is in joules. Actually, this electric potential energy is not due to the mechanical work in bringing the charge from one plate to another. But it is a chemical energy of the battery that transformed into potential energy. Now, we may find a relation between the energy and the electric field where the energy is stored. Consider a parallel plate capacitor of the plate area A and the plates separation d. Then, capacitance C of parallel plate capacitance is

$$C = \frac{\epsilon_0 A}{d} \qquad \text{...(6.30)}$$

Now, the work required to charge a capacitor to q is

$$U = \frac{1}{2}\frac{q^2}{C} \quad ...(6.31)$$

since $\quad q = \sigma A$

where σ is the surface charge density, Thus,

$$U = \frac{1}{2}\frac{\sigma^2 Ad}{\epsilon_0} \quad ...(6.32)$$

The electric field between the plates of the parallel plate capacitor is given by

$$E = \frac{\sigma}{\epsilon_0}$$

On substituting in the Eq. (6.32) we get

$$U = \frac{1}{2}\epsilon_0 E^2 \cdot Ad \quad ...(6.33)$$

If V is the volume of the space between the plates, $V = Ad$, Thus,

$$\frac{U}{V} = U_d = \frac{1}{2}\epsilon_0 E^2$$

or

$$\boxed{U_d = \frac{1}{2}\epsilon_0 E^2} \quad ...(6.34)$$

where U_d is the energy density. Thus, the energy stored per unit volume (energy density) depends on the square of the electric field. In general, the electrostatic energy is given by

$$U = \frac{1}{2}\epsilon_0 \int_V E^2 dV \quad ...(6.35)$$

Example 6.3. Consider two capacitors of capacitances $C_1 = 2\mu f$ and $C_2 = 4\mu f$ in series. The capacitor C_1 has voltage of 6 volts and C_2 of 12 volts across their plates. compute the charge and energy stored in each capacitor.

Solution: Given,

$$C_1 = 2\mu f$$
$$V_1 = 6 \text{ volts}$$

and

$$C_2 = 4\mu f$$
$$V_2 = 12 \text{ volts}$$

$\therefore$ Charge

$$q_1 = C_1 V_1 = 12\ \mu C$$

and $$q_2 = C_2V_2 = 48\ \mu C$$

Now,

$$U_1 = \frac{1}{2}C_1V_1^2 = \frac{1}{2}\times 2\times 10^{-6}\times 36$$

or $$U_1 = 3.6\times 10^{-5}\text{ joules}$$

and

$$U_2 = \frac{1}{2}C_2V_2^2 = \frac{1}{2}\times 4\times 10^{-6}\times 144$$

or $$U_2 = 2.88\times 10^{-4}\text{ joules.}$$

6.7. FORCE BETWEEN PLATES OF A CAPACITOR

Since potential energy is stored in the electric field between the plates of a capacitor, it allows us to calculate the force between the plates of the capacitor. For this, consider a parallel plate capacitor of plate area A and the separation between the plates d. Then, capacitance C is

$$C = \frac{\epsilon_0 A}{d} \quad \text{...(6.36)}$$

and associated stored energy is given by

$$U = \frac{q^2}{2C}$$

$$= \frac{q^2 d}{2\epsilon_0 A} = \left(\frac{q^2}{2\epsilon_0 A}\right)d \quad \text{...(6.37)}$$

Suppose that the electric force between the plates is F and the plates separation increases by an amount dx where dx is known as the virtual displacement, then,

$$dW = -F\,dx$$

or $$dU = -F\,dx \quad \text{...(6.38)}$$

with the analogy of Eq. (6.37), for a small displacement dx, we have the change in electric energy as

$$dU = \frac{q^2}{2\epsilon_0 A}dx \quad \text{...(6.39)}$$

on comparing the Eqs. (6.38) and (6.39), we get

$$\boxed{F = -\frac{q^2}{2\epsilon_0 A}} \quad \text{...(6.40)}$$

Here, negative sign indicates that there is an attractive force between the plates of a capacitor.

Example 6.4. A spring is connected to a parallel plate capacitor as shown in fig. 6.10. If the capacitor is charged to q_0, show that the expansion in the spring is given by

$$x = \frac{q^2 \sigma}{2 \epsilon_0 k}$$

where k is force constant.

Solution: Let σ be the charge density, $\sigma = \frac{q_0}{A}$.

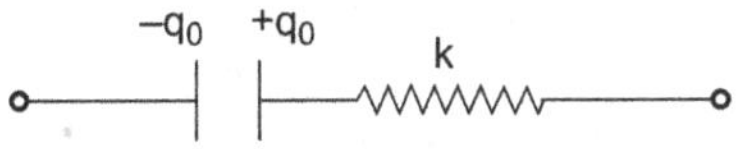

Fig. 6.10. A capacitor with a spring.

The restoring force acting on the right plate of the capacitor is,

$$F = -kx$$

The electric force due to right plate will be

$$F = q_0 E = q_0 \frac{\sigma}{2 \epsilon_0}$$

Since restoring force is balanced by the electric force, we have

$$\frac{q_0 \sigma}{2 \epsilon_0} = -kx$$

or
$$x = \frac{q_0 \sigma}{2 \epsilon_0 k}$$

6.8. ELECTRIC ENERGY STORED IN CONDUCTING SPHERE

Consider the conducting sphere of radius R as shown in fig. 6.11. The electric field for $r < R$ is zero.

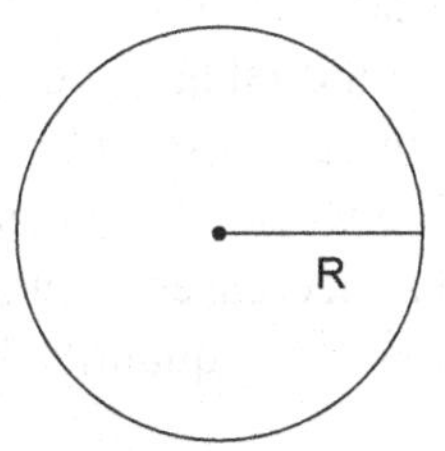

Fig. 6.11. Conducting sphere

and electric field at a distance r from the centre of the sphere is

$$E = \frac{1}{4\pi \epsilon_0} \frac{q}{r^2} \qquad ...(6.41)$$

The electrostatic energy is given by

$$U = \frac{1}{2} \epsilon_0 \int E^2 dV \qquad ...(6.42)$$

For $r > R$, consider a spherical shell of radius r and thickness dr, the volume of the shell is

$$dV = 4\pi r^2 dr \qquad ...(6.43)$$

Then,

$$U = \int_R^{\infty} \frac{1}{2} \epsilon_0 \left(\frac{q}{4\pi \epsilon_0 r^2} \right)^2 4\pi r^2 dr \qquad ...(6.44)$$

$$= \frac{q^2}{8\pi \epsilon_0} \int_R^{\infty} \frac{dr}{r^2}$$

$$\boxed{U = \frac{q^2}{8\pi \epsilon_0 R}} \qquad ...(6.45)$$

Since the potential at the surface of the sphere, is given by,

$$V = \frac{1}{4\pi \epsilon_0} \frac{q}{R} \qquad ...(6.46)$$

then, substituting Eq. (6.46) in the Eq. (6.45), we get

$$\boxed{U = \frac{1}{2} qV} \qquad ...(6.47)$$

6.9. DIELECTRIC MATERIALS

In the previous sections, we have discussed the electrostatic problems in the absence of the dielectric materials and all the discussions are taken with the air or vacuum. Now, we shall discuss the electrostatic problems with the dielectric materials. Dielectric material have no free charges and these are non-conducting materials such as mica, glass, rubber, wood or plastic etc. When a dielectric material is filled between the plates of a capacitor, the potential difference between the plates decreases. Thus, the capacitance C of the capacitor increases by a dimensionless quantity k which is called the dielectric constant of the material.

In the absence of the dielectric, a capacitor of capacitance C_0 has the charge q_0 as shown in fig. 6.12(a), the potential difference across the plate is given by

$$V_0 = \frac{q_0}{C_0} \quad ...(6.48)$$

Now, suppose that a dielectric material such as mica or glass is placed between the plates of a capacitor as shown in fig. 6.12.(b), the capacitance increases to a value given by

$$C = kC_0 \quad ...(6.49)$$

where k is known as dielectric constant of the material.

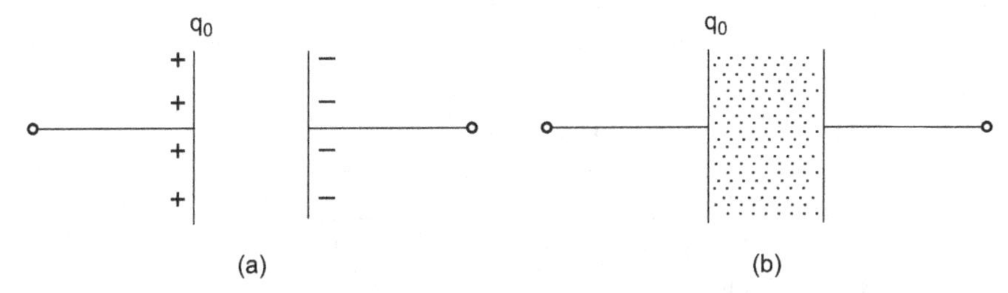

Fig. 6.12. Capacitor without dielectric (a) and with dielectric (b)

Therefore, the potential difference between the plates decreases to

$$V = \frac{V_0}{k} \quad ...(6.50)$$

The charge q_0 on the capacitor does not change, thus, we can write,

$$q_0 = C_0 V_0 = CV \quad ...(6.51)$$

or

$$\boxed{\frac{C}{C_o} = \frac{V_0}{V} = k} \quad ...(6.52)$$

If a dielectric slab is placed between the plates of a parallel plate capacitor, as shown in fig. 6.13, its capacitance is given by

$$C = \frac{\epsilon_0 A}{\left(\frac{d}{k}\right)}$$

or

$$\boxed{C = \frac{k \epsilon_0 A}{d}} \quad ...(6.53)$$

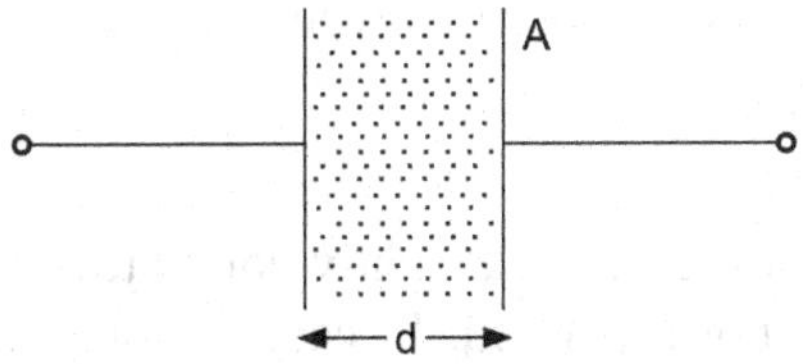

Fig. 6.13. Capacitor with dielectric slab.

where A is the area of plate and d is the separation between the plates. The dielectric constant k is just a number and is always greater than 1. That is, $k > 1$. The dielectric constant for various materials are listed in the table 6.1. Each dielectric has its own characteristic value of the electric field which is known as dielectric strength. By placing the a dielectric material between the plates of a capacitor, we have

(1) A maximum operating voltage for a capacitor.

(2) The capacitance of a capacitor increases by a factor k.

(3) It solves the mechanical problem between the conducting plates of a capacitor.

Table 6.1.

Material	Dielectric constant
Free space	1.00000
Dry Air	1.0006
Bakelite	4.9
Glass	~6
Mica	3 – 7.5
Paper	3.7
Water	80
Rubber	2.95
Transformer oil	2.0
Benzene	2.3

6.10. POLARISATION

In the solid materials, the atoms are arranged in a defined pattern and is known as crystal structure. In conductors, there are large number of free charge to move throughout the lattice. However, there are many solids in which the electrons are bound tightly to the atomic nuclei. Such solids are called the dielectrics. The behavior of the dielectrics depends on the electrical nature. The dielectric materials are unable to conduct the current. It contains the positive and negative charges in equal amount, it is electrically natural. In the materials, the binding is divided into two types.

(1) covalent binding.

(2) ionic binding.

In covalent binding, the atoms are tightly bound together, and the positive and negative atoms are not separated. In case of ionic crystals the positive ions may or may not be separated from the negative ions. But there are certain materials that have separation of the positive ions with the negative ions. Such materials possess permanent electric dipole moment. Thus, the polarisation at a point is defined as the vector sum of the electric dipole moments per unit volume of the material. We can explain the increase in the capacitance of a capacitor with dielectric on the basis of atomic or molecular point of view. When a conductor is placed in the electric field, there is a redistribution of charges on the surface of the conductor. Thus, the electric field inside the conductor is zero. Moreover, in case of dielectric, no charge is there to move, therefore a question arises that, how are the charges occur on the surface of the dielectric material. In this way, to answer the question, the dielectric materials are divided into two types as,

(1) Polar dielectric

(2) Non-polar dielectric

(1) Polar dielectrics: Such type of dielectrics have permanent electric dipole moment. The examples of polar dielectrics are HCl, HBr and H_2O etc. In polar dielectrics, the centre of negative charge does not coincide with the centre of positive charge as shown in fig. 6.14. The electric dipole moment of HCl is 3.40×10^{-30} coulomb-meter. In water molecule, the two hydrogen atoms are situated at an angle of 105° as shown in fig. 6.14(b).

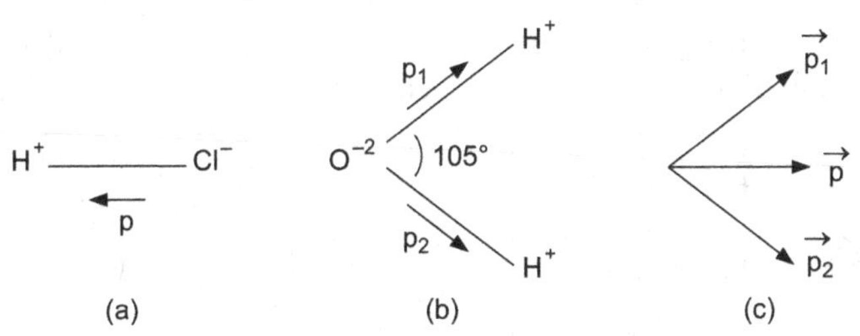

Fig. 6.14. Dipole moment of HCl and H_2O molecules.

For H_2O molecule, the resultant dipole-moment is

$$\vec{p} = \vec{p}_1 + \vec{p}_2 \qquad ...(6.54)$$

In the absence of external electric field, the orientations of these dipoles are random and the resultant electric dipole moment of the system becomes zero as shown in fig. 6.15(a). That is,

$$\sum_i p_i = 0 \qquad ...(6.55)$$

when these molecules are placed in the external electric field E_0, they tend to orient along the direction of the electric field as shown in fig. 6.15(b). The orientation of these polar molecules is due to the moment of force set up by the applied electric field. However, it can be seen that the orientation of these molecules along the direction of the electric field is not perfect. This is due to the thermal agitation. Thus, the orientation of the polar molecules creates a weak electric field in the direction opposite to the applied electric field.

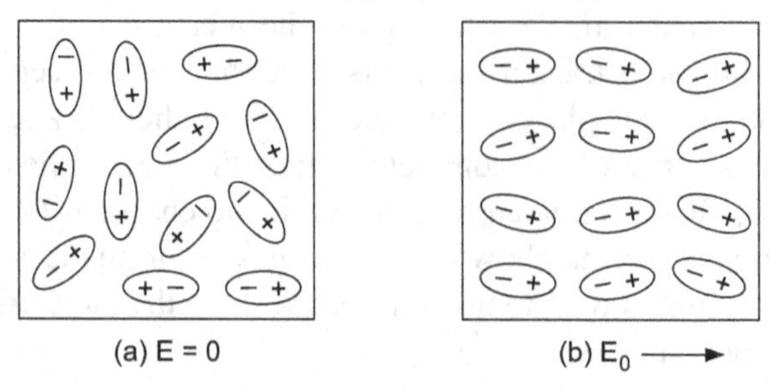

Fig. 6.15. The alignment of polar molecules without electric field and with electric field.

(2) Non-polar dielectrics: In non-polar dielectrics, the negative charge centre and the positive charge centre of a molecule coincide with each other as shown in fig. 6.16. (a) that is, they do not posses the permanent electric dipole moment. When the non-polar dielectric materials is placed in an external electric field, the material becomes polarised. Due to polarization, the positive and negative charges are then, separated and the electric dipole moments are induced as shown in fig. 6.16. (b). Such type of dipoles are called induced dipoles. The examples of non polar molecules are O_2, N_2, CO_2, CH_4 etc.

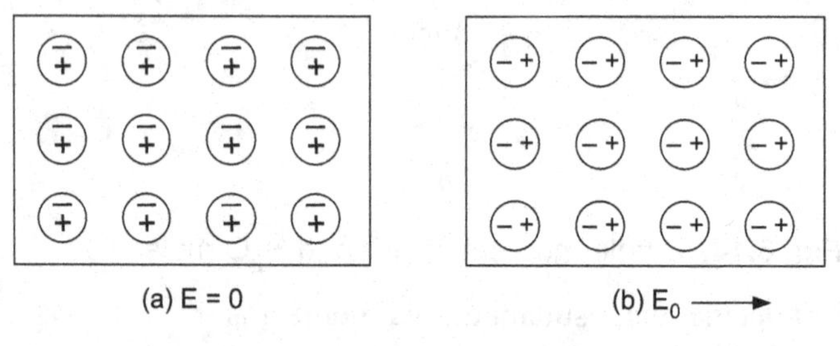

Fig. 6.16. Non-polar molecules (a) the positive and negative charge centres are coincide, when $E = 0$ (b) separation of charge centres in the field.

6.11. ELECTRIC FIELD IN DIELECTRICS

If a dielectric slab is placed between the charged plates of a parallel plate capacitor, the voltage across the plates decreases and as a result, an additional field is induced in the dielectric slab due to the polarization. This induced field points in the opposite direction to the applied electric field E_0 as shown in fig. 6.17. In this way, each molecule of the dielectric material becomes a tiny electric dipole. The charge σ_d is caused by the polarization and is called the bound charge, however, σ is the free charge density of the conducting plates of the capacitor. The charge σ_d does not leave its parent atom.

The polarization $\vec{P}$ of the dielectric material is the electric dipole moments per unit volume of the dielectric, and $\vec{p}$ is a vector quantity. Moreover, if p is the dipole moment of a molecule and N is the number of such molecules in a unit volume of the dielectric, then, polarization is

$$\vec{P} = Np \qquad ...(6.56)$$

Here, p is an average dipole moment of a molecule.

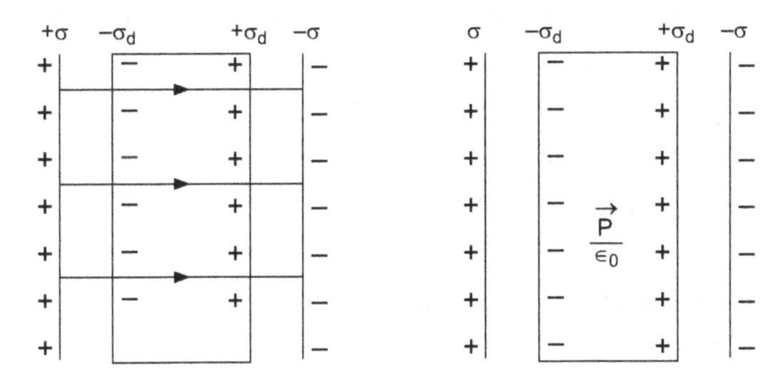

Fig. 6.17. Electric field in dielectric material.

If $\vec{P}(x, y, z)$ is the dipole moment per unit volume, the potential due to the dipole is given by

$$dV = \frac{1}{4\pi\epsilon_0} \frac{\vec{P} \cdot \vec{r}}{r^3} d^3r \qquad ...(6.57)$$

or

$$dV = \frac{1}{4\pi\epsilon_0} \vec{P} \cdot \nabla\left(\frac{1}{r}\right) d^3r \qquad ...(6.58)$$

where d^3r is a small volume. The net potential at any point may be found by integration of the Eq. (6.58). Thus,

$$V = \frac{1}{4\pi\epsilon_0} \int_V \vec{P} . \nabla\left(\frac{1}{r}\right) d^3r \qquad ...(6.59)$$

integrating the Eq. (6.59) by parts, we get

$$V = \frac{1}{4\pi\epsilon_0} \int_V \left(\nabla \cdot \frac{\vec{P}}{r}\right) d^3r - \frac{1}{4\pi\epsilon_0} \int_V \frac{\nabla \cdot \vec{P}}{r} d^3r \qquad ...(6.60)$$

For first term, using divergence theorem to convert volume integration into surface integration, we have

$$V = \frac{1}{4\pi\epsilon_0}\int_S \frac{\vec{P}\cdot\vec{dS}}{r} - \frac{1}{4\pi\epsilon_0}\int_V \frac{\nabla\cdot\vec{P}}{r}d^3r$$

or
$$V = \frac{1}{4\pi\epsilon_0}\int_S \frac{\vec{P}\cdot\hat{n}\,dS}{r} - \frac{1}{4\pi\epsilon_0}\int_V \frac{\nabla\cdot\vec{P}}{r}d^3r \qquad ...(6.61)$$

where S is the surface area which bounds the volume of the dielectric and the surface element $\vec{dS}$ points in outward direction. For free charge density, we may write

$$V = \frac{1}{4\pi\epsilon_0}\int_S \frac{\sigma\,dS}{r}$$

and
$$V = \frac{1}{4\pi\epsilon_0}\int_V \frac{\rho}{r}d^3r \qquad ...(6.62)$$

on comparison the Eqs (6.61) and (6.62), we get

Surface charge density is

$$\boxed{\sigma_d = \vec{P}\cdot\hat{n}} \qquad ...(6.63)$$

and volume charge density is given by

$$\boxed{\rho_d = -\nabla\cdot\vec{P}} \qquad ...(6.64)$$

where σ_d and ρ_d are bound charge densities and are distinguished from the free charge densities. Furthermore, if the surface covers all volume of the dielectric material, then $\nabla\cdot\vec{P}$ is sufficient to describe the source.

6.12. GAUSS'S LAW OF DIELECTRICS

The electric field due to dipoles can be evaluated from the bound charge density. Consider a parallel plate capacitor as shown in fig. 6.18 (a). If the dielectric material is not present between the plates, the electric field is given by

$$E_o = \frac{\sigma}{\epsilon_0} \qquad ...(6.65)$$

where σ is the free charge density. When a dielectric slab is placed between the plates, a charge q_p is induced over the surface of the dielectric, fig. 6.18 (b).

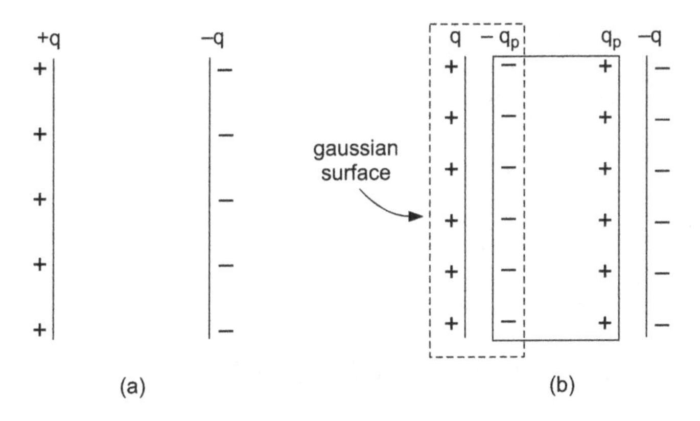

Fig. 6.18. (a) Capacitor without dielectric (b) with dielectric.

the net charge enclosed by the gaussian surface is given by

$$Q = q - q_p \qquad ...(6.66)$$

By Gauss's law,

$$\int_S \vec{E} \cdot \vec{dS} = \frac{Q}{\epsilon_0}$$

$$= \frac{q - q_p}{\epsilon_0} \qquad ...(6.67)$$

If A is the area of plate, then,

$$E = \frac{q - q_p}{\epsilon_0 A} \qquad ...(6.68)$$

on placing the dielectric slab, the potential is decreased by

$$V = \frac{V_0}{k} \qquad ...(6.69)$$

where k is known as the dielectric constant, and the electric field is

$$E = \frac{E_0}{k} \qquad ...(6.70)$$

on substituting the value of E_o from the Eq. (6.65) in the Eq. (6.70), we get

$$E = \frac{\sigma}{k \epsilon_0}$$

$$= \frac{q}{kA \epsilon_0} \qquad ...(6.71)$$

Now, equating the Eqs. (6.68) and (6.71), we have

$$\frac{q}{\epsilon_0 k A} = \frac{q - q_p}{\epsilon_0 A}$$

or

$$q_p = q\left(1 - \frac{1}{k}\right) \quad ...(6.72)$$

The surface charge density is

$$\sigma_d = \sigma\left(1 - \frac{1}{k}\right) \quad ...(6.73)$$

Therefore, from the Eqs. (6.67) and (6.72) we have

$$\boxed{\int_S \vec{E} \cdot \vec{dS} = \frac{q}{k\epsilon_0} = \frac{q}{\epsilon}} \quad ...(6.74)$$

where $\epsilon = \epsilon_o k$ is known as permittivity of the material medium.

or

$$\int_S \epsilon \vec{E} \cdot \vec{dS} = q$$

or

$$\boxed{\int_S \vec{D} \cdot \vec{dS} = q = q} \quad ...(6.75)$$

where $D = \epsilon E$ is called electric displacement vector.

6.13. POLARIZATION CURRENT DENSITY

If a dielectric slab is inserted into the plates of a capacitor, the polarization current occurs due to the motion of the bound charges. The polarization current is given by

$$I_d = \int_S \vec{J_d} \cdot \vec{dS} \quad ...(6.76)$$

since,

$$I_d = -\frac{\partial}{\partial t} \int_V \rho_d \, d^3r \quad ...(6.77)$$

The surface area bounds the volume d^3r. The rate of decrease of the bound charge at the surface S is equal to the rate of flow of charge through the surface. Thus,

$$-\frac{\partial}{\partial t} \int_V \rho_d \, d^3r = \int_S \vec{J_d} \cdot \vec{dS} \quad ...(6.78)$$

Using divergence theorem, we have

$$\int_S \vec{J_d} \cdot \vec{dS} = \int_V \nabla \cdot \vec{J_d} \, d^3r$$

Now

$$\int \nabla \cdot J_d \, d^3r = -\frac{\partial}{\partial t}\int_V \rho_d \, d^3r \qquad ...(6.79)$$

Substituting $\rho_d = -\nabla \cdot \vec{P}$, we get

$$\int_V \nabla \cdot \vec{J}_d \, d^3r = \frac{\partial}{\partial t}\int_V \nabla \cdot \vec{P} \, d^3r$$

or

$$\int \nabla \cdot J_d \, d^3r = \frac{\partial}{\partial t}\int_V \nabla \cdot \vec{P} \, d^3r \qquad ...(6.80)$$

For any arbitrary volume, we write

$$\boxed{\vec{J}_d = \frac{\partial \vec{P}}{\partial t}} \qquad ...(6.81)$$

we have used subscript '*d*' for dielectric.

6.14. LOCAL FIELD IN A DIELECTRIC MATERIAL

To evaluate the local field at a molecule or ion in the state of polarization, suppose that a dielectric slab is placed between the conductors as shown in fig. 6.19.

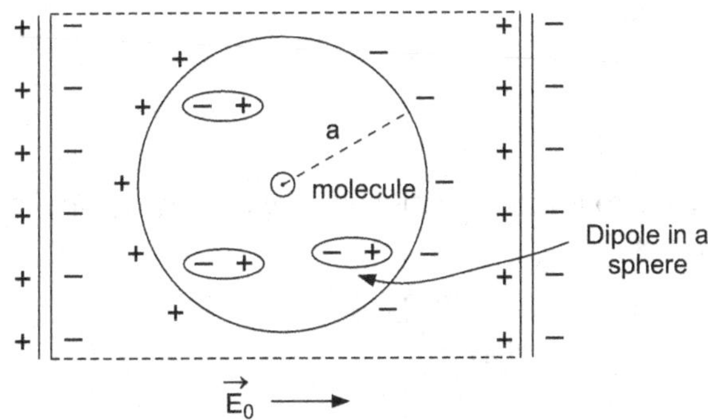

Fig. 6.19. Local field at the molecule in a sphere of radius *a*.

Since dielectric is placed in an external electric field, we can evaluate the field at a molecule at the centre of a sphere of radius *a*. This microscopic field is different from the macroscopic field governed by the Maxwell equations The electric field at the molecule is the vector sum of the external field and the electric field contributed by the dipoles within the dielectric material. Thus, the local field at the molecule is given by*

* Principle of Electricity and magnetism by Pugh, chapter-5.

$$E_{Loc} = E_0 + E_1 + E_2 + E_3 \qquad ...(6.82)$$

where

E_0 = Electric field due to free charge on the plates of a capacitor.

E_1 = Electric field due to the surface charge (induced charge or bound charge) on the dielectric material between the conductors.

E_2 = Field due to the polarization charge on the sphere of radius a

E_3 = Electric field due to individual dipoles in the sphere of radius a.

Now, we shall obtain the expressions for these electric fields as

(1) The electric field due to the charges on the surface of plates of the capacitor is

$$\vec{D} = \epsilon_0 \vec{E}_0 = \epsilon_0 \vec{E} + \vec{P} \qquad ...(6.83)$$

or

$$\vec{E}_o = \frac{\vec{D}}{\epsilon_0} = \vec{E} + \frac{\vec{P}}{\epsilon_0} \qquad ...(6.84)$$

(2) The induced charge on the surface of the dielectric material produces an electric field at the molecule is,

$$\vec{E}_1 = -\frac{\vec{D}}{\epsilon_0} = -\frac{\vec{P}}{\epsilon_0} \qquad ...(6.85)$$

(3) The field E_2 due to the polarization charge density σ_P on the sphere may be calculated as

$$dE_2 = \frac{\sigma_P \cos\theta \, dS}{4\pi \epsilon_0 a^2} \qquad ...(6.86)$$

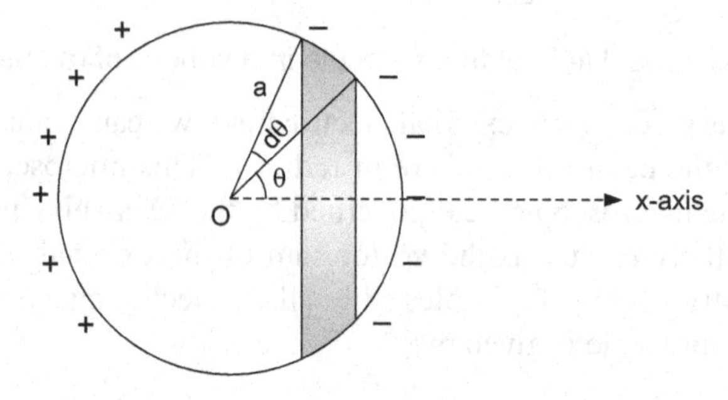

Fig. 6.20. Molecule in a spherical cavity of radius *a*.

dE_2 is the component of the electric field parallel to the x-axis, Fig. 6.20. The area of the element is

$$dS = 2\pi a \cdot a \sin\theta \, d\theta$$
$$= 2\pi a^2 \sin\theta \, d\theta \qquad ...(6.87)$$

Since the spherical cavity contains polarization charge density,

$$\sigma_p = \vec{P} \cos\theta \qquad ...(6.88)$$

Field due to polarization charge density is

$$E_2 = \int \frac{P \cdot 2\pi a^2 \cos^2\theta \sin\theta \, d\theta}{4\pi \epsilon_0 a^2}$$

$$= \frac{P}{2\epsilon_0} \int \cos^2\theta \sin\theta \, d\theta \qquad ...(6.89)$$

Integrating Eq. (6.89) in the limit $\theta = 0$ to $\theta = \pi$,

$$E_2 = \frac{P}{2\epsilon_0} \int_0^{\pi} \cos^2\theta \sin\theta \, d\theta$$

$$= \frac{P}{2\epsilon_0} \cdot \frac{2}{3} = \frac{P}{3\epsilon_0}$$

Thus,

$$E_2 = \frac{P}{3\epsilon_0} \qquad ...(6.90)$$

(4) E_3 is the field of the individual dipoles. We know that the potential due to the dipole at a distance r is

$$V = \frac{1}{4\pi\epsilon_0} \frac{\vec{p} \cdot \vec{r}}{r^3} \qquad ...(6.91)$$

The electric field due to the dipole is calculated as

$$E = -\nabla V$$

$$= \frac{1}{4\pi\epsilon_0 r^5} [3(\vec{p} \cdot \vec{r}) - \vec{p} r^2] \qquad ...(6.92)$$

Now, taking summation over all dipoles, we get

$$E_3 = \sum_i \frac{3(\vec{p_i} \cdot \vec{r_i}) r_i - p_i r_i^2}{4\pi \epsilon_0 r_i^5} \qquad ...(6.93)$$

or

$$E_3 = \frac{p}{4\pi \epsilon_0} \sum_i \frac{3x_i^2 r_i - r_i^2}{r_i^5} \qquad ...(6.94)$$

since all dipoles are oriented along x-axis, taking the symmetry over sphere, we have

$$\sum_i 3x_i^2 - r_i^2 = 0$$

$$\sum_i x_i y_i = 0 \qquad ...(6.95)$$

and $$\sum_i x_i z_i = 0$$

Hence,

$$\vec{E}_3 = 0 \qquad ...(6.96)$$

Thus, the local electric field at the molecule is given by

$$\vec{E}_{LOC} = \vec{E} + \frac{\vec{P}}{\epsilon_0} - \frac{\vec{P}}{\epsilon_0} + \frac{\vec{P}}{3\epsilon_0}$$

or $$\boxed{\vec{E}_{LOC} = \vec{E} + \frac{\vec{P}}{3\epsilon_0}} \qquad ...(6.97)$$

6.15. DIELECTRIC CONSTANT, ELECTRIC SUSCEPTIBILITY AND POLARIZABILITY

Consider a dielectric slab between the plates of a capacitor, then the surface charge density is

$$\sigma = \frac{q}{kA} + \sigma_P \qquad ...(6.98)$$

where

σ = free charge density.

k = dielectric constant.

A = area of plates of the capacitor.

and σ_P = polarization density.

Here, $\sigma = \frac{q}{A}$ and $\sigma_P = \frac{q'}{A} = \vec{P}$

Thus, the Eq. (6.98) can be written as

$$\hat{\sigma} = \epsilon_0 \vec{E} + \vec{P} \qquad ...(6.99)$$

where $E = \dfrac{q}{k\epsilon_0 A}$ is the macroscopic field.

we know that the electric displacement vector $\vec{D} = \hat{\sigma}$, we have

$$\vec{D} = \epsilon_0 \vec{E} + \vec{P} \quad ...(6.100)$$

for the free space,

$$\vec{P} = 0,$$

$$\vec{D} = \epsilon_0 \vec{E} \quad ...(6.101)$$

Moreover, for dielectric medium,

$$\vec{D} = \epsilon \vec{E} \quad ...(6.102)$$

where ϵ is the permittivity of the medium. On substituting the Eq. (6.102) in the Eq. (6.100), we get

$$\epsilon \vec{E} = \epsilon_0 \vec{E} - \vec{P} \quad ...(6.103)$$

or

$$\vec{P} = (\epsilon - \epsilon_0) \vec{E}$$

$$= (k\epsilon_0 - \epsilon_0) \vec{E}$$

or

$$\vec{P} = (k-1)\epsilon_0 \vec{E} \quad ...(6.104)$$

Now, we write

$$\frac{\vec{P}}{\epsilon_0 \vec{E}} = (k-1) \quad ...(6.105)$$

Here, we can define a dimension less quantity

$$\chi = \frac{\vec{P}}{\epsilon_0 \vec{E}} \quad ...(6.106)$$

where x (a Greek letter, χ-chi) is known as electric susceptibility.

or

$$\vec{P} = \chi \epsilon_0 \vec{E} \quad ...(6.107)$$

Thus, dipole moment per unit volume, polarization is directly related to the macroscopic field in the dielectric. Moreover, the electric polarizability is defined as the ratio of the induced dipole moment to the local electric field, it is denoted by α. Thus,

$$\alpha = \frac{p}{E_{LOC}}$$

or $$\vec{P} = \alpha \vec{E}_{LOC} \qquad ...(6.108)$$

or $$P \propto E_{LOC} \qquad ...(6.109)$$

Thus, the induced dipole moment is proportional to the E_{LOC}. Since, we know that

$$\vec{P} = N\vec{p} \qquad ...(6.110)$$

On substituting in the Eq. (6.109), we get

$$\boxed{\vec{P} = N\alpha \vec{E}_{LOC}} \qquad ...(6.111)$$

From the Eq. (6.11), we have

$$\vec{P} \propto \vec{E}_{LOC}$$

That is, polarization is proportional to the local electric field. Such dielectrics are known as linear dielectrics.

6.16. CLAUSIUS-MOSSOTTI EQUATION

Clausius-Mossotti equation is a relation between the electric polarizability and dielectric constant. We know that the local field at the molecule is given by

$$\vec{E}_{LOC} = \vec{E} + \frac{\vec{P}}{3\epsilon_0} \qquad ...(6.112)$$

and the dipole moment is given by

$$\vec{p} = \alpha \vec{E}_{LOC} \qquad ...(6.113)$$

where $\propto$ is electric polarizability. Moreover, the polarization is

$$\vec{P} = N\vec{p}$$

$$= N\alpha E_{LOC}$$

or $$\vec{P} = N\alpha\left(\vec{E} + \frac{\vec{P}}{3\epsilon_0}\right) \qquad ...(6.114)$$

Since, from the Eq. (6.104), we have

$$\vec{P} = \epsilon_0 (k-1)\vec{E} \qquad ...(6.115)$$

Substituting the Eq. (6.115) in eq. (6.114), we get

$$\epsilon_0 (k-1)\vec{E} = N\alpha\left[\vec{E} + \frac{\epsilon_0 (k-1)\vec{E}}{3\epsilon_0}\right]$$

or $$3\in_0(k-1) = N\alpha(k+2)$$

or $$\boxed{\alpha = \frac{3\in_0}{N}\frac{(k-1)}{(k+2)}} \quad ...(6.116)$$

This is known as famous clausius-Mossotti equation. If ρ is the density of the mass and N_A is the Avogadro's number, then

$$N = \frac{\rho}{M} N_A \quad ...(6.117)$$

where M is the molecular weight. Substituting the value of N from the Eq. (6.117) in the Eq. (6.116), we get

$$\boxed{\left(\frac{M}{\rho}\right)\left(\frac{(k-1)}{(k+2)}\right) = \frac{N_A}{3\in_0}\cdot\alpha} \quad ...(6.118)$$

Furthermore, if the number of molecules per unit volume approaches to $\left(\frac{3\in_0}{\alpha}\right)$. The dielectric constant $k \to \infty$. Since k is finite and small for the liquids and gases, the Eq. (6.118) provides the better results. This is not valid for crystalline solids, this is because that the dipole interactions are very complex in the solids.

6.17. LANGEVIN EQUATION OF POLAR-DIELECTRICS

The polarization is an important phenomenon in some applications of the physics and engineering. The polar dielectrics are those where the positive and negative ions are separated with a distance. That is, these dielectrics consist of permanent electric dipoles. In the absence of the electric field, the dipoles are aligned randomly throughout the dielectric and an external electric field causes the electric dipoles to orient in the direction of the field. However, the alignment of the dipoles is not complete due to the thermal agitation. In Langevin-Debye theory, the polarizability depends on temperature. Now, suppose that a system consists of N-dipoles in a unit volume. The potential energy for a dipole in a uniform electric field is given by

$$U = -\vec{p}\cdot\vec{E}$$
$$= -pE\cos\theta \quad ...(6.119)$$

According to Boltzmann's law, the number of electric dipoles per unit volume oriented in the direction between θ and $\theta + d\theta$ is

$$dN = Ce^{-U/KT}\sin\theta\, d\theta \quad ...(6.120)$$

where C is a constant

Thus, $$dN = C\, e^{pE\cos\theta/KT}\sin\theta\, d\theta \quad ...(6.121)$$

Since N is the total number of molecules in a unit volume, then,

$$N = \int_o^\pi dN \qquad \text{....(6.122)}$$

$$= C\int_o^\pi e^{pE\cos\theta/KT}\sin\theta\, d\theta$$

Let $u = \dfrac{pE}{KT}$,

$$N = C\int_o^\pi e^{u\cos\theta}\sin\theta\, d\theta \qquad \text{...(6.123)}$$

Now, polarization of the molecules whose dipole moments lie in the range θ and $\theta + d\theta$ is given by

$$\vec{dP} = \vec{p}\cos\theta\, dN$$

$$= C\vec{p}\, e^{u\cos\theta}\sin\theta\cos\theta \qquad \text{...(6.124)}$$

on substituting the value of C from the Eq. (6.123) in the Eq. (6.124), we get

$$\vec{dP} = \frac{N\vec{p}e^{u\cos\theta}\sin\theta\cos\theta\, d\theta}{\int_o^\pi e^{u\cos\theta}\sin\theta\, d\theta} \qquad \text{...(6.125)}$$

To compute the net polarization, we may integrate Eq. (6.125) from $\theta = 0$ to $\theta = \pi$. We have

$$\vec{P} = N\vec{p}\,\frac{\int_o^\pi e^{u\cos\theta}\sin\theta\cos\theta\, d\theta}{\int_o^\pi e^{u\cos\theta}\sin\theta\, d\theta} \qquad \text{...(6.126)}$$

Suppose that $u\cos\theta = y$

$u\sin\theta\, d\theta = -dy$

Thus, the equation (6.126) takes the form,

$$\vec{P} = \left(\frac{N\vec{p}}{u}\right)\frac{\int_{-u}^{u} y e^y\, dy}{\int_{-u}^{u} e^y\, dy} \qquad \text{...(6.127)}$$

$$= \left(\frac{N\vec{p}}{u}\right)\frac{u(e^u + e^{-u}) - (e^u - e^{-u})}{(e^u - e^{-u})}$$

$$= N\vec{p}\left[\frac{(e^u + e^{-u})}{(e^u - e^{-u})} - \frac{1}{u}\right]$$

or $$\vec{P} = N\vec{p}\left[\cot hu - \frac{1}{u}\right] \qquad ...(6.128)$$

or $$\vec{P} = N\vec{p}\ L(u) \qquad ...(6.129)$$

The Eq. (6.129) is known as Langevin equation and $L(u)$ is called the Langevin function. We can expand $L(u)$ as

$$L(u) = \frac{u}{3} - \frac{u^3}{45} + \ldots \qquad ...(6.130)$$

and inverse Langevin function is

$$L^{-1}(u) = 3u + \frac{9}{5}u^3 - \frac{297}{175}u^5 + \ldots \qquad ...(6.131)$$

The plot of the Langevin function $L(u)$ against u is shown in fig. 6.21. Now, we have following cases as

Case –1 For large u, that is, at low temperature $T << \frac{pE}{K}$,
we have, $L(u) \sim 1$

Thus, polarization is, then,

$$\vec{P} = N\vec{p} \qquad ...(6.132)$$

At the low temperature, all the dipoles are aligned parallel to the electric field and thus polarization $\vec{P}$ is maximum.

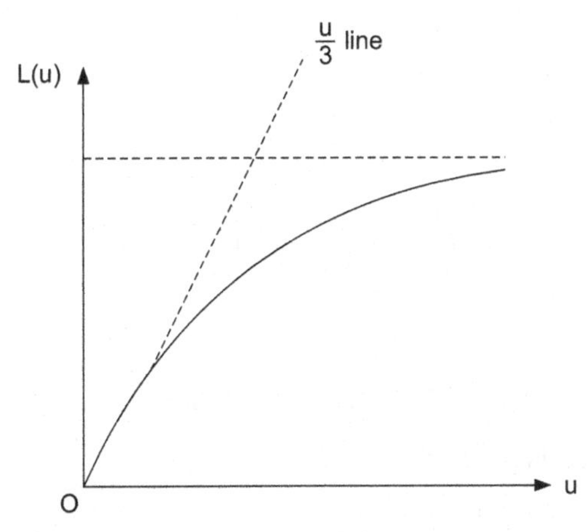

Fig. 6.21. Langevin function $L(u)$ against u.

Case-2: For high temperature, that is, $T >>> \frac{pE}{K}$,

$$\vec{P} = N\vec{p}\,\frac{u}{3}$$

or
$$P = \frac{Np^2E}{3KT} \qquad ...(6.133)$$

or
$$P \sim \frac{1}{T}$$

Thus, the polarization is directly proportional to the electric field, moreover if E is uniform, P is inversely proportion to the temperature. This dielectric is linear and the electric susceptibility is give by

$$\chi = \frac{\vec{P}}{\vec{E}} = \frac{Np^2}{3KT} \qquad ...(6.134)$$

or
$$\boxed{\chi = \frac{C}{T}} \qquad ...(6.135)$$

6.18. ENERGY STORED IN A DIELECTRIC

Consider a dielectric slab of thickness d inserted into a parallel plate capacitor of plate area A. Assuming that the dielectric slab is fitted completely between the plates as shown in fig. 6.22.

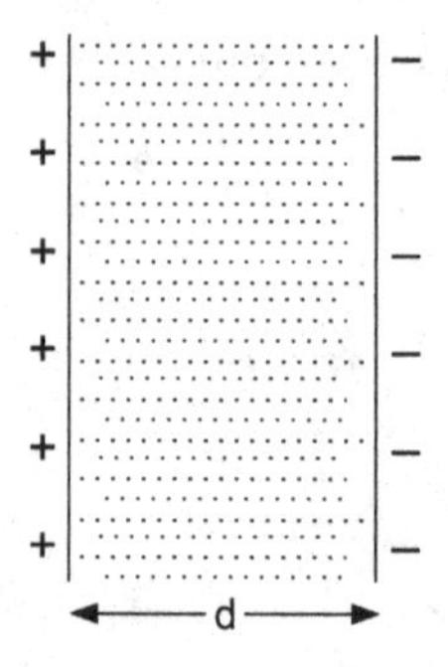

Fig. 6.22. Dielectric in a field $\vec{E}$

The energy stored is given by

$$U = \frac{1}{2}\in E^2 \cdot Ad \qquad ...(6.136)$$

since potential is,

$$V = Ed \qquad ...(6.137)$$

Substituting the value of E from the Eq. (6.137) in the Eq. (6.136), we get

$$U = \frac{1}{2} \in \cdot Ad \frac{V^2}{d^2}$$

$$= \frac{1}{2}\left(\frac{\in A}{d}\right)V^2 \quad \text{...(6.138)}$$

The capacity of the capacitor will be

$$C = \frac{\in A}{d} \quad \text{...(6.139)}$$

Substituting for C in the Eq. (6.138), we get

$$\boxed{U = \frac{1}{2} CV^2} \quad \text{...(6.140)}$$

From the Eq. (6.136), the energy density is

$$U_d = \frac{U}{Ad} = \frac{1}{2} \in E^2$$

or

$$U_d = \frac{1}{2}(\in \vec{E})\cdot \vec{E}$$

$$= \frac{\vec{E}\cdot\vec{D}}{2} \quad \text{...(6.141)}$$

Thus, we have

$$\boxed{U_d = \frac{D^2}{2\in}}, \quad D = \in E \quad \text{...(6.142)}$$

which is the expression for the energy density.

Example 6.5 Find the capacitance of a capacitor when it is filled with two dielectrics of dielectric constant k_1 and k_2 as shown in fig. 6.23.

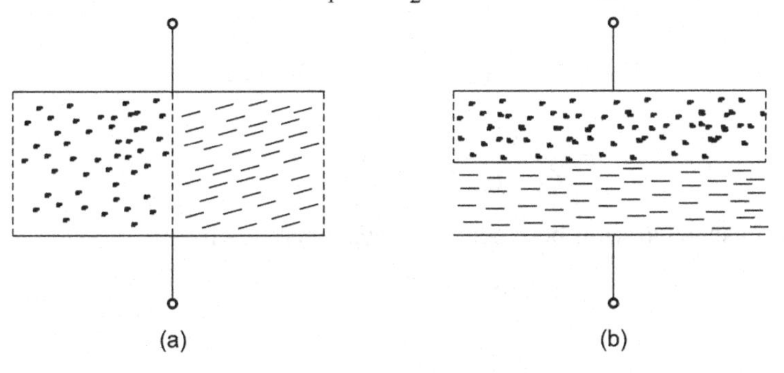

Fig. 6.23. Capacitors with different dielectrics.

Solution: In the fig. 6.23 (a), we have a parallel combination of the capacitors with dielectric constants k_1 and k_2. In this case, the plate are becomes half, Fig. 6.24.

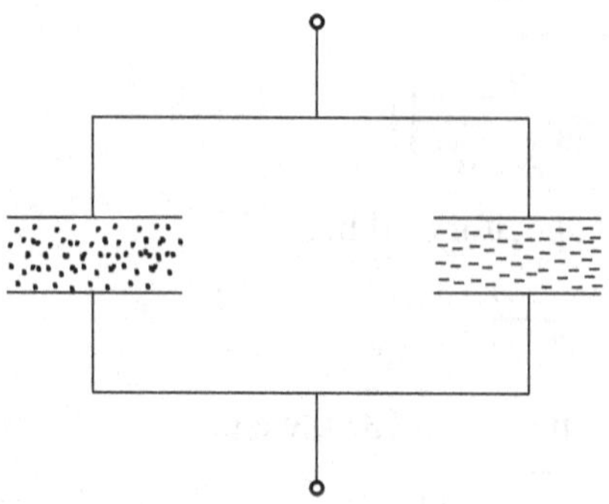

Fig. 6.24. Two capacitors with different dielectrics.

∴ Total capacitance is

$$C = C_1 + C_2$$

But,
$$C_1 = \frac{k_1 \in_0 A/2}{d}$$

and
$$C_2 = \frac{k_2 \in_0 A/2}{d}$$

where d is the plates separation and A is the plate area. Thus, on substitution for C_1 and C_2, we get

$$C = \frac{\in_0 (k_1 + k_2) A}{2d}$$

For Fig. 6.23(b), we have a series combination of C_1 and C_2, Fig. 6.25.

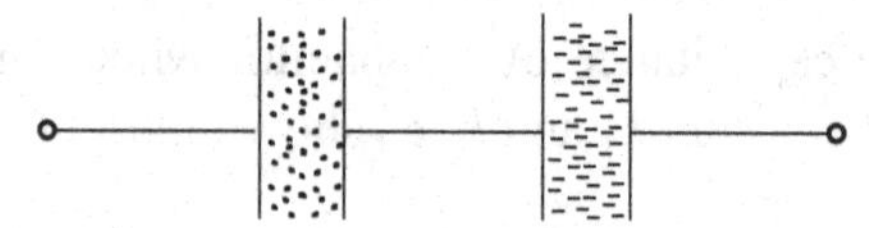

Fig. 6.25. Capacitors with dielectrics.

$$\therefore \qquad C = \frac{C_1 C_2}{C_1 + C_2}$$

In this case separation between the plates becomes half.

$$\therefore \qquad C_1 = \frac{\in_0 k_1 A}{d/2}$$

and $$C_2 = \frac{\in_0 k_2 A}{d/2}$$

on substituting in C, we get

$$C = \frac{2 \in_0 A}{d}\left(\frac{k_1 + k_2}{k_1 + k_2}\right).$$

Example 6.6. Two capacitors, where one capacitor is charged to a potential V_1 and other capacitor is uncharged, are connected in parallel. Then,

(1) Prove that when equilibrium is reached, the charge on the second capacitor is equal to the ratio of its capacitance to the sum of two capacitances of the capacitors multiplied by the initial charge.

(2) Show that the initial energy is greater than the final energy.

Solution: Let C_1 and C_2 be the capacitances of the capacitors and V be the common potential, then

$$V = \frac{\text{Total charge}}{\text{Total capacitances}}$$

$$= \frac{q_1 + q_2}{C_1 + C_2}$$

$$= \frac{C_1 V_2 + C_2 \cdot O}{C_1 + C_2} = \frac{C_1 V_1}{C_1 + C_2}$$

$\therefore$ Thus, the charge on the capacitor C_2 is

$$q_2 = C_2 V$$

$$= \frac{C_1 C_2}{C_1 + C_2} \cdot V_1 = \frac{C_1 q_1}{C_1 + C_2} \text{ (initial charge).}$$

Now, charge on the capacitor C_1 is

$$q'_1 = C_1 V$$

$$= \frac{C_1}{C_1 + C_2} \cdot C_1 V_1$$

$$= \frac{C_1}{C_2 + C_1} \cdot q_1$$

(2) The initial energy

$$U_i = \frac{1}{2} C_1 V_1^2$$

Final energy, when the uncharged capacitor is connected, will be

$$U_f = \frac{1}{2}(C_1 + C_2)V^2$$

$$= \frac{1}{2}(C_1 + C_2)\frac{C_1^2V_1^2}{(C_1 + C_2)^2}$$

$$= \frac{1}{2}\frac{C_1^2V_1^2}{C_1 + C_2}$$

$\therefore$

$$U_i - U_f = \frac{1}{2}C_1V_1^2 - \frac{1}{2}\frac{C_1^2V_1^2}{C_1 + C_2}$$

$$= \frac{1}{2}C_1V_1^2 - \frac{1}{2}\frac{C_1^2V_1^2}{C_1 + C_2}$$

$$= \frac{1}{2}\frac{C_1C_2}{C_1 + C_2}\cdot V_1^2$$

or $$U_i - U_f > 0$$

or $$U_i > U_f$$

Example 6.7. Consider a parallel plate capacitor of plate area A and plates separation d. A dielectric slab of thickness $t < d$, is placed between the plates as shown in fig. 6.26. Compute the capacitance.

Solution: Let C_0 and C be the capacitances of the capacitor without and with dielectric respectively. When dielectric is not present, the capacitance is

$$C_o = \frac{\epsilon_0 A}{d}$$

Now, to calculate the capacitance C of a capacitor with dielectric, we have to compute the potential difference between the plates.

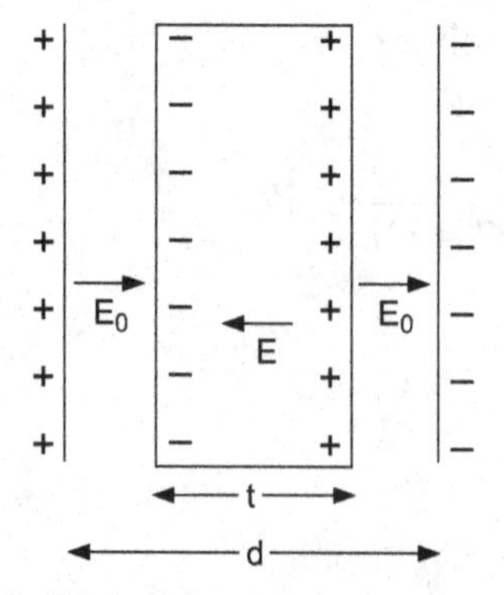

Fig. 6.26. Dielectric slab in an electric field.

The electric field between the plates are

$$E_o = \frac{q}{\epsilon_0 A}$$

and
$$E = \frac{E_0}{k} = \frac{q}{k \epsilon_0 A}$$

∴ the potential difference

$$\Delta V = E_0(d - t) + Et$$

$$= \frac{q}{\epsilon_0 A}(d - t) + \frac{q}{k \epsilon_0 A} \cdot t$$

$$= \frac{q}{\epsilon_0 A}\left[(d - t) + \frac{t}{k}\right]$$

or
$$\Delta V = \frac{q}{\epsilon_0 A}\left[d - t\left(1 - \frac{t}{k}\right)\right]$$

Thus, capacitance is,

$$C = \frac{q}{\Delta V}$$

or
$$C = \frac{\epsilon_0 A}{\left[d - t\left(1 - \frac{1}{k}\right)\right]}$$

From the above result, we have following points,

(1) As thickness t of the dielectric slab approaches zero, that is, $t \to 0$,

$$C = \frac{\epsilon_0 A}{d} = C_0$$

(2) As dielectric constant $k \to 1$,

$$C = \frac{\epsilon_0 A}{d} = C_0$$

(3) When $t = d$,

$$C = \frac{k \epsilon_0 A}{d} = \frac{k \epsilon_0 A}{t}$$

or
$$C = kC_0$$

Example 6.8. Consider two capacitors of equal capacitance C connected in parallel and this system is charged to a voltage V_1. Now, the system is disconnected from the voltage source and in one capacitor, a dielectric slab of dielectric constant k is placed. Then, compute

(a) The free charge transferred from one capacitor to another.

(b) Final potential V_2 across the capacitor.

Solution:

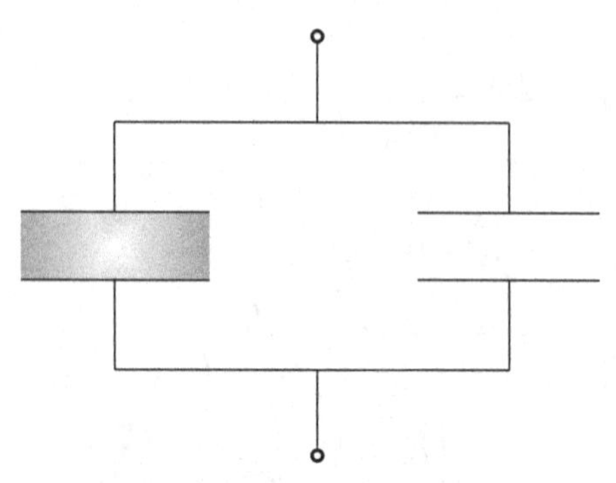

Fig. 6.27. Capacitors system.

Consider two capacitors as shown in fig. 6.27.

The potential V_2 is

$$V_2 = \frac{\text{Total charge}}{\text{Total capacitance}}$$

$$= \frac{C_1 V_1 + C_2 \cdot O}{C + kC} = \frac{CV_1}{C(1+k)}$$

or

$$V_2 = \frac{V_1}{(1+k)}$$

Now, charge on the capacitor is

$$q' = C_2 V_2$$

$$= \frac{kCV_1}{(1+k)}$$

or

$$q' = \frac{kCV_1}{(1+k)}.$$

6.19. BOUNDARY CONDITIONS AT THE INTERFACE OF TWO DIELECTRICS

The conditions that satisfied by the field at the interface separating two media are known as boundary conditions. Maxwell's field equations are used to determine these boundary conditions.

(1) Boundary Conditions for the Electric Displacement $\vec{D}$.

Consider two dielectric media separated by an interface, as shown in fig. 6.28.

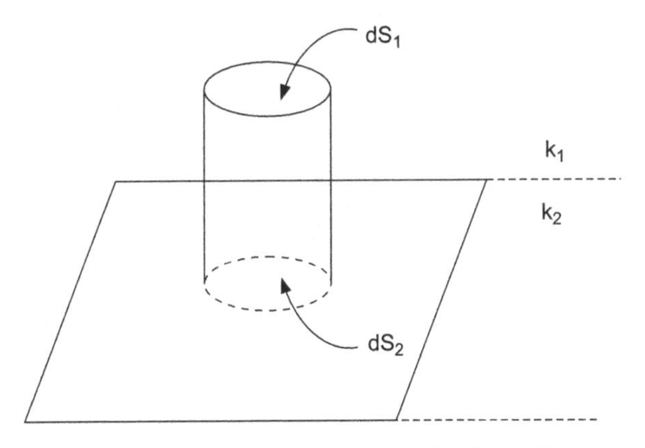

Fig. 6.28. Gaussian pillbox at the boundary.

we have Maxwell's equation as

$$\int_S \vec{D} \cdot \vec{dS} = q = \int \sigma \, dS \qquad ...(6.143)$$

or $\int_S \vec{D} \cdot \vec{dS_1} + \int_{S_2} \vec{D_2} \cdot \vec{dS_2} = \sigma$

or $$\vec{D}_{2n} - \vec{D}_{1n} = \sigma \qquad ...(6.144)$$

where σ is the surface charge density for free charge. From the Eq. (6.144), it is clear that the normal component of $\vec{D}$ is discontinuous at the boundary. If no free charge is present at the interface, $\sigma = 0$, thus,

$$\vec{D}_{1n} = \vec{D}_{2n}$$

or $$k_1 E_{1n} = k_2 E_{2n} \qquad ...(6.145)$$

Therefore, the normal component of $\vec{D}$ is continuous across the boundary. Now, the behavior of the tangential component of the electric field can be determined by the Maxwell's equation

$$\oint \vec{E} \cdot \vec{dl} = 0 \qquad ...(6.146)$$

we have a closed path at the interface, as shown in fig. 6.29. Now, applying the Eq. (6.146) to the path *abcda*. The paths *bc* and *da* are perpendicular to the interface, thus, the integrals are vanished, we have

$$\int_a^b E \cdot dl - \int_c^d E \cdot dl = 0$$

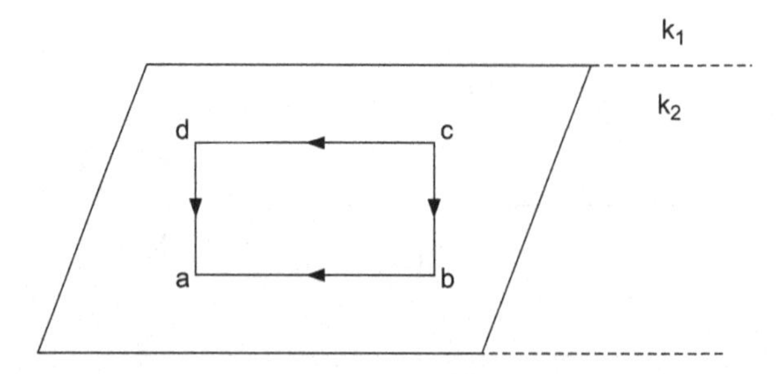

Fig. 6.29. Interface between two media.

or $$(E_{2t} - E_{1t})L = 0$$

or $$E_{1t} = E_{2t} \quad ...(6.147)$$

Thus, the tangential component of the electric field $\vec{E}$ is continuous across the boundary.

Example 6.9. Consider a parallel plate capacitor of length l and width b as shown in fig. 6.30. The plates separation is d. If a dielectric slab is inserted partially between the plates of the capacitor, Find

(a) Capacitance of the system

(b) energy stored in the system

(c) force acting on the slab due to the electric field in the capacitor.

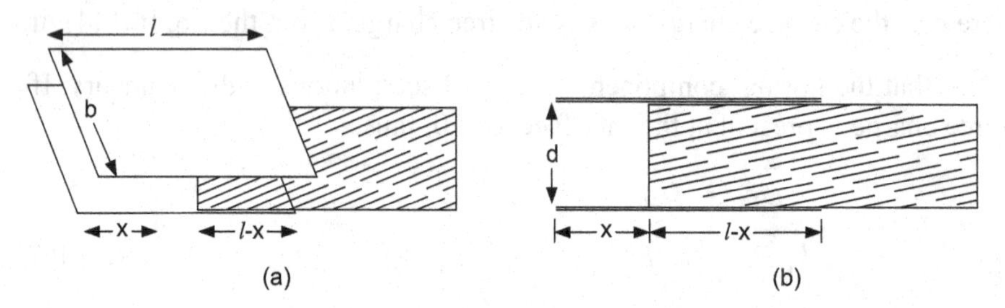

Fig. 6.30. A dielectric in a capacitor.

Solution: The electric field between the plates

$$E = \frac{V}{d}$$

or $$V = Ed$$

(a) The capacitance of the system is

$$C = C_1 + C_2$$

where $$C_1 = \text{Capacitance without dielectric}$$

$$= \frac{\epsilon_0 A}{d}$$

$$C_1 = \frac{\epsilon_0 bx}{d}$$

and $$C_2 = \text{capacitance with dielectric}$$

$$= \frac{\epsilon_0 k (l-x)b}{d}$$

Therefore,

$$C = \frac{\epsilon_0 b}{d}[x+(l-x)k]$$

(b) The energy of the system is

$$U = \frac{1}{2}CV^2$$

$$= \frac{\epsilon_0 b}{2d}[x+(l-x)k]V^2$$

or $$U = \frac{1}{2}\epsilon_0 E^2 bd[x+(l-x)k]$$

(c) The force acting on the slab due to the electric field across the capacitor is

$$F = -\frac{dU}{dx}$$

Now,

$$F = -\frac{1}{2}\epsilon_0 E^2 bd(k-1)$$

Thus,

$$F = \frac{1}{2}\epsilon_0 E^2 bd(k-1)$$

Aliter: Force can also be calculated as

$$Fdx = -dU + Vdq$$

or $$F = -\frac{dU}{dx} + V\frac{dq}{dx}$$

since $$\frac{dU}{dx} = -\frac{1}{2}\epsilon_0 E^2 bd(k-1)$$

Now, charge on the capacitor is

$$q = \sigma_1 A_1 + \sigma_2 A_2$$

where $$E = \frac{\sigma_1}{\epsilon_0} \text{ and } E = \frac{\sigma_2}{k\epsilon_0}$$

Thus, $q = \in_0 Exb + k\in_0 E(l-x)b$

$\therefore$ $\dfrac{dq}{dk} = -\in_0 E_b\,(k-1)$

and $V = Ed$

Thus,

$$F = -\frac{1}{2}\in_0 E^2 bd\,(k-1) + E^2 \in_0 bd\,(k-1)$$

or $$F = -\frac{1}{2}\in_0 bd\, E^2 (k-1)\,.$$

Example 6.10. A co-axial cylindrical capacitor of the inner radius a and outer radius b is half filled with a dielectric of dielectric constant k. The length of the capacitor is l. Compute the capacitance of the system.

Solution: We may compute the electric field using Gauss' law.

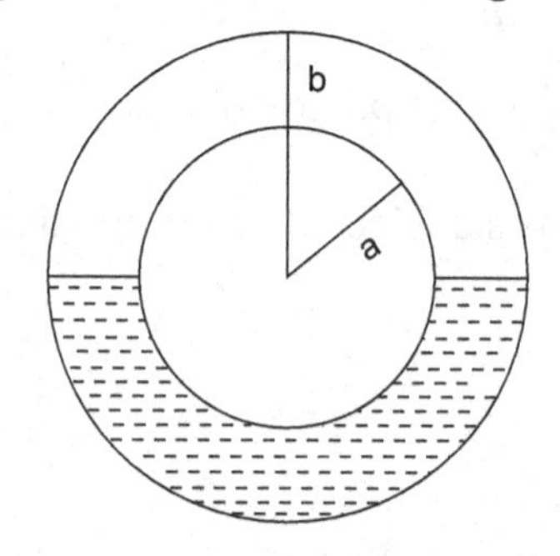

Fig. 6.31. Cylindrical capacitor.

The flux through the upper half, Fig. 6.31, is given by

$$\frac{1}{2}\oint_0 \vec{E}\cdot\vec{dS} = \frac{q}{\in_0}$$

or $$\frac{E\cdot 2\pi r l}{2} = \frac{q}{\in_0}$$

or $$E = \frac{q}{\pi \in_0 lr}$$

the potential difference V is

$$V = \int_a^b E\,dr$$

$$= \frac{q}{\pi \in_0 l}\ln\,(b/a)$$

∴ Capacitance

$$C_1 = \frac{q}{V}$$

$$C_1 = \frac{\pi \epsilon_0 l}{\ln (b/a)}$$

Similarly, the capacitance of lower half is

$$C_2 = \frac{\pi \epsilon_0 kl}{\ln (b/a)}$$

Total capacitance is

$$C = C_1 + C_2$$

or $$C = \frac{\pi \epsilon_0 l}{\ln (b/a)} (1+k)$$

Example 6.11. Find the capacitance of a half spherical shell field with dielectric of dielectric constant *k* as shown in fig. 6.32.

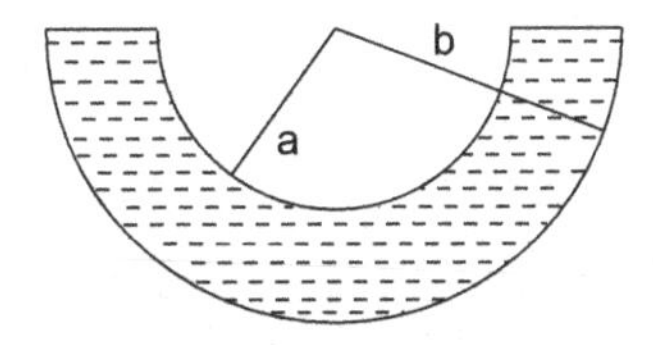

Fig. 6.32. A spherical shell with dielectric.

Solution: According to Gauss's law

$$\oint \vec{E} \cdot \vec{dS} = \frac{q}{\epsilon_0 k}$$

$$E \cdot 2\pi r^2 = \frac{q}{\epsilon_0 k}$$

or $$E = \frac{1}{2\pi \epsilon_0 k} \frac{q}{r^2}$$

∴ potential between the spherical conductor is

$$V = -\int_a^b \frac{q}{2\pi \epsilon_0 k} \frac{dr}{r^2}$$

$$V = \frac{q}{2\pi \epsilon_0 k} \frac{ab}{(b-a)}$$

Thus, capacitance is

$$C = \frac{q}{V} = \frac{2\pi \in_0 kab}{(b-a)}$$

Example 6.12. A parallel plate capacitor is field with two different dielectrics of dielectric constants k_1 and k_2 as shown in Fig. 6.33. Find the capacitance of the system.

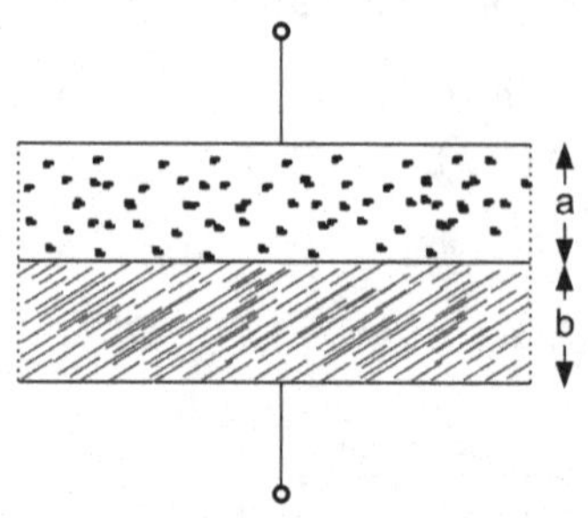

Fig. 6.33. A capacitor with dielectrics

Solution: Suppose that *a* and *b* are the thicknesses of the slabs. At the interface of two slabs,

$$\vec{D_1} = \vec{D_2}$$

or $$k_1 E_1 = k_2 E_2$$

where, $$E_1 = \frac{q}{\in_0 k_1 A}$$

and $$E_2 = \frac{q}{\in_0 k_2 A}$$

the potential difference

$$V = E_1 a + E_2 b$$

$$= \frac{q}{\in_0 A}\left(\frac{a}{k_1} + \frac{b}{k_2}\right)$$

∴ Thus, capacitance,

$$C = \frac{q}{V}$$

or $$C = \frac{\in_0 A}{\left(\frac{a}{k_1} + \frac{b}{k_2}\right)}$$

Example 6.13. Two capacitors of capacitances $C_1 = 2\mu f$ and $C_2 = 6\mu f$ are connected in series with an external voltage source of 200V. Then, compute the

(a) charge on the each capacitor.

(b) potential difference across each capacitor.

Solution: Let q be the common charge and V_1 and V_2 be the potential difference across C_1 and C_2 respectively, then,

(a) $$V_1 + V_2 = 200$$

or $$\frac{q}{C_1} + \frac{q}{C_2} = 200$$

or $$q = \frac{C_1 C_2}{C_1 + C_2}.200$$

$$= \frac{2 \times 6}{8} \times 200 \times 10^{-6}$$

or $$q = 3 \cdot 0 \times 10^{-4} \text{C}$$

(b) $$V_1 = \frac{q}{C_1} = \frac{3 \cdot 0 \times 10^{-4}}{2 \times 10^{-6}} = 150 \text{ V}$$

$$V_2 = \frac{q}{C_2} = \frac{3 \times 10^{-4}}{6 \times 10^{-6}} = 50 \text{ V}$$

Example 6.14. Two capacitors $C_1 = 4$ μf and $C_2 = 6$ μf are connected to a 12 V supply as shown in fig. 6.34. Compute

(1) equivalent capacitance of the circuit.

(2) voltage across the capacitor.

(3) charge on each capacitor.

(4) charge on equivalent capacitor.

(5) energy stored in the system.

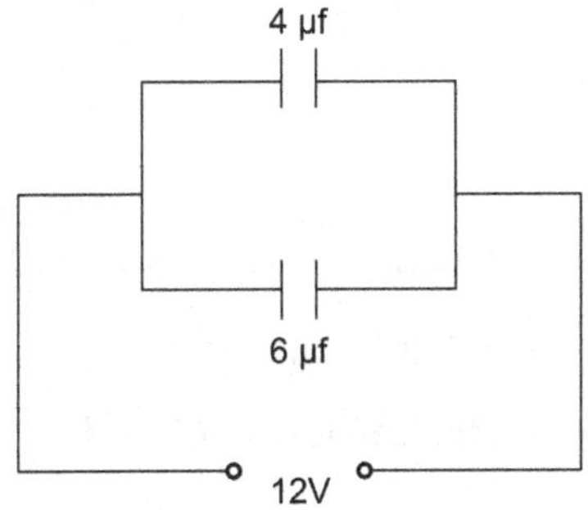

Fig. 6.34. Capacitors in parallel.

Solution: Both capacitors are in parallel,

(1) The equivalent capacitance is, $C = C_1 + C_2$

$$C = 4 + 6 = 10\mu f$$

(2) The voltage across each capacitor is 12V.

(3) Charge on $C_1 = 4\mu f$ is

$$q_1 = C_1 V_1$$
$$= 4 \times 12 = 48\ \mu C$$

and
$$q_1 = C_2 V_2$$
$$= 6 \times 12 = 72\ \mu C$$

(4) Charge on equivalent capacitor C is

$$q = CV$$
$$= 10\ \mu f \times 12$$
$$= 120\ \mu C$$

(5) Energy stored in the system is

$$U = \frac{1}{2} CV^2$$
$$= \frac{1}{2} \times 10 \times 10^{-6} \times 144$$
$$= 7.2 \times 10^{-4}\ J$$

Example 6.15. Two capacitors C_1 and C_2 are connected in parallel with a supply of V volts as shown in fig. 6.35. Show that the sum of energies stored in the individual capacitors is equal to the energy stored in the equivalent capacitor.

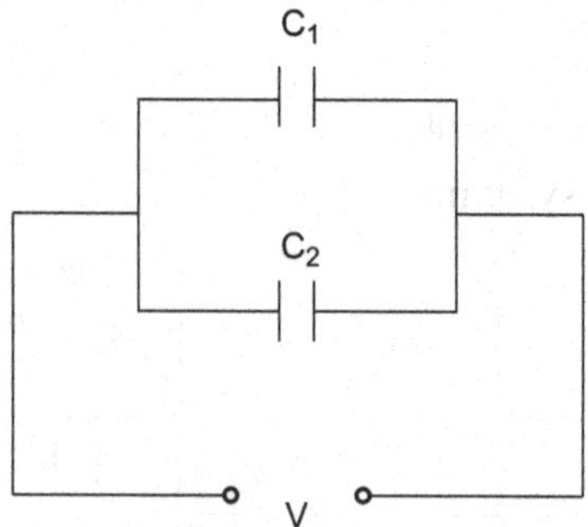

Fig. 6.35. Capacitors.

Solution: The energy stored in the capacitors C_1 and C_2

are
$$U_1 = \frac{1}{2} C_1 V^2$$

and $$U_2 = \frac{1}{2}C_2V^2$$

Thus,
$$U = U_1 + U_2$$
$$= \frac{1}{2}(C_1 + C_2)V^2$$
$$U = \frac{1}{2}CV^2$$

But, $$C = C_1 + C_2$$

Hence, $$U = \frac{1}{2}CV^2$$

Example 6.16. Two metallic spheres, each carrying equal and opposite charge of 10μC, are hanging with the weightless insulated threads. The distance between them is 1.0 cm and potential difference is 100 volts. Find the capacitance of the system.

Solution: Since, $$q = 10\mu\text{C}$$
$$V = 100 \text{ volts}$$

The capacitance is

$$C = \frac{q}{V} = \frac{10\times 10^{-6}}{100}$$

or $$C = 0.1\ \mu\text{C}$$

Example 6.17. Find the equivalent capacitance of the circuit shown in fig. 6.36.

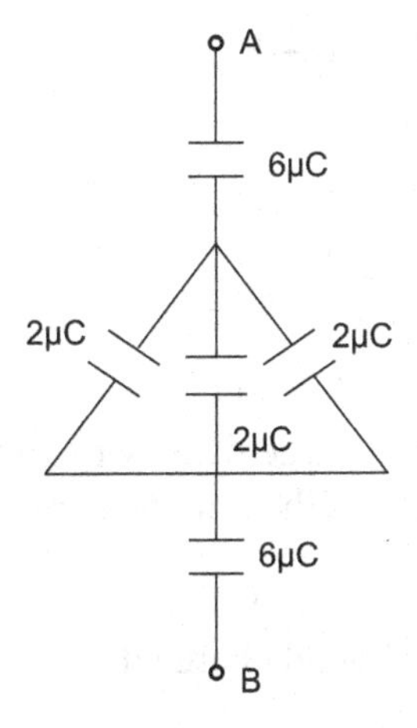

Fig. 6.36. System of capacitors.

Solution: The capacitors of capacitance 2μC are is parallel, the equivalent capacitance of these,

$$C' = 2 + 2 + 2 = 6\ \mu C$$

Now, C', 6μC and 6μC are in series, we have

$$\frac{1}{C} = \frac{1}{6} + \frac{1}{6} + \frac{1}{6} = \frac{3}{6}$$

$$C = \frac{6}{3} = 2\mu C.$$

Example 6.18. A slab of dielectric material with the dielectric constant $k = 3{\cdot}0$ is placed across an electric field of 10^7 V/m as shown in fig. 6.37.

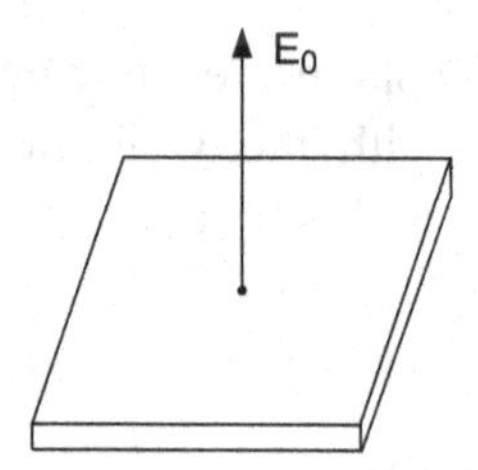

Fig. 6.37. Dielectric slab.

Compute the

(a) $\vec{E}$ (b) $\vec{D}$ (c) $\vec{P}$ (d) ρ_d (e) σ_d.

Solution: (a) The induced electric field is

$$E = \frac{E_0}{k} = \frac{10^7}{3} = 3.33 \times 10^6 \text{ V/m}$$

(b) $D = \epsilon_0 E_0 = \epsilon_0 kE$

$$= 8.85 \times 10^{-15} \text{ c/m}^2$$

(c) Now,

$$D = \epsilon_0 E + P$$
$$P = D - \epsilon_0 E$$
$$= 8.85 \times 10^{-5} - 2.95 \times 10^{-5}$$
$$P = 5.9 \times 10^{-5} \text{ c/m}^2$$

(d) Since there is no volume charge density of free charge, there should be no volume charge density of the bound charge. Then,

$$\rho_d = -\nabla \cdot \vec{P} = 0$$

(e) At the outer surface, the bound charge density

$$\sigma_d = \pm \vec{P} \cdot \hat{n}$$

$$= \pm P$$
$$= \pm 5.9 \times 10^{-5}\ \text{c/m}^2.$$

Example 6.19. If the earth is considered as a conducting sphere of radius R, compute the capacitance of the earth.

Solution: The potential of the earth is

$$V = \frac{q}{4\pi\epsilon_0 R}$$

and

$$C = \frac{q}{V} = 4\pi\epsilon_o R$$

Here, $R = 6.4 \times 10^6$ m

Thus, $C = 4\pi\epsilon_0 \times 6.4 \times 10^6$

$$= \frac{6.4 \times 10^6}{9 \times 10^9} = 0.71\ \text{mf}$$

or $C = 710\ \mu\text{F}.$

Example 6.20. There are four plates of area A and the separation between any two plates is d, as shown in fig. 6.38. Find the capacitance between A and B.

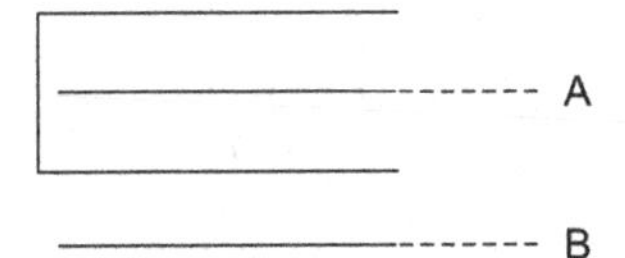

Fig. 6.38. Four conducting plates.

Solution:

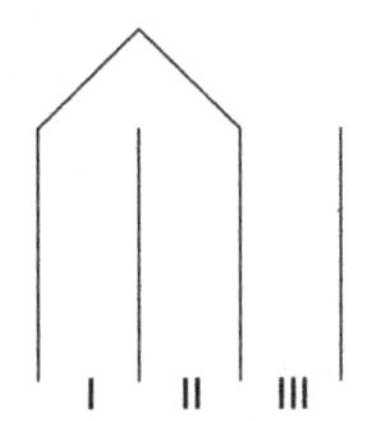

Fig. 6.39. Capacitors.

The I and II plates are in parallel, so we have $2C$, and III plate is in series with $2C$.

Thus, $$C' = \frac{2C \times C}{2C + C} = \frac{2}{3}C = \frac{2}{3} \cdot \frac{\epsilon_o A}{d}$$

Example 5.21. In the arrangement shown in fig. 6.40, $C_1 = 2\mu f$, $C_2 = 4\mu f$ and $C_3 = 3\mu f$. Then compute the

(a) total capacitance of the circuit

(b) charge on each capacitor

(c) the potential difference across each capacitor.

Solution:

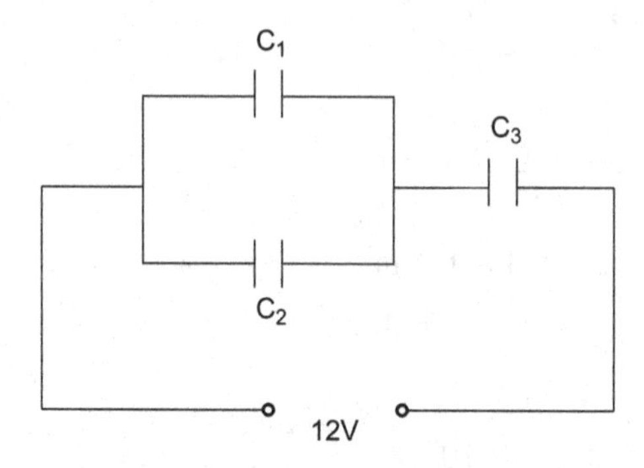

Fig. 6.40. Assembly of capacitors.

(a) The total capacitance can be evaluated as

$$\frac{1}{C} = \frac{1}{C_1 + C_2} + \frac{1}{C_3}$$

or $$C = \frac{(C_1 + C_2)\, C_3}{C_1 + C_2 + C_3} = \frac{6 \times 3}{9} = 2\mu f$$

(b) The total charge, $q = CV$

$$q = 12 \times 2$$
$$= 24\ \mu C$$

Now, charge on capacitor C_1 is

$$q_1 = \left(\frac{C_1}{C_1 + C_2}\right) q = \frac{2}{6} \times 24 = 8\ \mu C$$

and charge on capacitor C_2 is

$$q_3 = \left(\frac{C_2}{C_1 + C_2}\right) q$$

$$= \frac{4}{6} \times 24 = 16\mu C$$

It can also be calculated as

$$q_2 = q - q_1 = 24 - 8 = 16\ \mu C$$

Now, charge on C_3 will be same as supplied by battery.

Thus, $q_3 = q = 24\mu C$

(c) Potential difference across C_1 is

$$V_1 = \frac{q_1}{C_1} = \frac{8}{2} = 4V$$

Potential difference across C_2 is

$$V_2 = \frac{q_2}{C_2} = 4V$$

and potential difference across C_3 is

$$V_3 = \frac{q_3}{C_3} = \frac{24}{3} = 8V$$

It can also be calculated as

$$V_3 = V - V_1 = V - V_2 = 12 - 4 = 8 \text{ V}$$

EXERCISES

6.1. What is meant by the capacitance of a capacitor?

6.2. What is the difference between the permittivity and the dielectric constant of a medium?

6.3. Define electric displacement vector and electric susceptibility. What are their units?

6.4. Find the expression for the equivalent capacitance of three capacitors having capacities C_1, C_2 and C_3 connected (a) in series (b) in parallel.

6.5. Derive an expression for the capacitance of a conducting sphere of radius R.

6.6. Obtain an expression for the capacitance of a cylindrical capacitor having inner radius a and outer radius b.

6.7. Obtain an expression for the energy of a charged capacitor of capacitance C.

6.8. Differentiate between polar and non polar dielectrics and define the electric polarization.

6.9. State and explain Gauss's law in dielectrics.

6.10. Show that capacitance of a parallel plate capacitor increases with dielectric material.

6.11. Establish a relation between electric displacement vector and polarization vector.

6.12. Obtain an expression for the force between charges in a dielectric medium.

6.13. Compute the capacitance of the earth, regarding that it is a conducting sphere of radius R.

6.14. Define electric polarizability and prove the clasusius-mossotti equation

$$\frac{k-1}{k+2} = \frac{1}{3\epsilon_0}\sum_i N_i \alpha_i$$

where notations have their usual meanings.

6.15. For polar dielectrics, show that

$$P = Np\left(\coth\left(\frac{p\,E_{LOC}}{KT}\right) - \frac{KT}{p\,E_{LOC}}\right)$$

6.16. Define polarization current density and show that the polarization current density is equal to the time rate of change of polarization vector.

6.17. For bound charge densities, show that

$$\sigma_d = P \cdot \hat{n}$$

and $$\rho_d = -\nabla \cdot p$$

6.18. If a dielectric slab ($k = 2{\cdot}1$) is inserted into a charged capacitor having free charge density, $\sigma = 1\mu C/m^2$, Find electric displacement D, electric field E, polarization P and bound charge density σ_d.

6.19. A parallel-plate capacitor of area A and separation d is filled with three dielectrics, as shown in fig. 6.41 (a), (b). Find the capacitance.

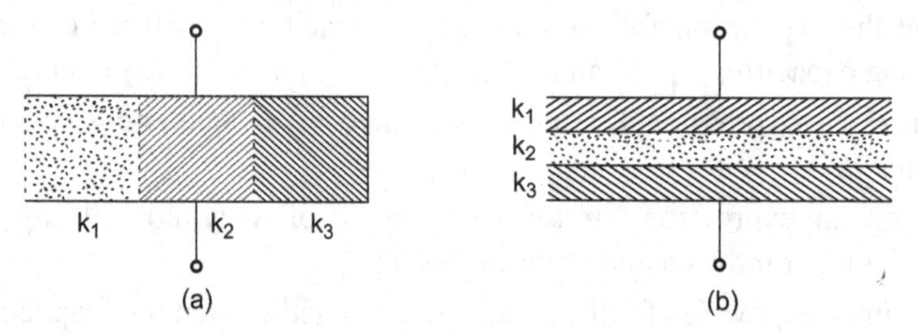

Fig. 6.41. A parallel plate capacitor with dielectrics.

6.20. A parallel plate capacitor has capacitance $C = 36\mu F$, $A = 100\ cm^2$ and space between plates is filled with a dielectric material ($k = 2.1$). When $\phi = 200$ V, Find

(a) the electric field in dielectric, E

(b) the free charge density, σ

(c) the induced charge density, σ_d

(d) the polarization, P

6.21. A parallel plate capacitor is half filled with the dielectric of dielectric constant k, as in fig. 6.42. Find the capacitance.

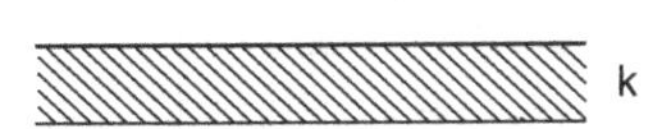

Fig. 6.42. Parallel plate capacitor.

6.22. Consider a thin conducting disk of radius R, Find the capacitance.

6.23. Find the equivalent capacitance of the arrangement shown in fig. 6.43.

Ans. $C = 3\ \mu f$

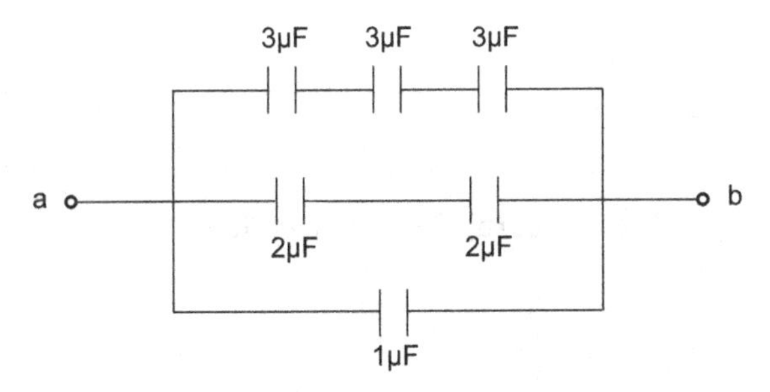

Fig. 6.43. System of capacitors.

6.24. A spherical shell has inner radius a and outer radius b. Two different dielectric are filled between a and b, as shown in fig. 6.44. Find the capacitance of the system.

$$\textbf{Ans. } C = \frac{4\pi \epsilon_0 k_1 k_2 abc}{k_1 a(b-c) + k_2 b(a-c)}$$

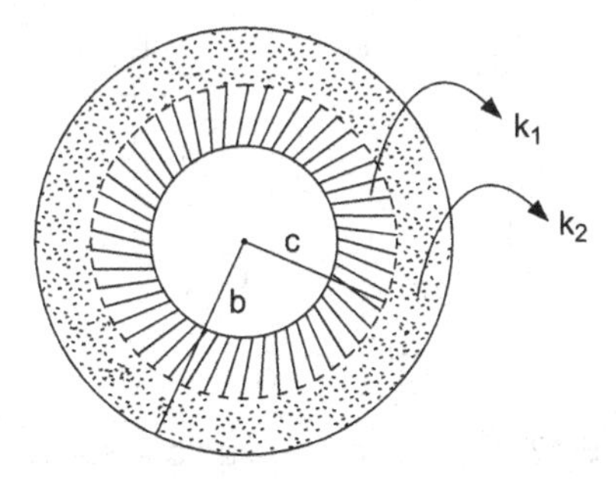

Fig. 6.44. Spherical shell with dielectirics.

6.25. Seven identical plates, each of area A are arranged, as shown in fig. 6.45. The distance between adjacent plates is d. Find the capacitance.

$$\textbf{Ans. } C = \frac{6 \epsilon_0 A}{d}$$

Fig. 6.45. Conducting plates.

6.26. Find the equivalent capacitances of the arrangements shown in fig. 6.46 (a) and (b).

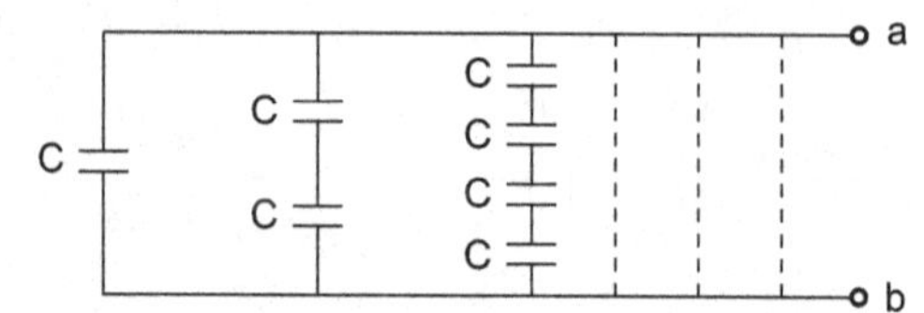

Ans. $C' = 2C$

Fig. 6.46. (a) Capacitor system.

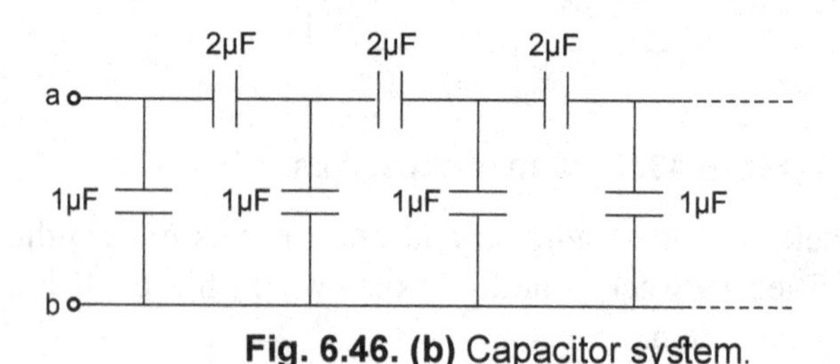

Ans. $C' = 2\mu F$.

Fig. 6.46. (b) Capacitor system.

6.27. find the equivalent capacitance between B and H, of the cubic arrangement of the capacitors, as shown in fig. 6.47.

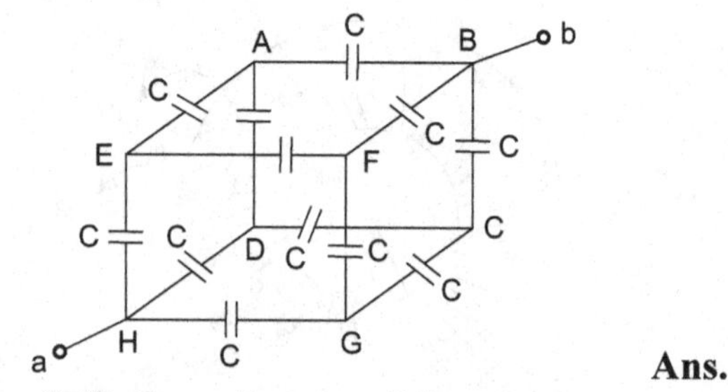

Ans. $C' = \dfrac{6C}{5}$

Fig. 6.47. Capacitors in cubic system.

6.28. Four capacitors are arranged, as shown in fig. 6.48. A 12V battery is connected between a and b. If $C_1 = 2\mu F$, $C_2 = 4\mu F$, $C_3 = 2\mu F$ and $C_4 = 4\mu F$, calculate the total charge drawn from the battery and charge on each capacitor when

(a) the switch s is closed (b) the switch s is open.

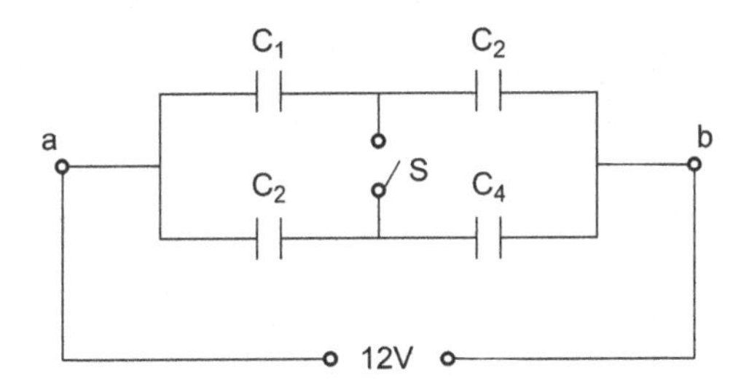

Fig. 6.48. Capacitor system.

6.29. A parallel plate capacitor of plate area A and plate separation d has capacitance C. Find the capacitance when an aluminium plate of thickness b is placed between the plates of capacitor.

Ans. $C' = \left(\frac{cd}{d-b}\right)$

6.30. Find the bound charge density in a sphere of radius R carrying polarization $\vec{P} = a\vec{r}$, where a is constant.

Hint:
$$\rho_d = -\nabla \cdot p$$
$$= -3a$$

6.31. A thin rod is placed along z axis from $z = 0$ to $z = L$. The rod is polarized along length and polarization $P = az^2 + b$. Find the bound charge densities.

Hint.
$$\sigma_d = p\,\hat{n}\Big|_{Z=L}$$

or
$$\sigma_d = az^2 + b\Big|_{Z=L}$$
$$= aL^2 + b$$

and
$$\sigma_d = -\nabla \cdot p$$
$$= -2aZ\Big|_{Z=L}$$

or
$$\sigma_d = -2aL$$

6.32. A point charge q is placed at the centre of the dielectric spherical shell of inner radius a and outer radius b, as shown in fig. 6.49. Find E, D and polarization P.

Ans. $E = \dfrac{q}{4\pi \epsilon_0 r^2}, r < a$

$$= \frac{q}{4\pi \in_0 k r^2}, a < r < b$$

$$= \frac{q}{4\pi \in_0 r^2} r < b$$

$$D = \frac{q}{4\pi r^2} a < r < b$$

$$P = \frac{(k-1)q}{4\pi k r^2} a < r < b$$

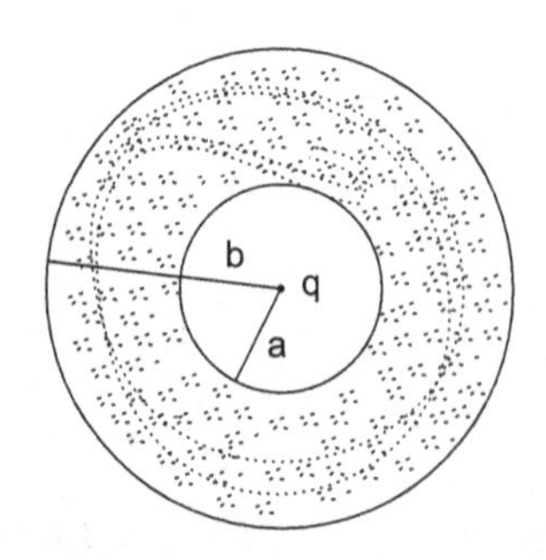

Fig. 6.49. Dielectric spherical shell.

6.33. Two capacitors, $C_1 = 3$ μF and $C_2 = 6$μF, are charged separately to same potential difference 12V. Now, positive plate of one is connected to negative plate of other and outermost connections are shorted together. Find the

(a) charge on each capacitor

(b) loss in electrostatic energy.

Hint:

$$q_1 = \frac{C_1(C_1 - C_2)V}{C_1 + C_2} = 12 \ \mu\text{C}$$

$$q_2 = \frac{C_2(C_1 - C_2)V}{C_1 + C_2} = 24 \ \mu\text{C}$$

$$dU = U_i - U_f = 5.76 \times 10^{-4} \text{ J}$$

■ ■ ■

Current, Resistance and Circuits

In the previous chapters, we have discussed the electric phenomena associated with the charge at rest. In the present chapter, we shall describe the situations having charge in motion. The electric current may exist where the charge is free to move. In different media, the current is caused by the different charge. In conductors the current is due to the motion of the free electrons. However, in semiconductors, the current is caused by the motion of the electrons and holes. The motion of positive and negative ions constitute a current in the electrolytes. Moreover, the displacement current is the result of the bound charge in the dielectrics. It is the important fact that when current flows through the resistance, there will be dissipation of the energy. In this chapter, we also analyze the electric circuits with the fundamental electrical theory and we shall discuss the relations among the electrical parameters viz, current, voltage, resistance and electric field etc.

7.1. ELECTRIC CURRENT

If there is a flow of charge continuously, it constitutes an electric current. The direct current is the average motion of the electrons in the same direction. We know that the charge moves from higher potential to the lower potential, that is, the current flows from higher potential to the lower potential that is, current flows from higher potential to the lower potential. Consider a portion of conductor connected to a voltage source as shown in fig. 7.1. In the conductor, the charge carriers are the free electrons and these charges move perpendicular to the area of cross-section A of the conductor.

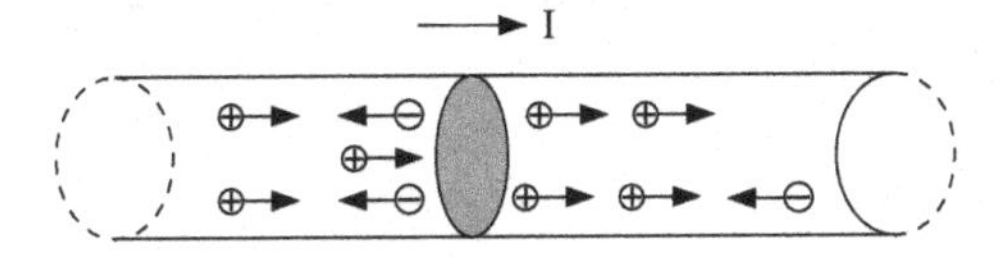

Fig. 7.1. Current in the conductor

The positive charges move from left to right whenever the electrons move from right to left. That is, the direction of the conventional current is from left to right as shown in fig. 7.1. Thus, the electric current is defined as the rate at which the net charge flows across the area per unit time. If a net charge dq flows through a area in a time interval dt, the current is given by

$$\boxed{I = \frac{dq}{dt}} \text{ coulomb/sec.} \quad ...(7.1)$$

current is a scalar quantity, however, the current has direction but it is not specified and does not obey law of vector addition. In S.I units, the unit of current is ampere (A) and

$$1 \text{ Ampere} = \frac{1 \text{ coulomb}}{1 \text{ second}}$$

the smaller units of ampere are given by

$$1 \text{ mA} = 10^{-3} \text{ A}$$
$$1 \text{ μA} = 10^{-6} \text{ A}$$

7.2. CURRENT DENSITY

The current density is defined as the flow of charge passing through a given area per unit time. Consider a conductor of cross-sectional area A and of length l, as shown in fig. 7.1, the current density is related to the current by the equation,

$$I = \int_S \vec{J} \cdot \vec{dS} \quad ...(7.2)$$

where $\vec{dS}$ is the area element. Suppose that n is the number of electrons per unit volume and the current is due to the drift of the charge carriers. Let v_d the mean drift velocity of the charge carriers. The total charge in the given section of length l of the conductor is given by

$$q = n\,e\,A\,l \quad ...(7.3)$$

Now, the distance covered by each charge carriers in time t will be

$$l = v_d\,t \quad ...(7.4)$$

substituting the value of l from Eq. (7.4) in the Eq. (7.3), we get

$$q = ne\,A\,v_d\,t$$

the rate at which the charges are flowing through the cross-section of a conductor is

$$I = \frac{q}{t}$$

or

$$I = n\,e\,A\,v_d \quad ...(7.5)$$

The current I depends on A. The current density which is independent of A is then,

$$\boxed{J = \frac{I}{A} = nev_d} \qquad \text{...(7.6)}$$

or in vector form,

$$\vec{J} = ne\vec{v}_d \qquad \text{...(7.7)}$$

The S.I. unit of the current density is Amp/m^2. The direction of the current density is opposite to the direction of flow of electrons, that is opposite to $\vec{v}_d$. Thus, the Eq. (7.7) can be written as

$$\vec{J} = -ne\vec{v}_d \qquad \text{...(7.8)}$$

Moreover, the charge per unit volume is given by

$$\rho = ne$$

Thus, from the Eq. (7.8), we write,

$$\vec{J} = \rho\vec{v}_d \qquad \text{...(7.9)}$$

Now, we want to obtain a relation between the current density and the electric field. We know that the force experienced by the electrons in the conductor is,

$$\vec{F} = -e\vec{E} \qquad \text{...(7.10)}$$

and the acceleration of the electron is given by

$$\vec{a} = \frac{\vec{F}}{m}$$

$$= \frac{-e\vec{F}}{m} \qquad \text{...(7.11)}$$

where m is the mass of the electron. During the motion, electrons collide each other. Suppose that v_0 is the velocity of the electron just after a collision and the velocity of the electron just before the next collision is given by, Fig. 7.2,

$$v = v_0 + at$$

$$= v_0 - \frac{eEt}{m} \qquad \text{...(7.12)}$$

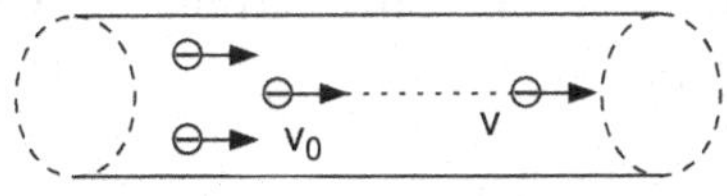

Fig. 7.2. Collisions of electrons

where t is the time interval between two collisions. In the absence of the electric field, the electrons in the conductor move in any direction, it is due to thermal agitation as shown in fig. 7.3. As a result, electrons constantly collide with the other atoms of the conductor. On the application of the electric field, they move with the drift velocity in the direction opposite to the electric field $\vec{E}$.

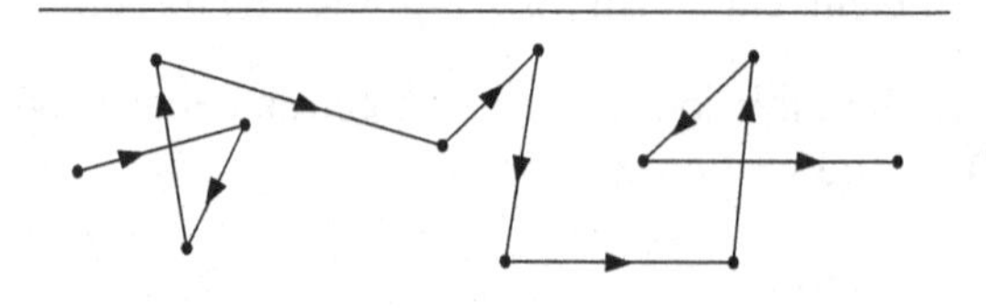

Fig. 7.3. Zig Zag path of the lectron without field.

Taking the average over v for all time intervals, we get

$$\langle v \rangle = \langle v_0 \rangle - \frac{eE}{m} \langle t \rangle \quad \text{...(7.13)}$$

when $\vec{E} = 0$, $\langle v_0 \rangle = 0$, and In the presence of the electric field, $\langle v \rangle$ represents the drift velocity of the electron. Moreover, $\langle t \rangle$ is known as mean free time and it is the time between two successive collisions and it is represented by τ(tau). Thus, we write,

$$\vec{v}_d = \frac{-e\vec{F}}{m}\tau \quad \text{...(7.14)}$$

Substituting the value of v_d from the Eq. (7.14) in the Eq. (7.8), we get

$$\vec{J} = -ne\left(-\frac{e\vec{E}\tau}{m}\right)$$

or

$$\vec{J} = \left(\frac{ne^2\tau}{m}\right)\vec{E} \quad \text{...(7.15)}$$

It is clear that the direction of the current density is same as that of the electric field $\vec{E}$. Hence the current density vector $\vec{J}$ is always parallel to the applied electric field $\vec{E}$. Further-more, the current is given by

$$I = JA$$

or

$$I = \left(\frac{ne^2 A\tau}{m}\right)E \quad \text{...(7.16)}$$

The Eq. (7.16) predicts that the direction of the conventional current is in the direction of the electric field E.

7.3. RESISTANCE AND OHM'S LAW

We know that the metallic conductors have a large number of free electrons. For example, the copper has 10^{29} electrons/m^3, but Ge, a semiconductor, has 10^{19} electrons/m^3. Since, the current in the conductor is given by

$$I = ne\,A\,v_d \qquad ...(7.17)$$

that is, the current is proportional to the average drift velocity which provides a linear relation between the current and the potential difference across the conductor. This linear relation between the current and voltage was first given by a German physicist, Georg Simon Ohm. Ohm was inspired by the work of Fourier which was on the rate of flow of heat through the conductor.

On the basis of the experimental observations, Ohm suggested a relation between the rate of flow of charge and the voltage across the conductor. Ohm's law states that, for a conductor at constant temperature, the current flowing through the conductor is proportional to the potential difference between the ends of the conductor. Thus, if V represents the potential difference across the conductor, the current through the conductor is given by

$$I = \frac{V}{R} \qquad ...(7.18)$$

or

$$\frac{V}{I} = R \text{ (constant)} \qquad ...(7.19)$$

where R is a proportionality constant and is known as the electrical resistance The S.I unit of R is ohm (Ω). Now,

$$1\Omega = \frac{1 \text{ Volt}}{1 \text{ Ampere}}$$

The conductors which obey Ohm's law are called ohmic conductors. This law is valid for all metals and some semiconductors under certain limitations. The limitations are

(1) The drift velocity v_d is always less than c, the speed of light.

(2) The material should have low resistivity.

(3) Temperature of the conductor.

On the otherhand, there are some materials which do not follow Ohm's law and these are called non-ohmic. If we plot current I versus potential difference V, we get a straight line as shown in fig. 7.4.

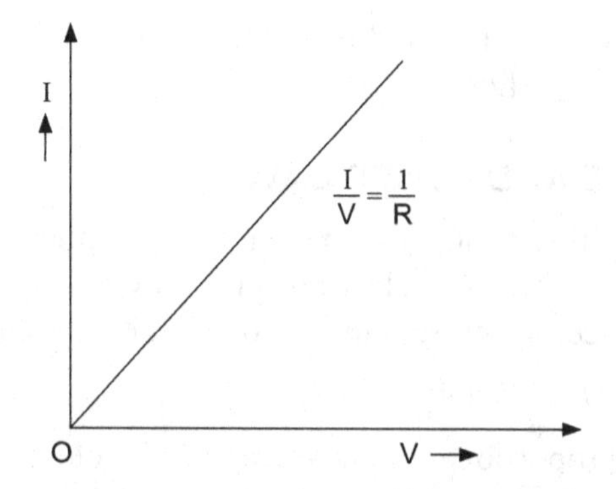

Fig. 7.4. Ohmic behavior of a conductor

Moreover, these are many elements which show non-ohmic behavior. The examples are semiconductor devices. If a graph between current and voltage is plotted for such devices, we do not have a straight line, but it is a non linear curve as shown in fig. 7.5. Thus, semiconductor devices are called non-linear devices.

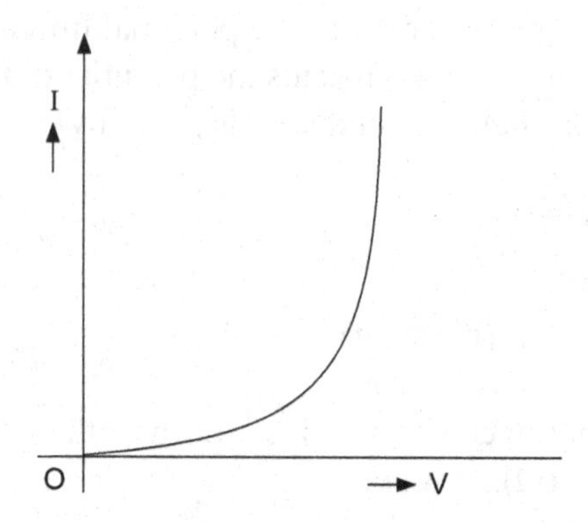

Fig. 7.5. Non-linear behavior of a semiconductor devices

Now, we develop a relation between the resistivity and resistance of a conductor. Consider a conductor of length l and cross-sectional area a as shown in fig. 7.6. If we apply an electric field E across the conductor, the charges drift along the electric field. The drift velocity of the charge is always less than the random velocity. The current density $\vec{J}$ depends linearly on the electric field across the conductor, thus,

$$\vec{J} = \sigma \vec{E} \quad ...(7.20)$$

where σ is called the conductivity of the conductor. The Eq. (7.20) is also known as Ohm's law.

On comparing the Eq. (7.15) and (7.20) we get

$$\sigma = \frac{ne^2\tau}{m} \qquad ...(7.21)$$

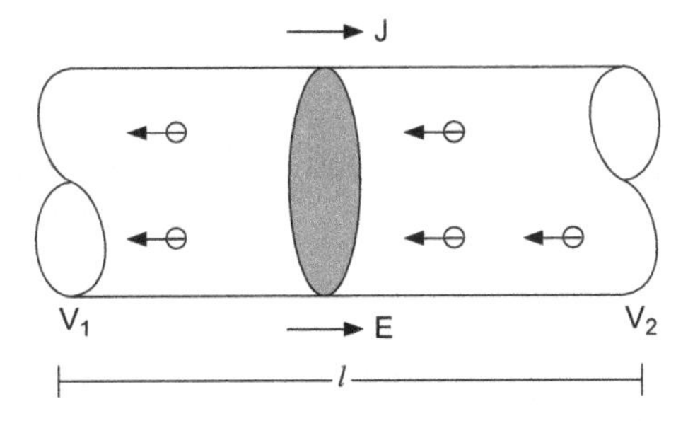

Fig. 7.6. A conductor of length *l* and cross-sectional area A.

Now, we are going to relate the resistance of a conductor to its size. The electric field across the conductor is

$$E = \frac{V}{\ell} \qquad ...(7.22)$$

where $V = V_2 - V_1$ is the potential difference, substituting the value of E from the Eq. (7.22) in the Eq. (7.20), we get

$$J = \frac{\sigma V}{\ell} \qquad ...(7.23)$$

The current flowing through the conductor is

$$I = JA$$

$$= \frac{\sigma A V}{\ell} \qquad ...(7.24)$$

If 2τ is the mean free time (a time between two successive collisions), the drift velocity is given by

$$v_d = \frac{1}{2} eE\left(\frac{2\tau}{m}\right)$$

$$= \frac{eE\tau}{m} = \frac{eV\tau}{ml} \qquad ...(7.25)$$

Thus, current is

$$I = neA\,v_d \qquad ...(7.26)$$

$$= \frac{ne^2 A\tau V}{ml}$$

The resistance of the conductor is,

$$R = \frac{V}{I} = \left(\frac{m}{ne^2\tau}\right)\frac{l}{A}$$

or

$$R = \frac{1}{\sigma}\frac{l}{A} \qquad ...(7.27)$$

we know that the reciprocal of the conductivity is called the resistivity and is given by

$$\rho = \frac{1}{\sigma} \qquad ...(7.28)$$

or

$$\rho = \frac{m}{ne^2\tau} \qquad ...(7.29)$$

Thus, the Eq. (7.27) can be written as

$$\boxed{R = \rho\,\frac{l}{A}} \qquad ...(7.30)$$

The Eq. (7.30) shows that the resistance of the conductor is proportional to its length and inversely proportional to its area of cross-section. On measuring the quantities V, I and l, R and ρ can be evaluated easily. The unit of resistivity is ohm-meter, or ohm-m. Let us see, how does the resistivity depend on the temperature. When the temperature of a conductor increases, the random velocity of the electrons in the conductor increases and as a result, more and more collisions of the electrons in the conductor occur. Thus, there is a increase in the resistivity of the conductor. The resistivity varies linearly with the temperature and is given by

$$\rho(T) = \rho_0[1 + \alpha\,(T - T_o)] \qquad ...(7.31)$$

where α is a temperature coefficient of resistivity, ρ is the resistivity at the temperature T and ρ_0, the resistivity at some reference temperature T_0. For 25°C, we write

$$\rho(T) = \rho_0(25°C)\,[1 + \alpha\,(T - 25°C)] \qquad ...(7.32)$$

At high temperatures,

$$\rho \propto T \qquad ...(7.33)$$

and at low temperatures, ρ varies as T^5.

At room temperature, ρ, σ and α for some substances are listed in the table 7.1. The variation of ρ(resistivity) with T for a conductor and a semiconductor are shown in fig. 7.7.

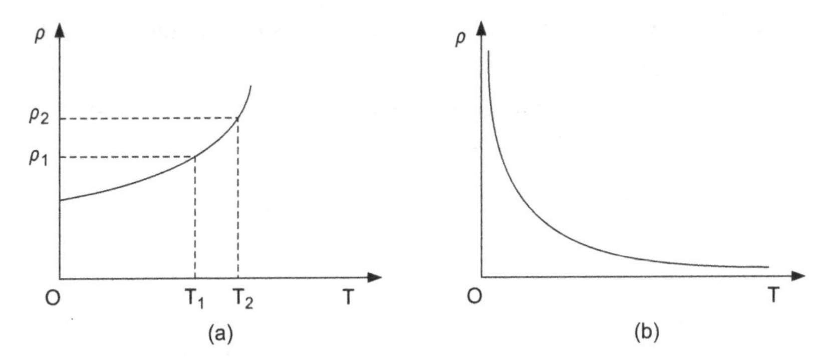

Fig. 7.7. Plot of ρ Vs T (a) conductor (b) semiconductor

From the fig. 7.7.(b), it is clear that the semiconductors consist of negative temperature coefficient of resistivity.

Table 7.1. Resistivity and conductivity of the different materials.

Substance	Resistivity (ρ) (Ω-m)	Conductivity (σ) (mho-m^{-1})	α (per °C)
Aluminum	2.8×10^{-8}	3.5×10^{7}	3.9×10^{-3}
Silver	1.6×10^{-8}	6.3×10^{7}	3.8×10^{-3}
Copper	1.7×10^{-8}	5.8×10^{7}	3.9×10^{-3}
Tungusten	5.5×10^{-8}	1.8×10^{7}	4.5×10^{-3}
Platinum	1.06×10^{-7}	1.0×10^{7}	3.9×10^{-3}
Iron	1.0×10^{-7}	1.0×10^{7}	5.0×10^{-3}
Manganin	4.4×10^{-7}	2.3×10^{6}	1.0×10^{-5}
Brass	7.0×10^{-8}	1.4×10^{7}	2.0×10^{-3}
Nichrome	1.0×10^{-6}	1.0×10^{6}	4.0×10^{-4}
Silicon	640	1.6×10^{-3}	-7.5×10^{-2}
Germanium	0.46	2.2	-4.8×10^{-2}
Teflon	10^{14}	10^{-14}	–
Glass	$10^{10} - 10^{14}$	$10^{-14} - 10^{-14}$	–
Blood	1.5	0.66	–
Polyethylene	$10^{8} - 10^{9}$	$10^{-9} - 10^{-8}$	–

From the Eq. (7.31) a may be expressed as

$$\alpha = \frac{1}{\rho_0} \frac{(\rho - \rho_0)}{(T - T_0)} \quad ...(7.34)$$

or

$$\alpha = \frac{1}{\rho_0} \left(\frac{\Delta \rho}{\Delta T} \right) \quad ...(7.35)$$

For metals, α is very low and positive and for super-conductor, $\rho = 0$, thus $\alpha = 0$.

Example 7.1. For 1Ω coil, compute the length of nichrome wire of cross-sectional are of 1×10^{-6} m^2 for a heater covl. ($\rho = 1 \times 10^{-6}$ Ω-m).

Solution:
$$R = \rho\frac{l}{A}$$

or
$$l = \frac{RA}{\rho}$$

$$= \frac{1\times1\times10^{-6}}{1\times10^{-6}} = 1 \text{ m.}$$

7.4. SUPERCONDUCTORS

The resistance always exists in the conductors that controls the current to flow. Now, we may have a class of materials that has no resistance. In 1911, Kamerlingh Onnes succeeded to measure the resistivity of the mercury at a very low temperature. The resistivity of the mercury drops suddenly to zero at the temperature $T = 4.18$ K. This phenomenon is known as the super-conductivity and the specimen used is called the superconductor. In the super conducting state, the total d.c. resistance of the material is equal to zero. The temperature at which this phenomenon exists is known as critical temperature (T_C). In case of mercury, $T_C = 4.18$ K as shown in fig 7.8.

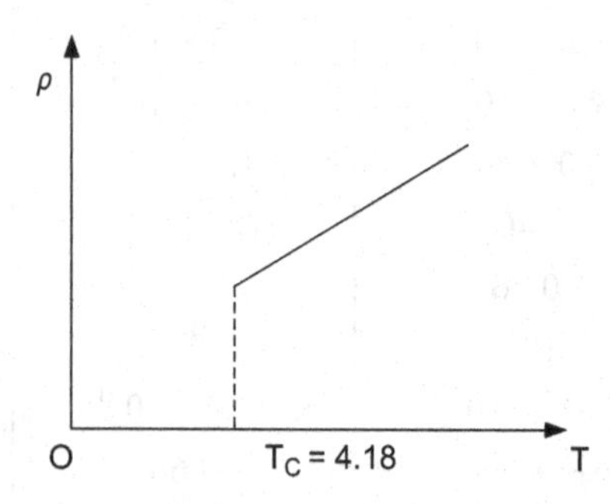

Fig. 7.8. Super-conducting state of mercury.

Below this temperature, the mercury in normal state undergoes a state of superconductivity after it. Thus, we have a phase transition from one state to another at this temperature. It is also known as transition temperature. The superconducting state is also known as ordered state.

Example 7.2. Show that the total charge density at the junction of two conductors is given by

$$\sigma = J\epsilon_o(\rho_2 - \rho_1)$$

where ρ_1 and ρ_2 are the resistivity of two conductors, and J is current density.

Solution: Consider the two conductors having resistivities ρ_1 and ρ_2 as shown in fig. 7.9.

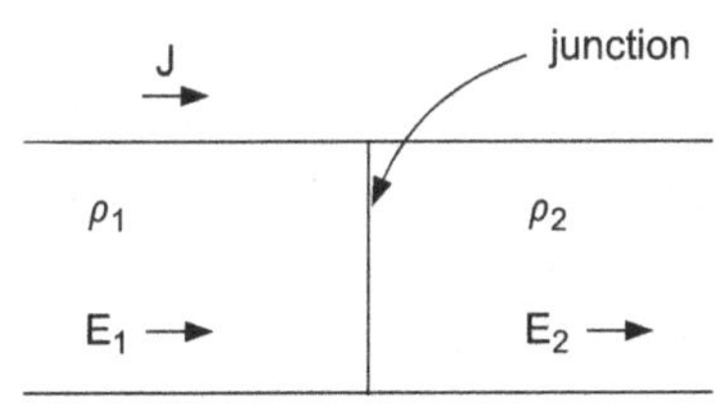

Fig. 7.9. Two conductors of different resistivity.

At the interface, by Maxwell's equation

$$\rho_C = \nabla\cdot\vec{D} = \nabla\cdot(\epsilon_0\vec{E}) = \epsilon_0\nabla\cdot(\rho\vec{J})$$

$$= \epsilon_0 J\,\nabla\cdot(\rho) \qquad (\nabla\cdot J = 0 \text{ for steady state})$$

or $$\rho_C = \epsilon_0 J\frac{d\rho}{dn}$$

since $\nabla(\rho)$ is normal to the interface, we have

now, the charge density at the junction is

$$\sigma = \int\rho_C\,dn = \epsilon_0 J\int_{\rho_1}^{\rho_2} d(\rho)$$

or $$\sigma = \epsilon_0 J(\rho_2 - \rho_1)$$

7.5. CIRCUITS CONTAINING RESISTORS IN SERIES AND PARALLEL

We wish to find the equivalent resistance of the circuit containing the resistors in series and in parallel.

(a) Resistors in series: Consider three resistors of the resistances R_1, R_2 and R_3 connected in series as shown in fig. 7.10.

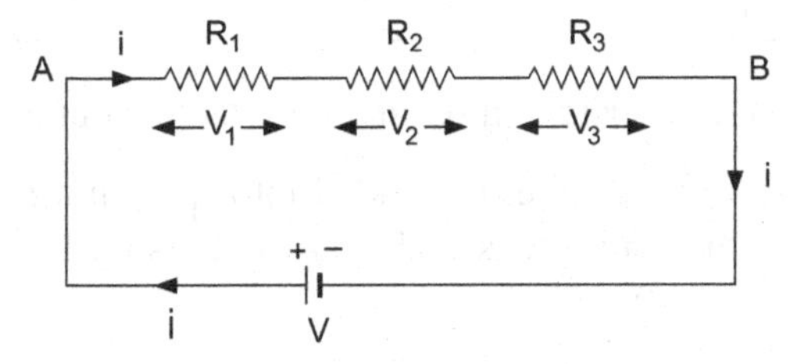

Fig. 7.10. Resistors in series with a source.

The potential difference between the points A and B is equal to the voltage of the battery, and it is equal to V. The potential difference across each resistor is given by

$$\begin{aligned} V_1 &= i R_1 \\ V_2 &= i R_2 \\ V_3 &= i R_3 \end{aligned} \qquad \text{...(7.36)}$$

The potential difference across points A and B can be obtained by adding three voltage drops. Thus,

$$V = V_1 + V_2 + V_3 \qquad \text{...(7.37)}$$

$$= iR_1 + iR_2 + iR_3$$

or

$$\frac{V}{i} = (R_1 + R_2 + R_3) \qquad \text{...(7.38)}$$

Therefore, the equivalent resistance is given by

$$R = \frac{V}{i}$$

or

$$\boxed{R = R_1 + R_2 + R_3} \qquad \text{...(7.39)}$$

If n resistors are connected in series, the equivalent resistance is

$$R = R_1 + R_2 + \ldots + R_n$$

or

$$\boxed{R = \sum_{i=1}^{n} R_i} \qquad \text{...(7.40)}$$

(b) Resistors in Parallel. Fig. 7.11 shows three resistors of resistances R_1, R_2 and R_3 are connected in parallel with a d.c. source of voltage V.

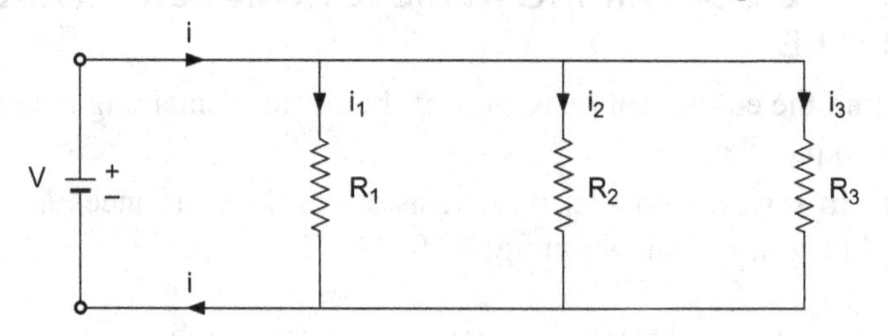

Fig. 7.11. Resistors in parallel with a d.c. source.

The current i, from the source, is divided into i_1, i_2 and i_3. The potential difference V will remain same across each resistor, thus we have

$$i_1 = \frac{V}{R_1}$$

$$i_2 = \frac{V}{R_2} \qquad ...(7.41)$$

and $$i_2 = \frac{V}{R_3}$$

Now, the total current is the sum of individual currents passing through each resistor, then,

$$i = i_1 + i_2 + i_3 \qquad ...(7.42)$$

$$= \frac{V}{R_1} + \frac{V}{R_2} + \frac{V}{R_3}$$

and $$\frac{i}{V} = \frac{1}{R_1} + \frac{1}{R_2} + \frac{1}{R_3} \qquad ...(7.43)$$

The equivalent resistance is

$$\boxed{\frac{1}{R} = \frac{1}{R_1} + \frac{1}{R_2} + \frac{1}{R_3}} \qquad ...(7.44)$$

If n resistors are connected in parallel, then

$$\frac{1}{R} = \frac{1}{R_1} + \frac{1}{R_2} + ... + \frac{1}{R_3}$$

or $$\boxed{\frac{1}{R} = \sum_{i=1}^{n} \frac{1}{R_i}} \qquad ...(7.45)$$

7.6. ELECTROMOTIVE FORCE AND SINGLE LOOP CIRCUIT

A circuit is a combination of passive elements (e.g. resistor, capacitor and inductor) and a source of energy which is used to maintain a constant current in the circuit. This energy source is called the electromotive force (emf). The source of emf are classified as the constant voltage source and the constant current source. The example of emf sources are cells, battery and solar cell etc. The emf is represented by ε. The purpose of a battery in a circuit is to maintain a constant voltage or current in the circuit, then

$$\varepsilon = V_0 \qquad ...(7.46)$$

where V_0 is a constant potential difference between the two terminals of a battery. In a battery, the positive terminal is at higher potential while negative terminal is at lower potential. Thus, the emf is equal to work done in carrying a unit charge from lower potential to higher potential. That is, the electrons move from negative terminal to positive terminal of a energy source. The unit of emf is volt. Since we know, for a close circuit,

$$\Delta V = \oint \vec{E} \cdot \vec{dl} = 0 \qquad ...(7.47)$$

This implies that it maintains a circuit at a constant potential or voltage. From the Eq. (7.47) it is clear that the electric field can not be used as the energy source and it can not provide the energy to the moving charge in a circuit.

The basic difference between emf and potential is that when the potential difference does the work in moving a charge from one point to another, the emf provides the energy in this process.

Single loop circuit: Now, consider a single loop circuit containing a resistor R and an emf as shown in fig. 7.12.

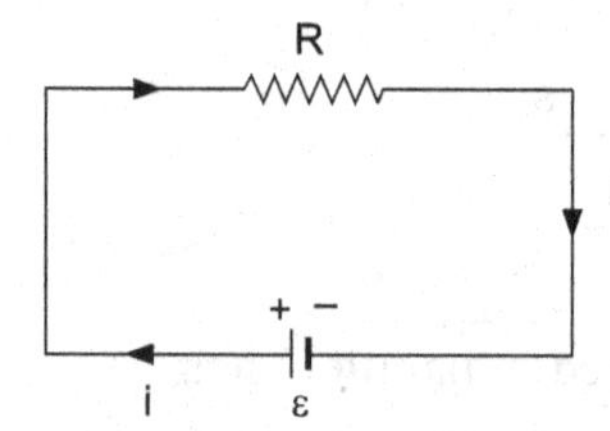

Fig. 7.12. Circuit containing an emf source and a resistor

The emf of the source is equal to the potential difference between the positive and negative terminals of the source. Here, we have assumed that the internal resistance of the source is zero, and there is only an energy dissipative element R, the resistance. Then,

$$\varepsilon = iR \qquad ...(7.48)$$

or

$$i = \frac{\varepsilon}{R} \qquad ...(7.49)$$

which is known as single loop equation.

In case where the source of emf is localized, that is, in a real system, the losses in the circuit is divided into two parts, viz,

(a) losses within the emf source.

(b) losses in the external circuit.

Now, consider a circuit containing a resistor R and a source of emf with internal resistance r as shown in fig. 7.13. Then,

$$\varepsilon = iR + ir \qquad ...(7.50)$$

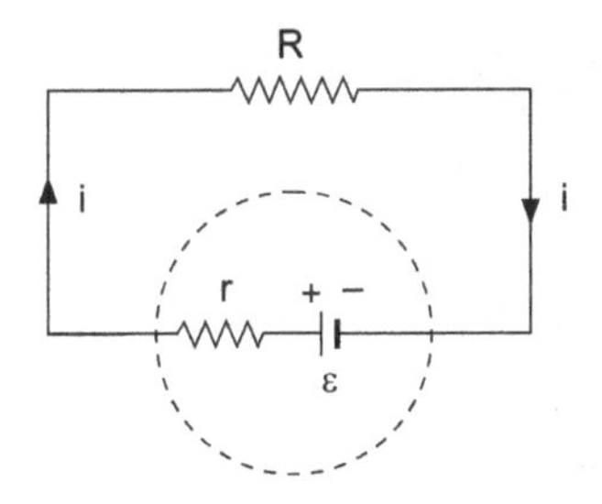

Fig. 7.13. Circuit with internal resistance of the source.

or $$i = \frac{\varepsilon}{R+r} \qquad ...(7.51)$$

Moreover the energy conservation law is

$$\varepsilon i = i^2 R + i^2 r$$

In this way, the current flowing in the circuit depends on both resistance R and r.

7.7. ENERGY AND ENERGY DISSIPATION IN A RESISTOR

In an electrical circuit, the energy is transferred from a source of emf to the load by means of the current. Consider a circuit containing a source of emf and a resistor of resistance R as shown in fig. 7.14.

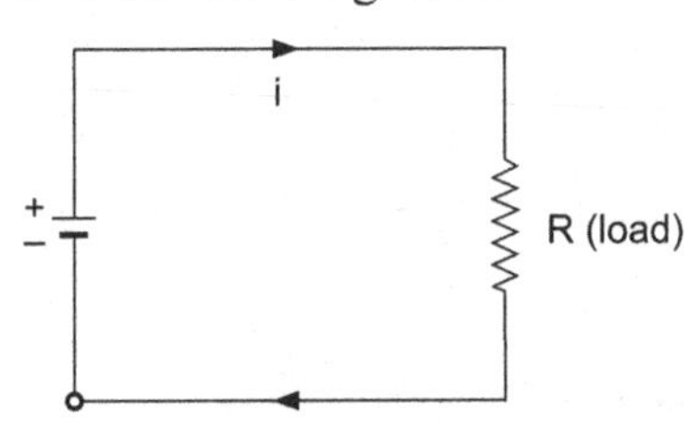

Fig. 7.14. A circuit containing load resistance.

When a current flows through a resistor, it is heated. It means that the electrical energy is transformed into heat energy. When a small charge dq is moving from higher potential to lower potential, the work done is

$$dW = V\,dq \qquad ...(7.52)$$

where V is a constant potential difference across the resistor R. Now, the power is defined as the rate of change of work done and is given by

$$P = \frac{dW}{dt}$$

or $$P = V\frac{dq}{dt} \qquad ...(7.53)$$

But $$i = \frac{dq}{dt},$$

$$P = Vi \qquad ...(7.54)$$

since, $V = iR$, Thus,

$$P = i^2 R$$

or $$\boxed{P = \frac{V^2}{R}} \qquad ...(7.55)$$

The unit of electrical power is Amp-volt.

Therefore,

$$1 \text{ Amp-volt} = 1 \text{ J/sec}$$
$$= 1 \text{ watt } (W)$$

The electrons are accelerated by the applied emf and they loss their energy in the collision with the atoms. For a given resistor, large the surface area, larger the energy dissipation.

Example 7.3. Compute the resistance between *A* and *B* of an infinitely long ladder of resistors as shown in fig. 7.15.

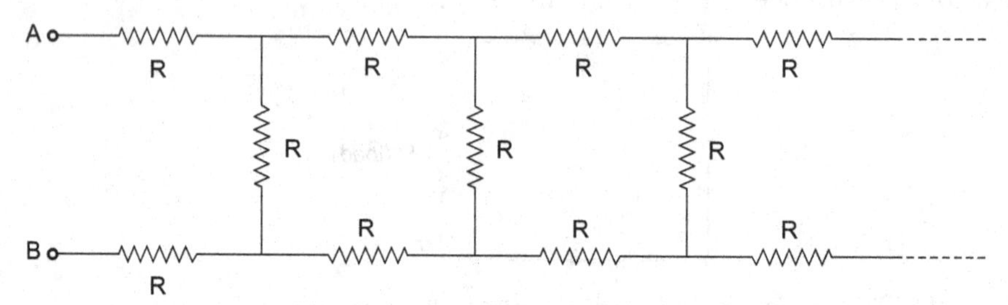

Fig. 7.15. Resistance system.

Solution: Suppose that the series is terminated by equivalent resistance R' as shown in fig. 7.16.

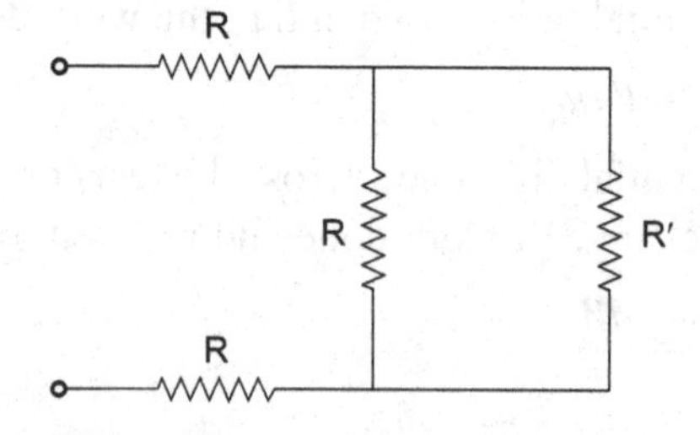

Fig. 7.16. Combination of resistances.

The resistance between A and B can be found as

$$R' = R + R + \frac{RR'}{R + R'}$$

or $$R'^2 - 2RR' - 2R^2 = 0$$

On solving this equation we get

$$R' = R(1 + \sqrt{3}).$$

7.8. COLOR CODE FOR CARBON RESISTOR

Every carbon resistor has color bands on it. These color bands may be read to compute the value of the given resistor as shown in fig. 7.17.

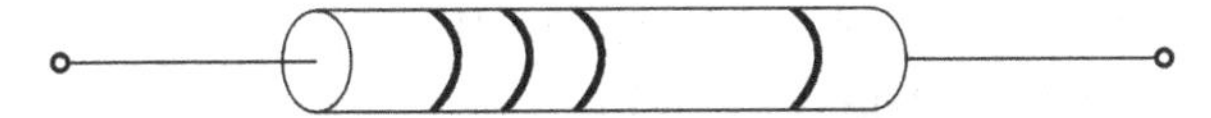

Fig. 7.17. Color bands on a resistor.

The codes for these color bands are listed in the table 7.2. We have following points for coding the colors bands on the resistor.

(a) the first strip indicates the first digit.

(b) the second strip indicates the second digit.

(c) the third band/strip indicates the number of zeros.

(d) Fourth band denotes the tolerance.

Table 7.2. Color code for resistors.

Color	Code
Black	0
Brown	1
Red	2
Orange	3
Yellow	4
Green	5
Blue	6
Violet	7
Gray	8
White	9
Gold	±5%
Silver	±10%
No color	±20%

Example 7.4. If a resistor has four color strips of colors Red, Red and orange and fourth strip is of Gold, compute the resistance of the resistor.

Solution: We have following bands on a resistor as shown in fig. 7.18.

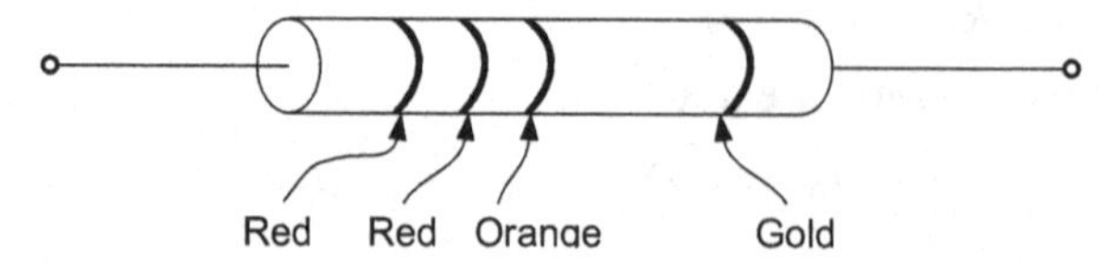

Fig. 7.18. Computation of color bands on a resistor.

(a) First band denotes a digit '2'.

(b) Second band indicates a digit '2'.

(c) Third digit shows number of zeros, i.e. 10^3.

(d) Tolerance is +5% as strip is of Gold.

Thus,

$$R = 22 \times 10^3 \pm 5\%$$

or

$$R = 22K \pm 5\%$$

7.9. KIRCHHOFF'S LAWS FOR ELECTRIC NETWORKS

For a given network with sources of emf and the circuit elements, we can compute the currents and the voltage drops across the various branches of the network. If the currents in various branches are known, the voltage drops can be computed by Ohm's law. To analyze any network, we have two rules called Kirchhoff's Laws.

(a) Junction Theorem: This states that the algebraic sum of the currents flowing into and out of any junction of a network is zero. That is

$$\boxed{\sum i = 0} \quad ...(7.56)$$

Now, consider five currents at a node as shown in fig. 7.19.

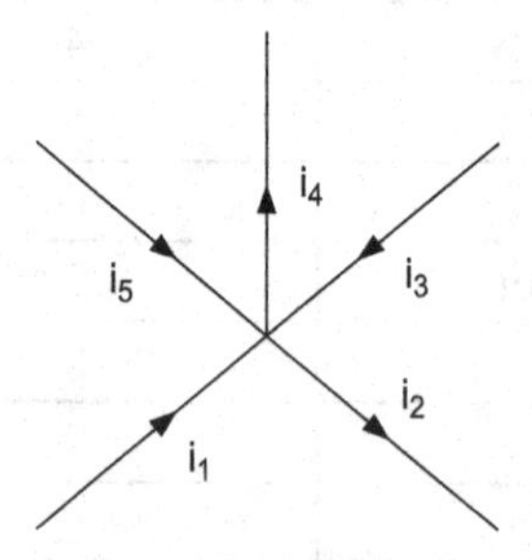

Fig. 7.19. Kirchhoff's node rule

Now, according to junction theorem, we may write,

$$-i_1 + i_2 - i_3 + i_4 - i_5 = 0 \quad ...(7.57)$$

or

$$i_2 + i_4 = i_1 + i_3 + i_5 \quad ...(7.58)$$

Here, we have assumed those currents as positive which directed away from the junction and those currents as negative which are directed into the junction.

Thus, the Eq. (7.58) takes the form

$$\left(\sum i\right)_{in} = \left(\sum i\right)_{out} \quad ...(7.59)$$

(b) Loop or Mesh Theorem: It states that the sum of the voltage drops is equal to the sum of the emf sources. That is, In a single mesh, we have

$$\boxed{\sum iR = \sum \varepsilon} \quad ...(7.60)$$

The voltage drop across a resistor depends on the choice of the direction of the travel of current. We may obtain the same equation for the voltage drops in a mesh whether the mesh is traversed clockwise or counter-clockwise. For example, consider a network as shown in fig. 7.20.

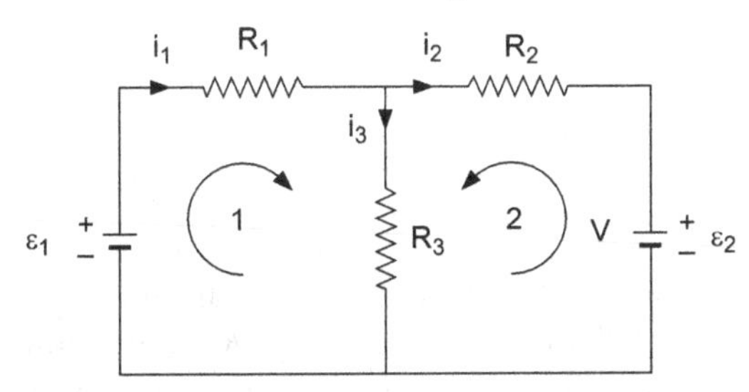

Fig. 7.20. An electrical network.

Here, we have three possible currents in different branches of the network. Now, we shall apply Kirchhoff's voltage law or loop theorem to this network.

For mesh–1:

$$i_1R_1 + i_2R_2 - \varepsilon_1 = 0 \quad ...(7.61)$$

For mesh–2:

$$-i_2R_2 + i_3R_3 - \varepsilon_2 = 0 \quad ...(7.62)$$

For mesh–3:

$$i_1R_1 + i_2R_2 = \varepsilon_2 - \varepsilon_1 \quad ...(7.63)$$

we get three equations for the three currents i_1, i_2 and i_3 and on solving these equations, we can determine the values of i_1, i_2 and i_3.

7.10. MATRIX AND DETERMINANT METHOD FOR SOLVING MESH EQUATIONS

The mesh equations may be written as

$$[R]\,[I] = [\varepsilon] \qquad ...(7.64)$$

where $[R]$ = Resistance matrix

$[I]$ = Current matrix

and $[\varepsilon]$ = Emf sources matrix

Here, $[R]$ may be a rectangular or a square matrix of m rows and n columns and $[I]$ and $[\varepsilon]$ are the column matrices. Now consider a network containing three loops, then matrices $[R]$, $[I]$ and $[E]$ can be written in form of the matrices as

$$[R] = \begin{pmatrix} R_{11} & R_{12} & R_{13} \\ R_{21} & R_{22} & R_{23} \\ R_{31} & R_{32} & R_{33} \end{pmatrix} \qquad ...(7.65)$$

$$[I] = \begin{pmatrix} i_1 \\ i_2 \\ i_3 \end{pmatrix} \qquad ...(7.66)$$

and

$$[\varepsilon] = \begin{pmatrix} \varepsilon_1 \\ \varepsilon_2 \\ \varepsilon_3 \end{pmatrix} \qquad ...(7.67)$$

The resistance R_{11} is the resistance of the first mesh, R_{22} of second mesh and R_{33} is of third mesh. The resistance R_{12} and R_{21} are common resistances for meshes 1 and 2. Similarly R_{23} and R_{32} are common for meshes 2 and 3 etc. These common resistance terms are positive, if there is a current passing through them in same direction and the resistance terms will be negative if current flows in opposite direction. Moreover, if there are n meshes, the dimensions of the resistance matrix is $n \times n$. To understand the concept, consider a network containing resistances R_1, R_2 and R_3 with the energy sources ε_1 and ε_2 as shown in fig. 7.21.

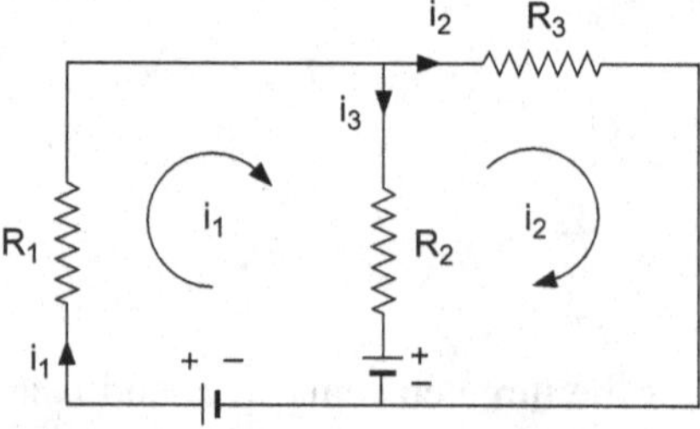

Fig. 7.21. Two loop network

According to Kirchhoff's loop rule,

$$i_1(R_1 + R_2) - i_2R_2 = \varepsilon_1 - \varepsilon_2 \qquad ...(7.68)$$

and

$$-i_1 R_2 + (R_2 + R_3)\, i_2 = \varepsilon_2 \qquad ...(7.69)$$

In matrix form,

$$\begin{bmatrix} R_1 + R_2 & -R_2 \\ -R_2 & R_2 + R_3 \end{bmatrix}\begin{bmatrix} i_1 \\ i_2 \end{bmatrix} = \begin{bmatrix} \varepsilon_1 - \varepsilon_2 \\ \varepsilon_2 \end{bmatrix} \qquad ...(7.70)$$

If

$$\Delta = \begin{vmatrix} R_1 + R_2 & -R_2 \\ -R_2 & R_2 + R_3 \end{vmatrix} = \det |R| \qquad ...(7.71)$$

and

$$\Delta_1 = \begin{vmatrix} \varepsilon_1 - \varepsilon_2 & -R_2 \\ -\varepsilon_2 & R_2 + R_3 \end{vmatrix} \qquad ...(7.72)$$

$$\Delta_2 = \begin{vmatrix} R_1 + R_2 & \varepsilon_1 - \varepsilon_2 \\ -R_2 & \varepsilon_2 \end{vmatrix} \qquad ...(7.73)$$

Then, the values of the currents i_1 and i_2 can be obtained as

$$i_1 = \frac{\Delta_1}{\Delta} \qquad ...(7.74)$$

and

$$i_2 = \frac{\Delta_2}{\Delta} \qquad ...(7.75)$$

Example 7.5. Compute the current in each mesh of network shown in fig. 7.22.

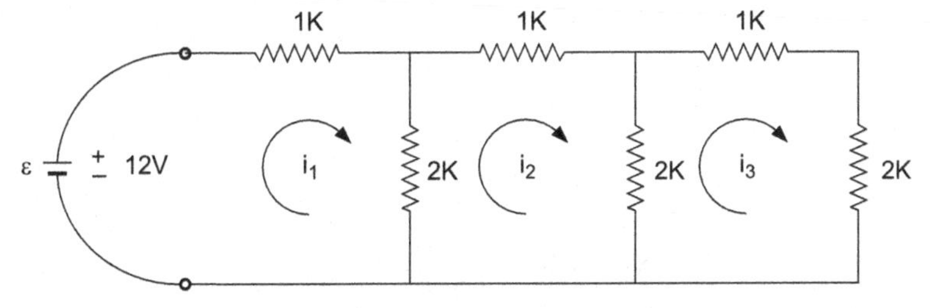

Fig. 7.22. Electrical network.

Solution: We have matrix equation

$$[R][I] = [\varepsilon]$$

or

$$\begin{pmatrix} R_{11} & R_{12} & R_{13} \\ R_{21} & R_{22} & R_{23} \\ R_{31} & R_{32} & R_{33} \end{pmatrix}\begin{pmatrix} i_1 \\ i_2 \\ i_3 \end{pmatrix} = \begin{pmatrix} \varepsilon_1 \\ \varepsilon_2 \\ \varepsilon_3 \end{pmatrix}$$

or

$$\begin{pmatrix} 3 & -2 & 0 \\ -2 & 5 & -2 \\ 0 & -2 & 5 \end{pmatrix}\begin{pmatrix} i_1 \\ i_2 \\ i_3 \end{pmatrix} = \begin{pmatrix} 12 \\ 0 \\ 0 \end{pmatrix}$$

Here

$$R_{11} = 1 + 2 = 3K$$
$$R_{22} = 2 + 1 + 2 = 5K$$
$$R_{33} = 2 + 1 + 2 = 5K$$

The common resistance between meshes 1 and 2 is $2K$ and similarly $2K$ is common to meshes 2 and 3. We take common resistances as negative.

Now

$$\Delta = \begin{vmatrix} 3 & -2 & 0 \\ -2 & 5 & -2 \\ 0 & -2 & 5 \end{vmatrix}$$

$$\Delta_1 = \begin{vmatrix} 12 & -2 & 0 \\ 0 & 5 & -2 \\ 0 & -2 & 5 \end{vmatrix}$$

$$\Delta_2 = \begin{vmatrix} 3 & 12 & 0 \\ -2 & 0 & -2 \\ 0 & 0 & 5 \end{vmatrix}$$

and

$$\Delta_3 = \begin{vmatrix} 3 & -2 & 12 \\ -2 & 5 & 0 \\ 0 & -2 & 0 \end{vmatrix}$$

Now,

$$i_1 = \frac{\Delta_1}{\Delta} = \frac{252}{43} = 5.89 \text{ A}$$

$$i_2 = \frac{\Delta_2}{\Delta} = \frac{120}{43} = 2.8 \text{ A}$$

$$i_3 = \frac{\Delta_3}{\Delta} = \frac{48}{43} = 1.1 \text{ A}$$

6.11. THE RC SERIES CIRCUIT

In RC circuit, the current and voltage are time dependent. Using RC circuit, we can produce a time varying signal.

(a) Charging of a Capacitor: Consider a circuit containing a resistor and a capacitor in series with a source of emf V as shown in fig. 7.23.

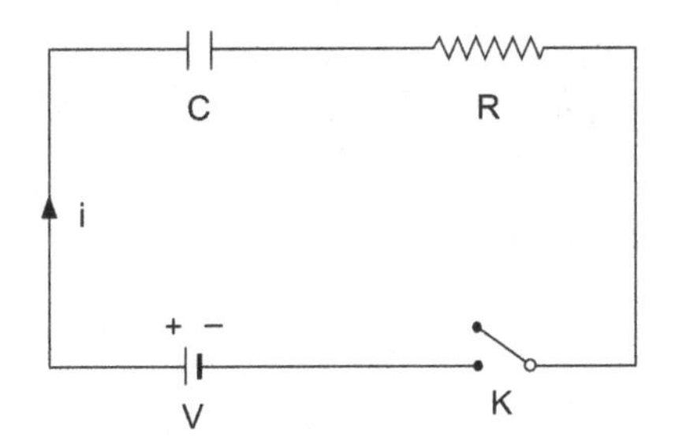

Fig. 7.23. Charging of a Capacitor.

To start charging, the key K is closed at $t = 0$, then, a transient current starts to flow. At $t = 0$, the current is given by

$$i_0 = \frac{V}{R} \quad ...(7.76)$$

As a result, the capacitor starts charging and it attains $+q$ and $-q$ charges on its plates. Thus, the potential difference across the capacitor is given by

$$V_C = \frac{q}{C} \quad ...(7.77)$$

Now, applying Kirchhoff's loop theorem, we have

$$V = iR + \frac{q}{C} \quad ...(7.78)$$

since $$i = \frac{dq}{dt},$$

we have

$$\frac{dq}{dt}R = V - \frac{q}{C}$$

or $$\frac{dq}{dt} = \frac{V}{R} - \frac{q}{RC} \quad ...(7.79)$$

or $$\frac{-dq}{(q - CV)} = \frac{dt}{RC} \quad ...(7.80)$$

Since current and voltage are time dependent, at $t = 0$ the charge on the capacitor is zero, thus current in the circuit is maximum. As the charge on the plates of the capacitor increases, the current in the circuit decreases and it becomes zero when capacitor is charged maximum. After that there is no charging of the capacitor. When $i = 0$, $q = q_0$ (maximum charge). Now integrating the Eq. (7.80), we have

$$\int_0^q \frac{dq}{(q-CV)} = -\frac{1}{RC}\int_0^t dt \qquad ...(7.81)$$

or $$\ln(q - CV) - \ln(-CV) = \frac{-t}{RC}$$

or $$\ln\left(\frac{q-CV}{-CV}\right) = \frac{-t}{RC}$$

or $$\left(\frac{q-CV}{-CV}\right) = e^{-t/RC}$$

or $$q - CV = -CV\,e^{-t/RC}$$

or $$q = CV(1 - e^{-t/RC}) \qquad ...(7.82)$$

Assuming $CV = q_0$ (maximum charge stored in the capacitor), then

$$q = q_0(1 - e^{-t/RC})$$

or $$\boxed{q = q_0(1 - e^{-t/\tau})} \qquad ...(7.83)$$

where $\tau = RC$ is known as the time constant of a circuit. The variation of the current can be obtained by differentiating the Eq. (7.82) with respect to t. Thus, we have

$$i = \frac{dq}{dt} = \frac{V}{R}e^{-t/RC}$$

or $$i = i_0\,e^{-t/\tau} \qquad ...(7.84)$$

where $i_0 = \frac{V}{R}$. The Eq. (7.84) shows that the current decreases with time and $i = 0$, when the potential different across the plates of the capacitor is equal to V. The Eq (7.83) shows that the capacitor is charged exponentially as shown in fig. 7.24.

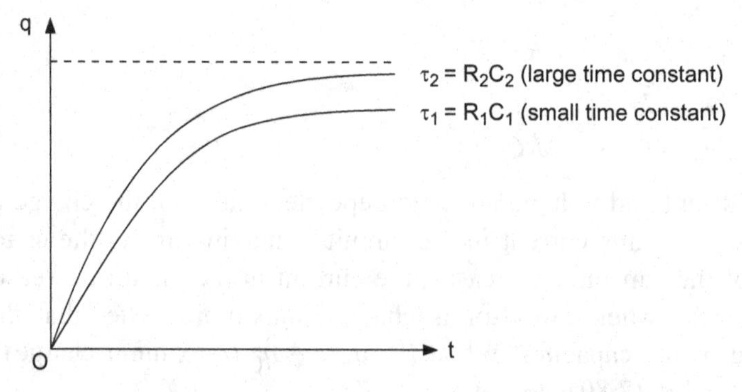

Fig. 7.24. The variation of *q* with respect to time *t*.

The equation (7.84) predicts that the current decreases exponentially with time as shown in fig. 7.25.

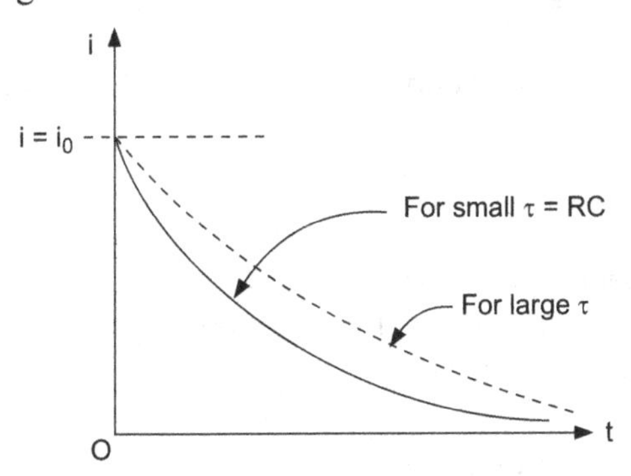

Fig. 7.25. Variation of current with time during charging of a capacitor

Now we have following cases,

(1) When $t = 0$, $e^0 = 1$ and $i = i_0$, Thus $q = 0$

(2) When $t = \infty$, $e^{-\infty} = 0$ and $i = 0$, thus the charge on the capacitor is

$$q = q_0 \qquad \text{...(7.85)}$$

(3) When $t = RC$, then

$$q = q_0(1-e^{-1}) = q_0\left(1-\frac{1}{e}\right)$$

or $$q = q_0\left(1-\frac{1}{2.712}\right) = q_0(1-0.368)$$

or $$q = 0.64q_0$$

and $$i = \frac{i_0}{e} \qquad \text{...(7.86)}$$

Thus τ is the time in which i decreases to $\frac{1}{e}$ times of its maximum value.

(b) Discharging of a Capacitor: Now, we shall obtain an expression for discharging of a capacitor. Since the capacitor is fully charged to q_0 and we open the key K, then the positive charge begins to flow through the negative plate of the capacitor as shown in fig. 7.26.

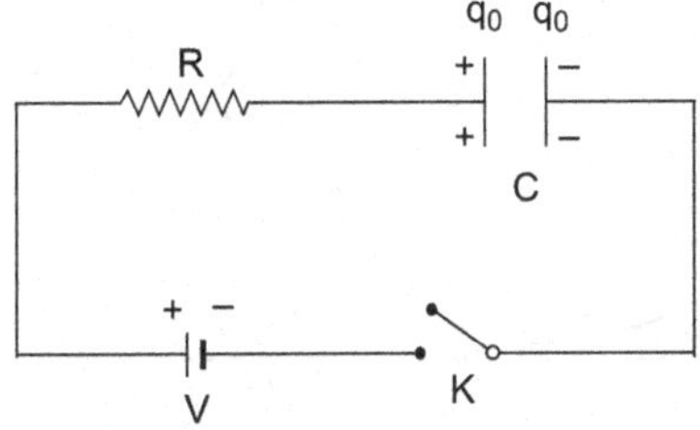

Fig. 7.26. Discharging of a capacitor.

At $t = 0$, $q = q_0$ and the potential difference across the capacitor is

$$V_C = \frac{q_0}{C} \qquad ...(7.87)$$

In this case, the direction of current is opposite and applying loop theorem, we have

$$\frac{-q}{C} + Ri = 0 \qquad ...(7.88)$$

Since charge capacitor acts as a source, the potential difference across the capacitor decreases. Thus, the current is given by

$$i = -\frac{dq}{dt}$$

The Eq. (7.88) takes the form,

$$\frac{-q}{C} - R\frac{dq}{dt} = 0$$

or

$$\frac{dq}{q} = -\frac{1}{RC}dt \qquad ...(7.89)$$

we integrate the Eq. (7.89), we get

$$\int_{q_0}^{q} \frac{dq}{q} = -\frac{1}{RC}\int_0^t dt$$

$$\ln q - \ln q_0 = -t/RC$$

or

$$\ln \frac{q}{q_0} = -\frac{t}{RC}$$

or

$$\boxed{q = q_0\, e^{-t/RC}} \qquad ...(7.90)$$

Further more, $\quad i = i_0\, e^{-t/RC} = i_0\, e^{-t/\tau} \qquad ...(7.91)$

In the time constant $\tau = RC$, the capacitor discharged 64% of the total charge. The charge and current decreases exponentially with time as shown in fig. 7.27.

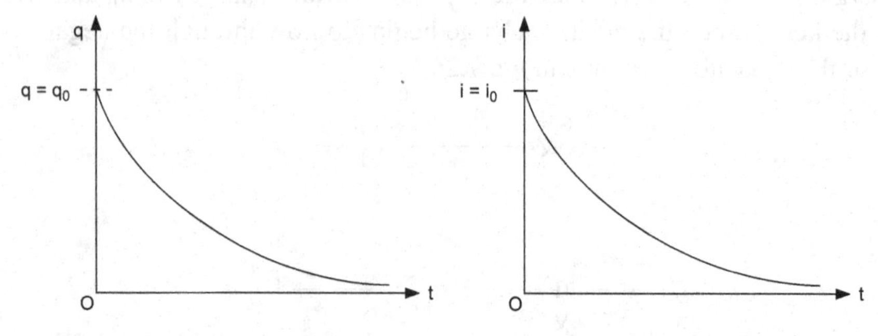

Fig. 7.27. The variation of charge (a) and current (b) with the time

Example 7.6. In a circuit Fig. 7.28, containing $R_1 = 3\Omega$, $R_2 = 6\Omega$ and $C = 6\mu f$. Compute

(1) time constant when key K is open.

(2) time constant when key K is closed.

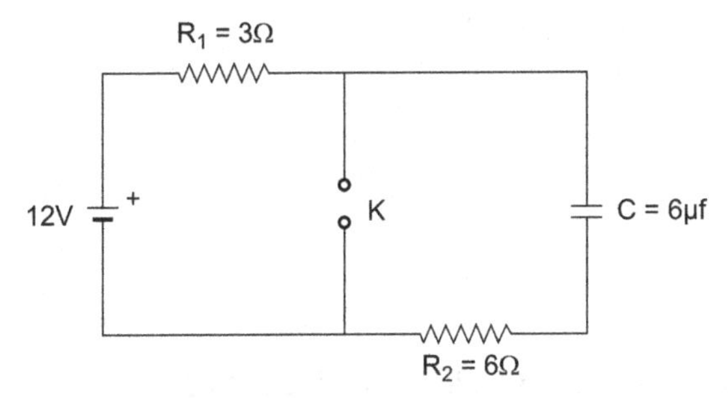

Fig. 7.28. Electrical network.

Solution: (1) When K is open,

$$\begin{aligned}\tau &= (R_1 + R_2)\, C \\ &= 9\Omega \times 6\mu f \\ &= 5.4 \times 10^{-5} \text{ sec.}\end{aligned}$$

(2) When K is closed

$$\begin{aligned}\tau &= R_1 C \\ &= 6 \times 6 \times 10^{-6} \\ &= 3{\cdot}6 \times 10^{-5} \text{ sec.}\end{aligned}$$

Example 7.7. A cube is formed with twelve identical resistors, each of resistance R as shown in fig. 7.29.

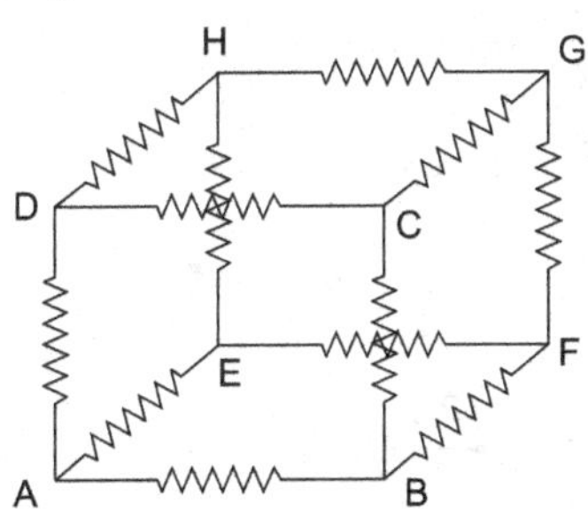

Fig. 7.29. Network of resistors.

Compute the equivalent resistance

(a) between points A and G.

(b) between points A and F

Solution: (a) Let i be the current entering from the point A as shown in fig. 7.30.

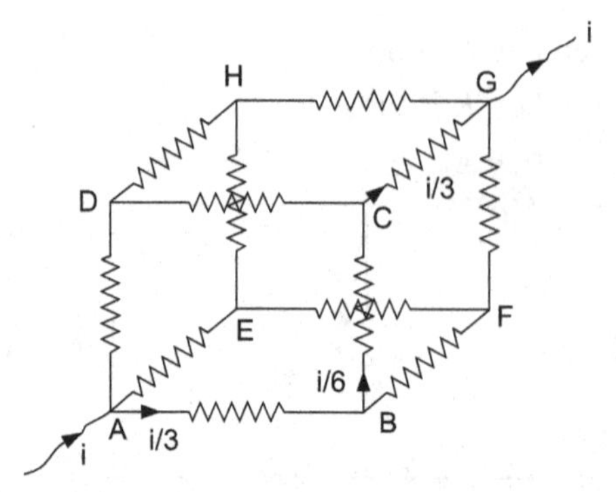

Fig. 7.30. Network of resistors.

We assume that V_{AG} is the potential difference between the point A and G, and choosing the path $ABCG$ for computing the equivalent resistance between A and G.

$$V_{AG} = V_{AB} + V_{BC} + V_{CG}$$

$$= \frac{i}{3}R + \frac{i}{6}R + \frac{i}{3}R$$

$$= \frac{5}{6}iR$$

or $$R_{AG} = \frac{V_{AG}}{i} = \frac{5}{6}R.$$

(b)

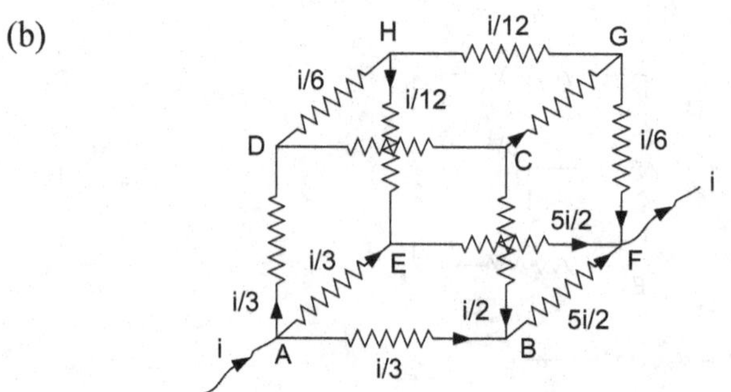

Fig. 7.31. Network of resistors.

In Fig. 7.31, considering the path $A \to B \to F$,

$$V_{AF} = V_{AB} + V_{BF}$$

$$= \frac{iR}{3} + \frac{5iR}{12}$$

$$= \frac{3}{4} iR$$

$$R_{eq} = \frac{V_{AF}}{i} = \frac{3}{4} R.$$

7.12. POTENTIO-METER

Potentiometer is an instrument commonly used for measuring the potential difference of the energy source. A voltmeter can be used to measure the potential difference, but it consists of a finite resistance which causes it to draw a current from the energy source. The beauty of the potentiometer is that it measures the potential difference without drawing any current from the source of emf. In this way, it is called the **infinite resistance voltmeter**. A simple arrangement of the potentiometer is shown in fig. 7.32. The potentiometer works on the principle of the comparing of the unknown potential difference across the battery with a standard source of emf.

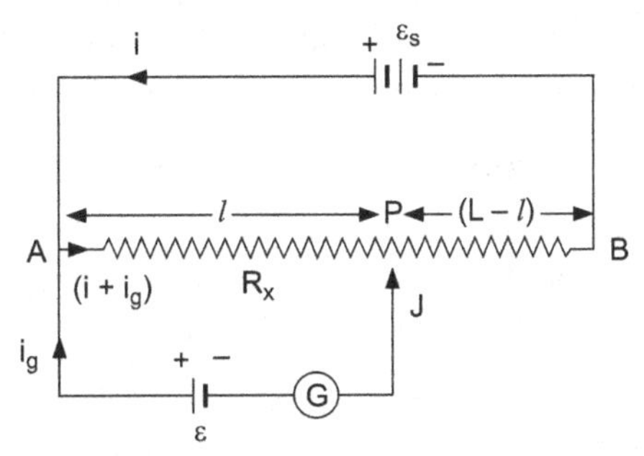

Fig. 7.32. Potentiometer

It consists of a standard (known) voltage source ε_S which is connected across A and B. Between the points A and B a resistance wire of resistance R is connected and this wire is of uniform thickness called the slide wire. Moreover, a unknown source of emf ε is connected to a sliding jockey J through a galvenometer G. The Jockey J moves on the sliding wire back and forth, and a null or zero deflection position is obtained by sliding the jockey on the slide wire AB. At the null point P(suppose). There is a no current passing in the branch containing galvenometer G.

Now, according to loop theorem,

$$\varepsilon_S = R_x(i + i_g) + i(R - R_x) \quad ...(7.92)$$

and

$$\varepsilon = (i_g + i)R_x \quad ...(7.93)$$

To obtain the value of unknown emf the jockey is set at the null position P at which $i_g = 0$

Thus, we have

$$\left.\begin{aligned} \varepsilon_S &= iR \\ \text{and} \qquad \varepsilon &= iR_x \end{aligned}\right\} \qquad \text{...(7.94)}$$

on dividing these equations we get

$$\frac{\varepsilon}{\varepsilon_S} = \frac{R_x}{R} \qquad \text{...(7.95)}$$

we know that the resistance is proportional to the length thus, we may write the Eq. (7.95) as

$$\frac{\varepsilon}{\varepsilon_S} = \frac{l}{L}$$

or

$$\boxed{\varepsilon = \left(\frac{l}{L}\right)\varepsilon_S} \qquad \text{...(7.96)}$$

where L is the length of wire AB. The potentiometer is a variable resistor having a resistance wire, a known emf and a source of unknown emf. We can compare two unknown emfs also using potentiometer.

Difference between a Potentiometer and a Rheostat

The potentiometer and rheostat both are known as the variable resistors. Since we know that a variable resistor is used as a voltage divider, we can differentiate between a potentiometer and a rheostat.

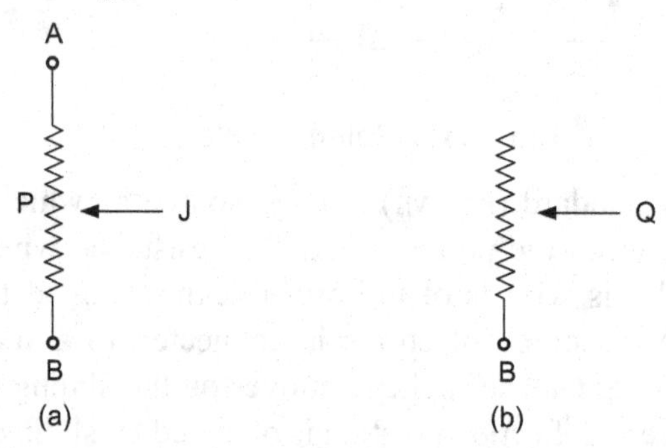

Fig. 7.33. (a) Potentiometer (b) rheostat

Suppose that a variable resistor divides the voltage between the points A and B as shown in fig. 7.33. Fig 7.33(a) a wire bound potentiometer consisting of a coil of resistance wire with a central slider. It selects the desired length of wire to offer a considerable resistance and a fraction of the potential difference. In this way, the potentiometer is a three terminals device. Now, we may have another variable resistor called rheostat, fig. 7.33(b). In the rheostat, current flows through a variable part of the resistor and it is used to control the current.

7.13. WHEATSTONE'S BRIDGE

Wheatstone's bridge is an arrangement of the resistors in a circuit used for measuring the unknown resistance accurately. In 1843, an English scientist Charles Wheatstone gave an excellent method of measuring the unknown resistance very accurately. In wheatstone bridge, Fig. 7.34, a galvenometer is used in the arm *BD* with a key *K*. When the key *K* is closed, no current passes through the arm *BD*, and bridge is said to be in balanced condition. Thus the potentials at the point *B* and *D* are same, that is, $V_B = V_D$.

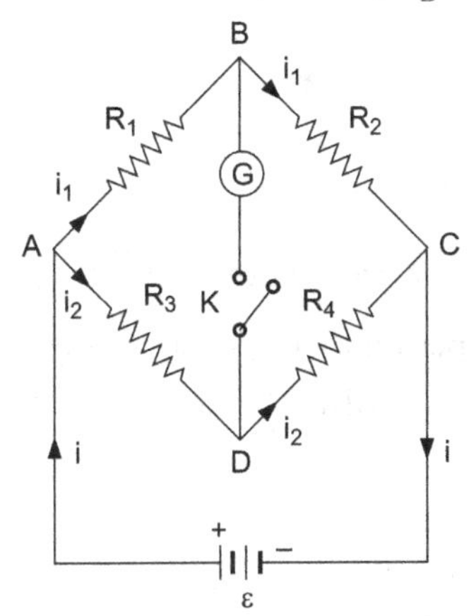

Fig. 7.34. Wheatstone bridge.

Now, when *K* is closed, applying Kirchhoff's loop law in the loops *ABDA* and *BCDB*, we have

$$i_1R_1 - i_2R_3 = 0$$

or
$$i_1R_1 = i_2R_3 \quad \text{...(7.97)}$$

and
$$i_1R_2 - i_2R_4 = 0$$

or
$$i_1R_2 = i_2R_4 \quad \text{...(7.98)}$$

on dividing the Eq. (7.97) by the Eq. (7.98), we get

$$\boxed{\frac{R_1}{R_2} = \frac{R_3}{R_4}} \quad \text{...(7.99)}$$

Suppose that R_1 is unknown resistance and the R_2, R_3 and R_4 all are known resistances, then R_1 can be calculated as

$$R_1 = R_2\left(\frac{R_3}{R_4}\right) \quad \text{...(7.100)}$$

Wheatstone's bridge provides the high accuracy of the measurement and the accuracy depends on the sensitivity of the galvanometer *G*.

Example 7.8. A ten wire potentiometer is connected to a battery of 3V. The potentiometer has resistance per unit length 1.5 Ω/m, and a laclanche cell of 1.5V is also connected to the jockey through galvenometer as shown in fig. 7.35. If null point occurs at a distance of 6 m, compute the internal resistance r of the battery.

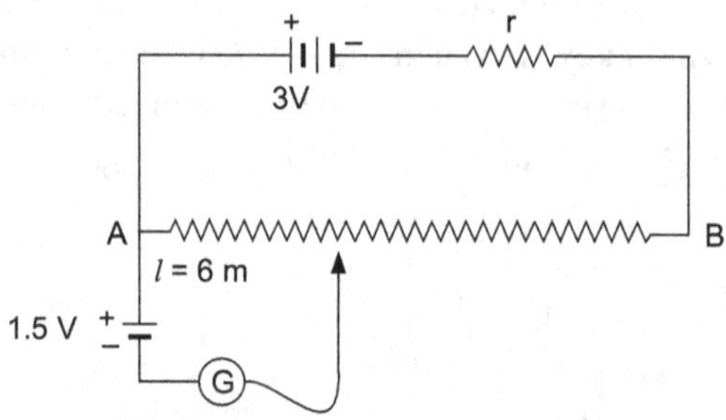

Fig. 7.35. A potentiometer.

Solution: Given resistance per unit length = 1.5 Ω/m

The resistance of 6 m length of wire = 6 × 1.5 = 9 Ω

and the resistance of 10 m wire = 1.5 × 10 = 15 Ω

Applying Kirchhoff's law,

$$V = i(r + R)$$

where R = resistance of 10 m wire.

and $V' = R_x\, i$

where R_X = resistance of 6 m wire.

Now

$$i = \frac{V'}{R_X} = \frac{1.5}{6 \times 1.5} = 0.166\ A$$

Thus,

$$3 = i(r + 10 \times 1.5)$$

$$3 = 0.166\,(r + 15)$$

or $r = 3\ \Omega$.

Example 7.9. Compute the current in each resistor shown in fig. 7.36.

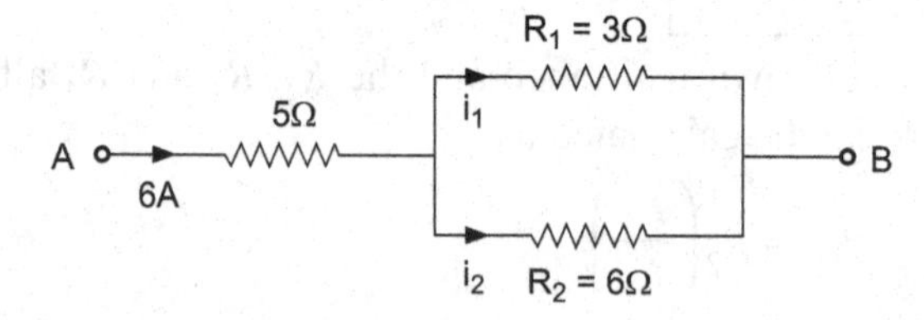

Fig. 7.36. An electrical network.

Solution: The current passing through 5 Ω is 6A. Since the voltage drops across the resistors R_1 and R_2 must be same, Thus,

$$i_1R_1 = i_2R_2$$

and $$i = i_1 + i_2$$

on solving these equations, we get

$$i_1 = \left(\frac{R_2}{R_1 + R_2}\right)i$$

and $$i_2 = \left(\frac{R_1}{R_1 + R_2}\right)i$$

Thus,

$$i_1 = \left(\frac{6}{9}\right)\cdot 6 = 4A$$

and $$i_2 = \frac{3}{9}\times 6 = 2A$$

Example 7.10. In the circuit shown in fig. 7.37, the internal resistance of the battery is R_i, compute the maximum power delivered to the load R.

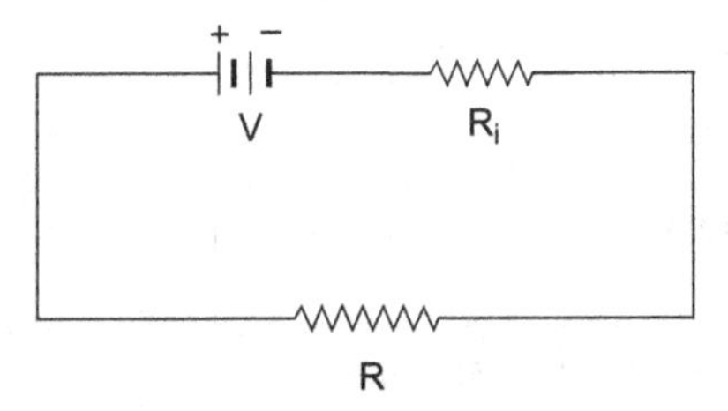

Fig. 7.37. Electrical network.

Solution: Current from the source V is

$$i = \frac{V}{R + R_i}$$

The power dissipation in the load R is

$$P = i^2 R$$

$$P = R\left(\frac{V}{R + R_i}\right)^2$$

or $$P = \frac{V^2 R}{(R + R_i)^2}$$

To obtain the maximum power dissipation in the load R, differentiating P w.r.to R,

$$\frac{dP}{dR} = \left[\frac{1}{(R+R_i)^2} - \frac{2R}{(R+R_i)^3}\right]V^2$$

$$= V^2 \frac{(R_i - R)}{(R+R_i)^3}$$

For maximum power,

$$\frac{dP}{dR} = 0$$

Thus, we get $R = R_i$

Now, maximum power dissipated in R is

$$P_m = \frac{V^2 R}{(2R)^2}$$

or

$$P_m = \frac{1}{4}\frac{V^2}{R}$$

Example 7.11. In the circuit shown in fig. 7.38, compute the current flowing in each resistor.

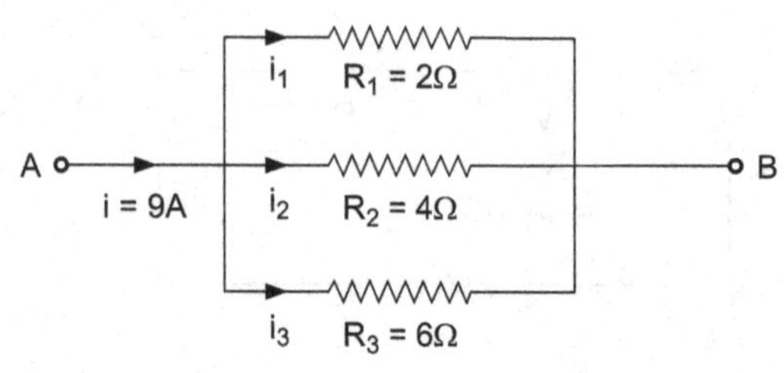

Fig. 7.38. Combination of resistors.

Solution: Let i_1, i_2 and i_3 be the currents passing through the resistors R_1, R_2 and R_3 respectively. Then

$$i = i_1 + i_2 + i_3$$

The equivalent resistance of the circuit is

$$R = \frac{R_1 R_2 R_3}{R_1 R_2 + R_2 R_3 + R_1 R_3}$$

since the voltage drops across the resistors R_1, R_2 and R_3 must be same, then,

$$i_1 R_1 = i_2 R_2 = i_3 R_3$$

Therefore,

$$i_1 = \frac{R_2 R_3 \, i}{R_1 R_2 + R_2 R_3 + R_3 R_1}$$

$$i_2 = \frac{R_1R_3\,i}{R_1R_2 + R_2R_3 + R_3R_1}$$

and
$$i_3 = \frac{R_1R_2\,i}{R_1R_2 + R_2R_3 + R_3R_1}$$

so,
$$i_1 = 4.90 \text{ A}$$
$$i_2 = 2.45 \text{ A}$$
$$i_3 = 1.65 \text{ A}$$

Example 7.12. A rod of length 2 long has diameter 0.5 m. The potential difference between its ends is 50 V and a current of 5 A flows into it. Then compute

(1) current density J

(2) electric field across the rod

(3) resistivity of the material of the rod

Solution: Given

$$i = 5 \text{ A}$$
$$l = 2 \text{ m}$$
$$V = 50 \text{ V}$$

Cross-sectional are $A = \pi r^2$

$$= 1.96 \times 10^{-5} \text{ m}^2$$

(1) The current density

$$J = i/A$$
$$= \frac{5}{1.96 \times 10^{-5}}$$
$$= 2.55 \times 10^5 \text{ A/m}^2$$

(2) The electric field,

$$E = \frac{V}{\ell} = \frac{50}{2} = 25 \text{ V/m}$$

(3) $\rho = \dfrac{E}{J} = \dfrac{25}{2.55 \times 10^5} = 9.8 \times 10^{-5}\ \Omega\text{-m.}$

Example 7.13. The current passing through a rod is given by

$$i(t) = 3 + 5t^2$$

where t is the time. Compute

(a) the amount of the charge passing through the rod in the time interval from $t = 0s$ to $t = 2s$.

(b) current in the interval $t = 0s$ to $t = 2s$.

Solution:

(a) $$i = \frac{dq}{dt}$$

$\therefore$ $$q = \int i\, dt$$

$$q = \int_o^2 (3 + 5t^2)\, dt$$

$$= \left[3t + \frac{5}{3}t^3\right]_o^2$$

$$= 19.33 \text{ C}$$

(b) $$i = \frac{q}{t} = \frac{19.33}{2} = 9.66 \text{ A}$$

Example 7.14. A 2000 watt radiant heater is to be used with 200 V. Then compute the

(1) current

(2) resistance of the heating coil.

(3) how much heat is produced in one hour by the heater.

Solution: (1) $$P = Vi$$

or $$i = \frac{P}{V} = \frac{2000}{200} = 10A.$$

(2) $$P = \frac{V^2}{R}$$

or $$R = \frac{V^2}{P} = \frac{2000 \times 10}{2000} = 10\,\Omega$$

(3) $$\text{Heat} = Pt$$

$$= 2000 \times 60 \times 60$$

$$= 7.2 \times 10^6 \text{ J}$$

Example 7.15. Three electric bulbs of 40 W, 60 W and 100 W are connected in series with 200 V mains. Compute the

(1) current in each bulb.

(2) potential difference across each bulb.

(3) energy produced by each bulb and which bulb does glow brightly.

Solution: Suppose that R_1, R_2 and R_3 are the resistances of the bulbs 40W, 60W and 100W respectively.

$$R_1 = \frac{V^2}{P_1} = \frac{(200)^2}{40} = 1\,\text{K}\Omega$$

$$R_2 = \frac{V^2}{P_2} = \frac{(200)^2}{60} = 0.66\,\text{K}\Omega$$

and
$$R_3 = \frac{V^2}{P_3} = \frac{(200)^2}{100} = 0.40\,\text{K}\Omega$$

since bulbs are connected in series,

$$R = R_1 + R_2 + R_3$$
$$= 2.066\ \text{K}\Omega$$

$$\therefore \qquad \text{current } i = \frac{V}{R} = \frac{200}{2.066} = 96.8\,\text{mA}$$

(2) potential difference across the each bulb is

$$V_1 = iR_1$$
$$= 96.8\ V$$
$$V_2 = iR_2 = 64V$$
$$V_3 = iR_3 = 38.7V$$

(3) The energy produced by each bulb is

$$P_1 = i^2R_1 = 9.37\text{ J}$$
$$P_2 = i^2R_2 = 6.18\text{ J}$$

and
$$P_3 = i^2R_3 = 3.75\text{ J}$$

Since the bulbs are connected in series, the current in all bulbs will be same. Again, the brightness is proportional to the power dissipation, thus, 40W bulb glows more brightly.

Example 7.16. A metallic conductor of the length l has diameter $2a$. If a potential difference of V volts is applied across the conductor, compute the

(1) electric field across the conductor.

(2) resistance of the conductor.

(3) current through the conductor.

(4) current density.

Solution: (a) The electric field across the conductor is

$$E = \frac{V}{l}$$

(b) resistance of the conductor is

$$R = \rho\frac{l}{A}$$

Since $$A = \pi a^2$$

$\therefore$ $$R = \frac{\rho l}{\pi a^2}$$

where ρ is the resistivity.

(c) current passing through the conductor is

$$i = \frac{V}{R}$$

or $$i = \frac{V\pi a^2}{\rho l}$$

(d) The current density,

$$J = \frac{i}{A} = \frac{V}{RA}$$

or $$J = \frac{V}{\rho l}$$

Example 7.17. Consider a circuit as shown in fig. 7.39.

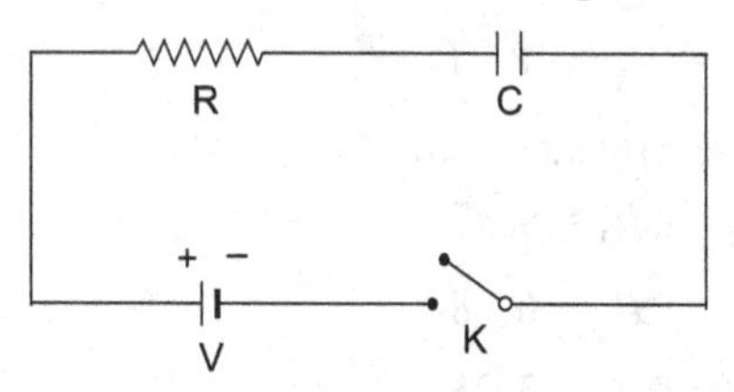

Fig. 7.39. Charging of a capacitor.

compute the

(a) energy supplied by the battery.

(b) energy across the resistor

(c) energy stored in the capacitor.

Solution: The power supplied by the battery is equal to the rate of change of the energy.

$$P = \frac{dU}{dt} = Vi$$

$$= V i_0 e^{-t/RC}$$

since $$i_0 = \frac{V}{R}$$

$$P = \frac{V^2}{R} e^{-t/RC}$$

the total energy supplied by the battery is

$$U = \int dU$$

$$= \int P\,dt = \int_{t=0}^{\infty} \frac{V^2}{R} e^{-t/RC} dt$$

or $$U = V^2C$$

But the energy stored in the capacitor is

$$U_C = \frac{1}{2}CV^2$$

Thus, the energy delivered to the resistor is

$$U_C = CV^2 - \frac{1}{2}CV^2 = \frac{1}{2}CV^2$$

It can be proved as

$$P_R = \frac{dU}{dt} = i^2R = \frac{V^2}{R^2} \cdot R\, e^{-2t/RC} = \frac{V^2}{R} e^{-2t/RC}$$

The, total energy across the resistor is

$$U_R = \int dU = \int_o^{\infty} \frac{V^2}{R} e^{-2t/RC} dt = \frac{1}{2}CV^2$$

Thus, the total energy of the battery is equally distributed across C and R.

EXERCISES

7.1. What is the drift velocity and how do you measure it, distinguish between drift velocity and random velocity of the electrons.

7.2. A wire has length 5 m and diameter 3 mm. Compute the resistance of wire, when its resistivity is 1.5×10^{-8} Ω-m.

7.3. Prove that $R = \dfrac{\rho l}{A}$.

where l and A are length and area of cross-section of a conductor.

7.4. Compute the potential difference across 1 m of a wire carrying a current of 10A and it has a resistance of 50 mΩ/m.

7.5. Show that the free charge density at the junction of two conductors is given by

$$\sigma = J\left(\frac{\epsilon_2}{\sigma_2} - \frac{\epsilon_1}{\sigma_1}\right)$$

where σ_1 and σ_2 are the conductivity of two conductors and J is the current density.

7.6. The 10 LED of 20W light every night for 10 hours for a month. Compute the cost of lighting if the rate is 3 R_S/kwh.

7.7. Compute the resistance of a copper rod of length 1 m and cross-sectional area 5×10^{-4} m^2.

7.8. The n resistors of resistance R are connected in series with an emf of V volt. If i is the current passing through the circuit, prove that the equivalent resistance is

$$R_e = nR$$

7.9. In a RC circuit, $R = 2.2$ K and $C = 1.5$ mf. If an emf of 12 V is applied, compute the

(1) charge on the capacitor.

(2) $\tau = RC$, time constant

(3) current i, and

(4) i when $t = \tau$

7.10. Compute the equivalent resistance of the circuits, shown in fig. 7.40.

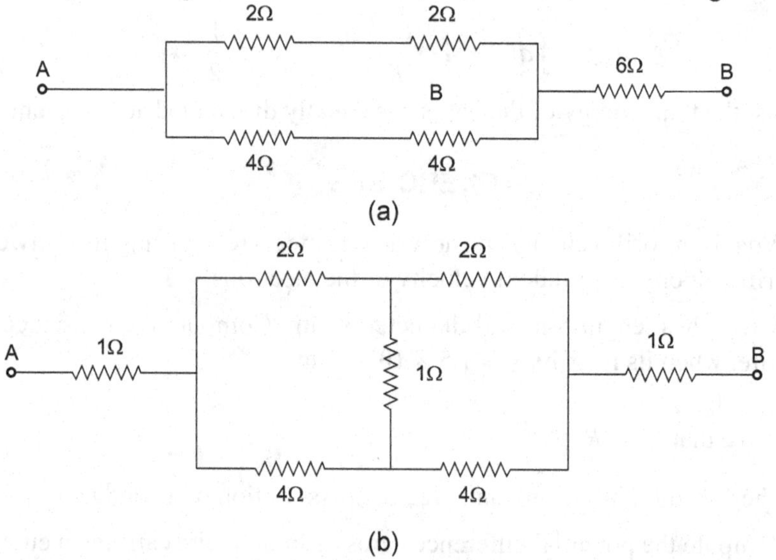

Fig. 7.40. Networks of resistors.

7.11. A *RC* circuit is shown in fig. 7.41.

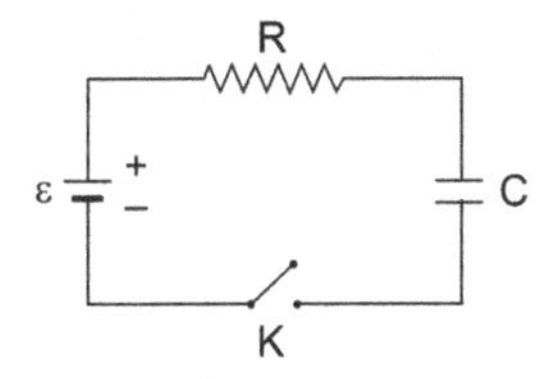

Fig. 7.41. C with R.

when *K* is closed, prove that the current is

$$i = -\frac{\varepsilon}{R} e^{-t/RC}$$

7.12. Compute the net resistance between *A* and *B*, Fig. 7.42.

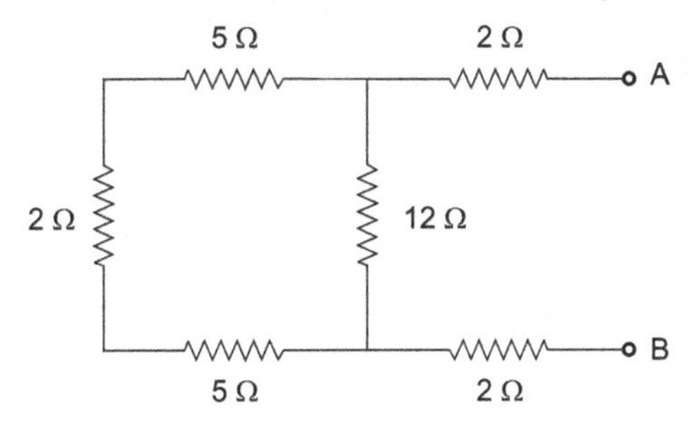

Fig. 7.42. Resistive network.

7.13. For a metallic conductor, show that the current density is

$$\vec{J} = \frac{ne^2\tau}{m}\vec{E}$$

where τ = time of electron between two collisions, $\vec{E}$ is the electric field, *n* is number of charge per unit volume.

7.14. Two 20W LED are connected with a supply of 100V. Compute the total power consumed by the LED when (a) both are in series (b) both are in parallel.

■ ■ ■

Magnetic Fields and Materials

The natural magnets, originally called lodestones were discovered many centuries ago. It was found that magnetic behavior of materials came into picture with the ancient city of Magnesia in Asia Minor. In 1820, Danish physicist Hans Christan Oersted showed that the electric current gives rise magnetic force. A current carrying conductor is associated with the magnetic field and causing a deflection in the compass needle. When a magnetized needle is suspend freely, it always in the north and the south directions. In this chapter we shall study the magnetic field and magnetic interactions.

8.1. MAGNETS AND THE MAGNETIC FIELD

The magnet attracts the pieces of iron, nails and other magnetic materials. This property of attracting materials is called the magnetism. A bar magnet or any shaped magnet has two magnetic poles marked as north $N(+)$ and south $S(-)$. When a magnet is suspended in air freely, the north and south poles are decided by pointing the magnet towards north and the south directions. Now, what does about magnetic interaction between like poles and also in un-like poles. The like poles repel each other and unlike poles attract each other as shown in fig. 8.1. Hence,

unlike poles attract and like poles repel

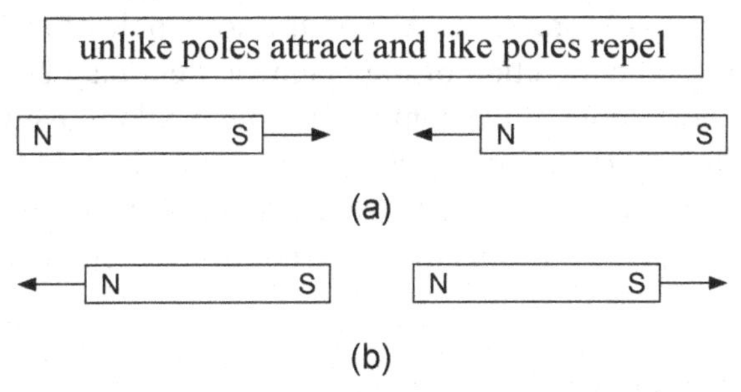

Fig. 8.1. Unlike poles attract (a) and like poles repel (b)

It is not possible to isolate the north pole and south pole of a magnet. If we cut a bar magnet into two pieces, two new magnets are obtained as shown in fig. 8.2.

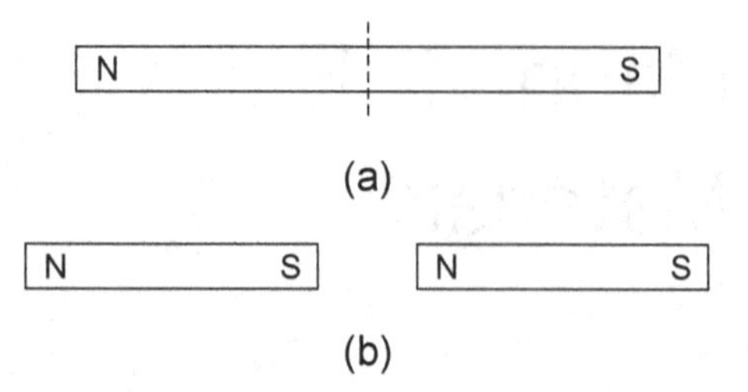

Fig. 8.2. Two pieces of a magnet means monopoles does not exist.

The new magnets have both north and the south poles. This process of cutting magnet continues until a tiny particle is obtained. This tiny particle of magnet has two poles also. This means that the magnetic monopoles do not exist. This is unlike the electric interaction, where electric monopole exists.

Since a static charge produces an electric field around it, the space around a magnet in which magnetic effect occurs, is called the magnetic field. The magnetic field is represented by the magnetic lines of force. The concentration of the magnetic field lines or number of magnetic field lines per unit cross-sectional area is proportional to the magnetic field strength. Consider a bar magnet shown in fig. 8.3.

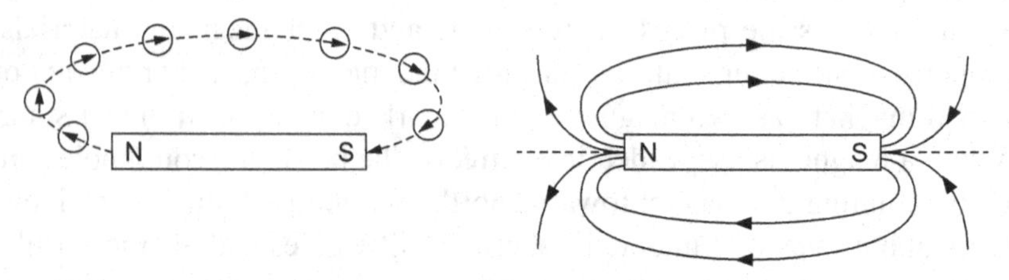

Fig. 8.3. Magnetic lines of force.

The magnetic lines of force emerge from the north pole of the magnet and enter the south pole. The direction of the magnetic field may be taken as the direction which a north pole of the compass needle would move when placed in a magnetic field. The direction of the magnetic field is represented by dot $\bigcirc$ or a cross $\otimes$. The symbol $\bigcirc$ means that field direction is coming out from the plane of paper, while $\otimes$ means the direction of magnetic field into plane of paper.

8.2. MAGNETIC FLUX

Like the electric field $\vec{E}$, the magnetic field is represented by $\vec{B}$. This is also called magnetic induction or magnetic flux density. The direction of the magnetic field B is the tangent to a line of force at any point. The magnetic lines of force as drawn in such a way that the number of field lines crossing per unit cross-sectional area is proportional to the magnetic field B at any point. The magnetic flux ϕ_B passing through a surface area S is shown in fig 8.4.

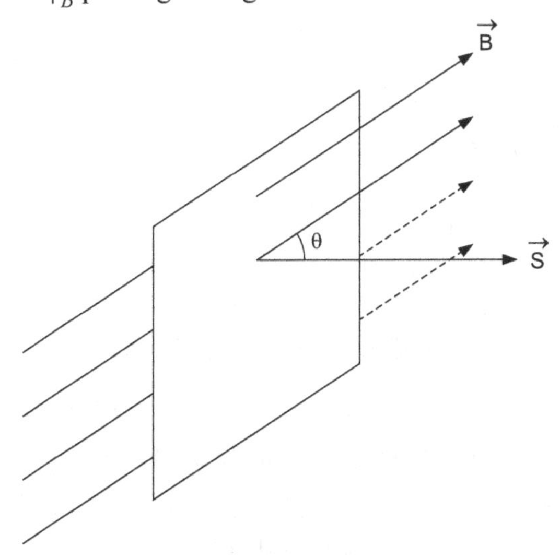

Fig. 8.4. Magnetic Flux pass through the surface S.

If the normal vector $\hat{n}$ to the surface area $\vec{S}$ makes an angle θ with the direction of the magnetic field $\vec{B}$, then,

$$\phi_B = \vec{B} \cdot \vec{S} \quad \text{...(8.1)}$$

$$\boxed{\phi_B = BS \cos \theta} \quad \text{...(8.2)}$$

Here, we have considered a constant magnetic field $\vec{B}$. If $\vec{B}$ is normal to the surface area $\vec{S}$, $\theta = 0$, hence

$$\phi_B = BS \quad \text{...(8.3)}$$

Again, if $\vec{B}$ is perpendicular to the normal $\hat{n}$,
Then, $\theta = 90°$, hence $\phi_B = 0$

Moreover, if the magnetic field $\vec{B}$ is non-uniform, Then the equation (8.1) may be written as

$$\boxed{\phi_B = \int_S \vec{B} \cdot \vec{dS}} \quad \text{...(8.4)}$$

The SI unit of area S is m^2, and B is weber m^{-2}. The S.I unit of magnetic flux is the Weber (Wb) in the honor of Wilhelm Weber.

Example 8.1. A circular coil of area 0.1 m^2 is placed perpendicular to the direction of the magnetic field ($B = 0{\cdot}5$ Weber m^{-2}). Calculate the flux passing through the coil.

Solution: Here,

$$S = 0{\cdot}1 \text{ m}^2$$
$$B = 0.5 \text{ Wm}^{-2}$$
$$\therefore \quad \phi_B = BS \cos\theta$$
$$= 0{\cdot}5 \times 0{\cdot}1 \times \cos 0$$
$$\phi_B = 0.05 \text{ Wb}$$

8.3. MAGNETIC FORCE ON A MOVING CHARGE

Suppose that a positive charge q is moving in a uniform magnetic field $\vec{B}$ with a velocity $\vec{v}$ in the x-y plane and the magnetic field is along x-axis as shown in fig. 8.5.

Experimentally, It has been observed that the magnitude of the force $\vec{F}$ on the moving charge,

(a) is proportional to the magnitude of charge q.

(b) is proportional to the component of velocity v perpendicular to the magnetic field B.

(c) is proportional to sine of angle between $\vec{v}$ and $\vec{B}$.

(d) is proportional to the magnetic field $\vec{B}$. Here the constant of proportionality is unity.

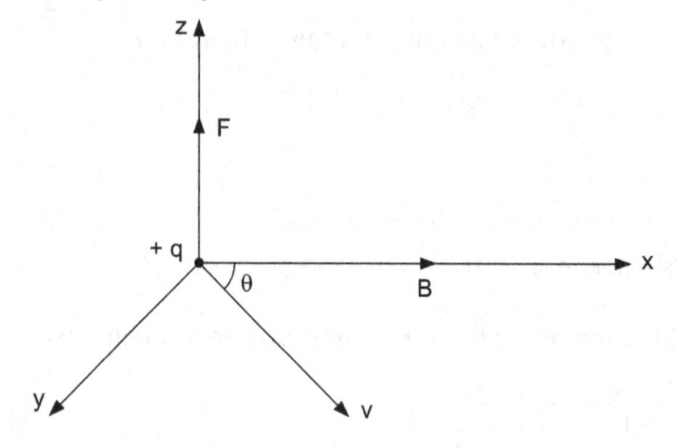

Fig. 8.5. A magnetic force experienced by a moving charge

Hence,

$$\boxed{F = qvB \sin\theta} \qquad \text{...(8.5)}$$

This force $\vec{F}$ is always perpendicular to the plane containing $\vec{v}$ and $\vec{B}$. Thus, the force $\vec{F}$ is defined by the cross product of $\vec{v}$ and $\vec{B}$. Hence, we may write,

$$\boxed{\vec{F} = q(\vec{v} \times \vec{B})} \quad ...(8.6)$$

From the equation (8.6) it is observed that

(1) the force is maximum when the charge is moving perpendicular to the magnetic field $\vec{B}$.
that is, for $\theta = 90°$, $\sin\theta = 1$

$$F_{max} = qvB \quad \text{(charge opts circular path)}$$

(2) the magnetic force F is zero if the charge q is moving parallel to the magnetic field $\vec{B}$. i.e.,

$$\text{for } \theta = 0°, \sin\theta = 0$$

$$F = 0 \quad \text{(charge moves in a straight line)}$$

The direction of force $\vec{F}$ is given by the right hand rule. According to this rule, if the fingers of the right-hand are stretched along the direction of velocity $\vec{v}$ of the charge and then bend the fingers towards magnetic field $\vec{B}$, the thumb will point in the direction of the force $\vec{F}$, as shown in fig. 8.6.

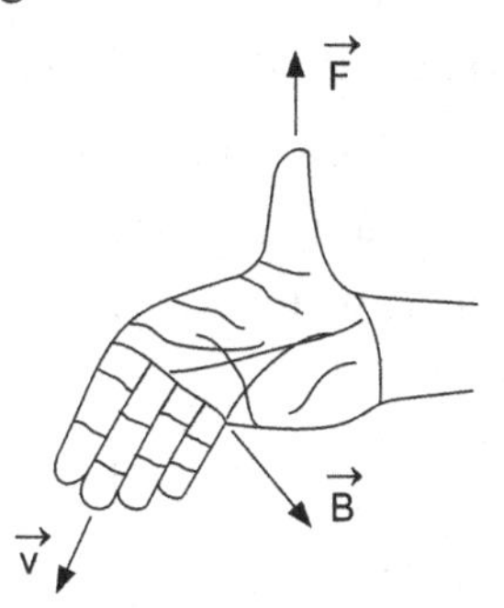

Fig. 8.6. Right hand rule for determining the direction of force $\vec{F}$.

(3) The force F is always perpendicular to the direction of moving charge. Thus, force does not work, hence force F does not change the magnitude of $\vec{v}$. But it changes the direction of $\vec{v}$. The force F is called the deflecting force. If a charge enters the magnetic field at an angle (θ) and $0 < \theta < 90°$, then, the two components of velocity viz $v\cos\theta$ and

$v \sin\theta$ cause to move charge q into helical or spiral path as shown in fig. 8.7.

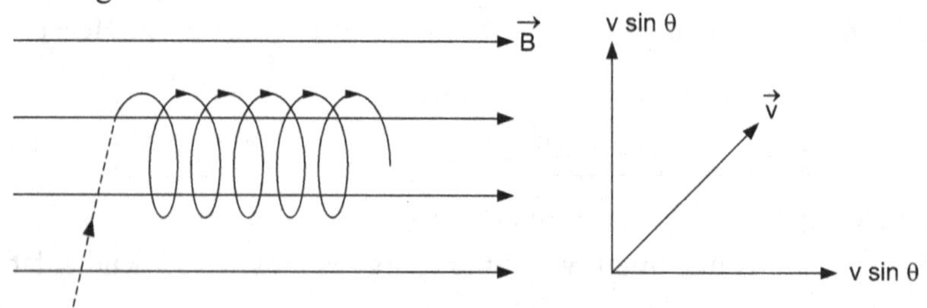

Fig. 8.7. A helical path of charge q.

Now, from the equation (8.5), we have

$$B = \frac{F}{vq\sin\theta} \quad \text{...(8.7)}$$

For $\theta = 90°$

$$B = \frac{F}{vq} \quad \text{...(8.8)}$$

$$B = \frac{\text{Newton}}{\text{meter sec}^{-1} \times \text{coulomb}}$$

$$B = \frac{\text{Newton}}{\text{Ampere-meter}}$$

Thus, the magnetic field B is defined as the force exerts on a unit charge moving with a unit velocity when v is perpendicular to magnetic induction B. The SI units of $\vec{B}$ are Weber m^{-2}, Newton-ampere^{-1} m^{-1} and Tesla T.

$$1T = \frac{\text{Newton}}{\text{Ampere meter}} = \frac{\text{Weber}}{\text{meter}}$$

and

$$\boxed{1T = \frac{\text{Wb}}{\text{meter}} = 10^4 \text{ gauss}}$$

When a charged particle is moving with a velocity $\vec{v}$ in a region, where the electric field $\vec{E}$ and the magnetic field $\vec{B}$ are present, the total force $\vec{F}$ is the vector sum of the electric force qE and the magnetic force $q(v \times B)$:

$$\left.\begin{aligned} \vec{F} &= \vec{F}_e + \vec{F}_m \\ \vec{F} &= q\vec{E} + q(\vec{v} \times \vec{B}) \end{aligned}\right\} \quad \text{...(8.9)}$$

$$\therefore \quad \boxed{\vec{F} = q[\vec{E} + (\vec{v} \times \vec{B})]} \quad ...(8.10)$$

The Force $\vec{F}$ is called the Lorentz force.

8.4. MOTION OF A CHARGE IN A UNIFORM MAGNETIC FIELD

Consider a charged particle moving in a magnetic field (having same magnitude and direction at all points). Let $+q$ be the charge and m be the mass of the charged particle moving with velocity $\vec{v}$ in a direction perpendicular to the magnetic field B as shown in fig. 8.8.

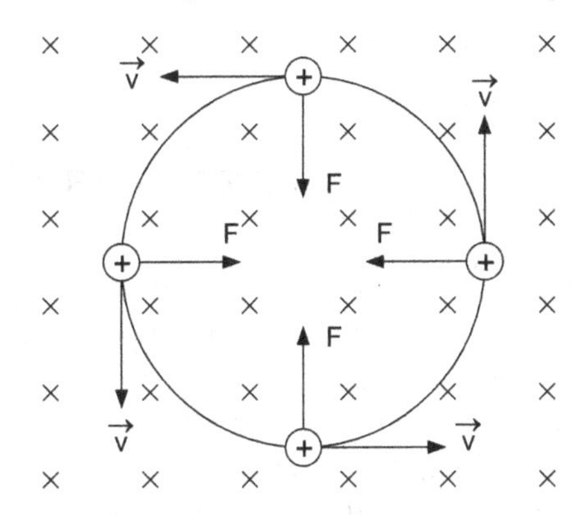

Fig. 8.8. A charged particle moving perpendicular to magnetic field in a circular path.

The magnetic force on the charge is given by

$$F = q\,v\,B \sin 90°$$

$$F = q\,v\,B \quad ...(8.11)$$

This force F is always perpendicular to the velocity $\vec{v}$ of the particle. The force F acts as a deflecting force, it can change the direction of velocity $\vec{v}$ without changing its magnitude. The charge q travels in a circle of radius r, the centripetal force is equal to $q\,v\,B$. Then, we may write,

$$q\,v\,B = \frac{mv^2}{r} \quad \text{(Balancing of forces)} \quad ...(8.12)$$

$$\therefore \quad \boxed{r = \frac{mv}{qB}} \quad ...(8.13)$$

Since $P = mv$ (momentum of the charged particle)

Then, equation (8.12) can be written as

$$\boxed{r = \frac{p}{qB}} \quad ...(8.14)$$

The radius of the circle r is directly proportional to p and inversely proportional to B. If the moving charged particle has larger energy, r will be larger. The angular speed of the particle is

$$\omega = 2\pi f = \frac{v}{r} \quad ...(8.15)$$

$$\omega = \frac{qB}{m} \quad ...(8.16)$$

ω is the cyclotron frequency

Now, $$\boxed{f = \frac{qB}{2\pi m}} \quad ...(8.17)$$

8.5. MAGNETIC FORCE ON A CURRENT CARRYING CONDUCTOR

The free charge experiences a force when it moves in a magnetic field. Here, we will see what happens, when charges are moving in a conductor. Consider a conductor of length l, carrying a current i placed in a magnetic field. The force on the conductor is the average force acting on all charge carriers moving with a drift velocity v_d. If n is the number of charge carriers, with each charge q, then, the total charge in the segment dl of the conductor of cross-sectional area A is given by

$$Q = q(n\,A\,dl) \quad ...(8.18)$$

Fig. 8.9 shows a current carrying conductor in a magnetic field.

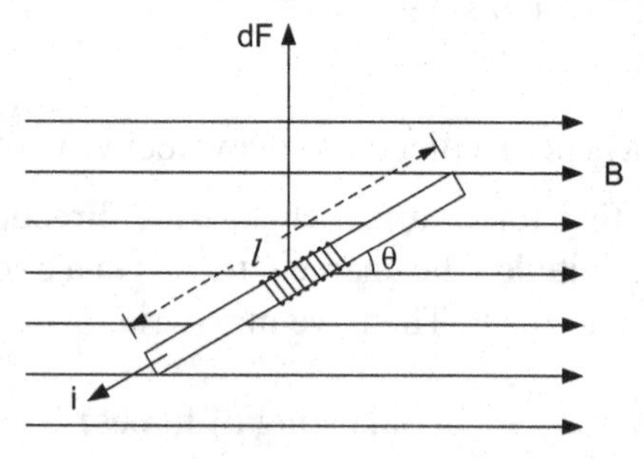

Fig. 8.9. Current carrying conductor in a magnetic field.

The force on a segment dl will be

$$\vec{dF} = Q(\vec{v_d} \times \vec{B}) \quad ...(8.19)$$

$$\vec{dF} = qnAdl\,(\vec{v_d} \times \vec{B}) \quad ...(8.20)$$

But current through the conductor is given by

$$i = nqv_dA$$

The equation (8.20) becomes

$$\boxed{\vec{dF} = i(\vec{dl} \times \vec{B})} \qquad ...(8.21)$$

the total magnetic force on the conductor of length l is

$$\vec{F} = \int i(\vec{dl} \times \vec{B}) \qquad ...(8.22)$$

$$= i\left[\int \vec{dl}\right] \times \vec{B}$$

$$\boxed{\vec{F} = i(\vec{l} \times \vec{B})} \qquad ...(8.23)$$

The direction of the magnetic force is perpendicular to the plane containing vector $\vec{l}$ and vector $\vec{B}$.

The magnitude of the force is

$$\boxed{|F| = i\,l\,\underline{B}\sin\theta} \qquad ...(8.24)$$

where θ is the angle between length $\vec{l}$ of the conductor and the magnetic field $\vec{B}$.

Now, if conductor is placed in the magnetic field parallel to the field direction, then, $\theta = 0°$

$$F = i\,l\,B\sin 0 = 0$$

therefore, no force is experienced by the conductor.

If $\theta = 90°$, the force on the conductor is

$$F = i\,l\,B\sin 0$$

$$F = i\,l\,B$$

this mean that a maximum force is experienced by the conductor.

Example 8.2. A conductor of 2 m long is placed perpendicular to the direction of magnetic field (B = 4 Wbm^{-2}), as shown in fig. 8.10. The direction of magnetic field is along x-axis, and conductor is placed along y axis.

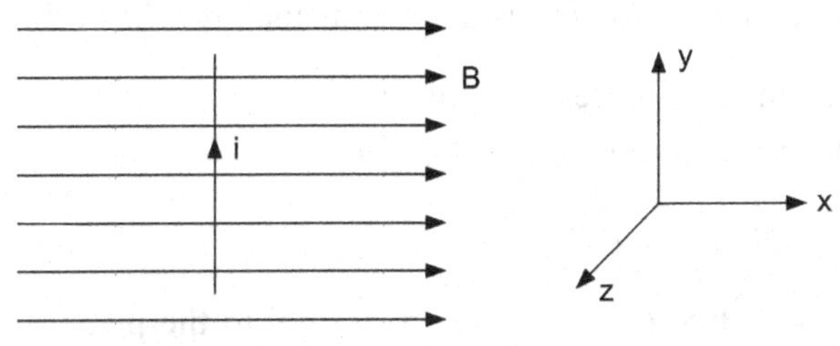

Fig. 8.10. A conductor in a magnetic field.

If a current of 5A is flowing through the conductor, find the force on the conductor and its direction.

Solution:

$$\vec{B} = 4\hat{i}$$

$$\vec{l} = 2\hat{j}$$

Then, Force is

$$\vec{F} = i(\vec{\ell} \times \vec{B})$$

$$= 8 \times (\hat{j} \times \hat{i})$$

$$= -8\hat{k}$$

The direction of force is along $-z$ axis.

8.6. MAGNETIC DIPOLE MOMENT

It is very interesting to give an idea of the magnetic dipole moment before going to discuss the torque on a current loop. A bar magnet or a current loop shows a similar pattern of magnetic lines of force. The magnetic lines of force emerge from the north pole and enter the south pole of the magnet. Since we have two poles, this arrangement is called magnetic dipole.

Consider a loop having number of turns n and area of cross-section A as shown in fig. 8.11 (a).

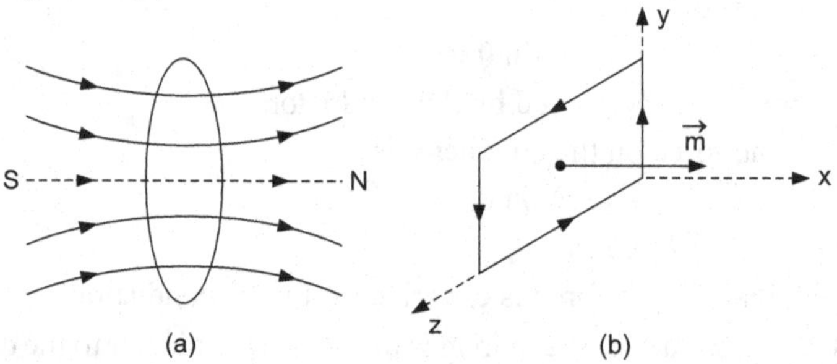

Fig. 8.11. Current loops

The magnetic dipole moment is defined as the product of the current in a loop and its area; and it is represented by $\vec{m}$.

Thus, dipole moment of a current loop is

$$\boxed{\vec{m} = n\,i\,\vec{A}} \qquad ...(8.25)$$

The direction of dipole moment $\vec{m}$ is normal to the plane of the loop as shown in fig. 8.11(b).

8.7. MAGNETIC TORQUE ON A CURRENT LOOP

Consider a rectangular loop of length a and width b placed in a uniform magnetic field $\vec{B}$ as shown in fig. 8.12. Suppose that an electric current i is flowing in the loop. This current loop experiences a torque due to the magnetic force. These forces are acting on the vertical arms of the loop and are equal and opposite. The magnitude of force is

$$F = i\,a\,B \quad \text{...(8.26)}$$

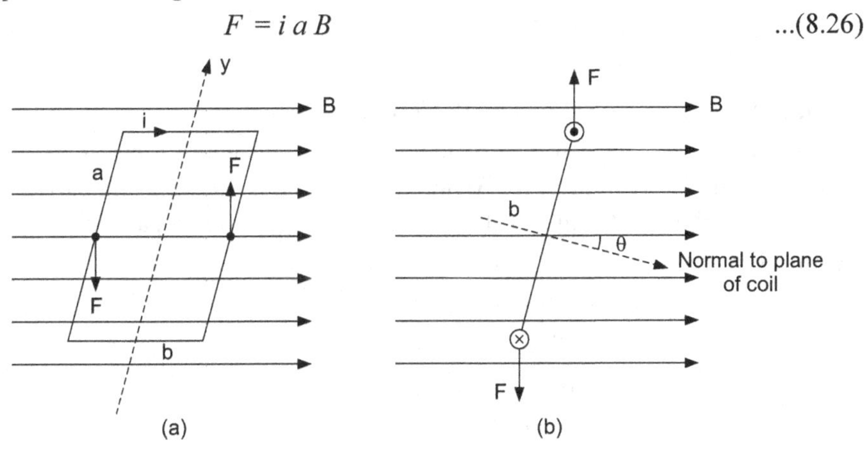

Fig. 8.12. (a) Current carrying rectangular loop (b) the normal to the plane of loop makes an angle θ with $\vec{B}$.

The total force acting on the loop is zero. These opposite and equal forces acting on the loop create a couple which tends to rotate the loop countercloke-wise. The total torque on the loop is

$$\tau = i\,a\,B \cdot \frac{b}{2}\sin\theta + i\,a\,B \cdot \frac{b}{2}\sin\theta$$

$$\tau = i\,a\,b\,B\sin\theta$$

$$\tau = i\,A\,B\sin\theta \quad \text{...(8.27)}$$

where $A = ab$ is the area of the loop.

If there are n term in the loop, the total magnetic torque will be

$$\boxed{\tau = N\,i\,A\,B\sin\theta} \quad \text{...(8.28)}$$

But $m = n\,i\,A$, then the equation (8.28) can be written as

$$\boxed{\tau = m\,B\sin\theta} \quad \text{...(8.29)}$$

Since the magnetic dipole moment $\vec{m}$ and the magnetic field $\vec{B}$ are both vector quantities, then we write equation (8.29) as

$$\boxed{\vec{\tau} = \vec{m} \times \vec{B}} \quad \text{...(8.30)}$$

It should be clear that the direction of the torque is along the axis of rotation. When the plane of loop is perpendicular to the magnetic field, $\theta = 0$ and $\tau = 0$ i.e. the position of equilibrium. In this situation, the sum of forces on the loop is zero.

When the plane of loop is parallel to the magnetic field $\vec{B}$, this means that normal to the plane of loop is perpendicular to $\vec{B}$, then,

$$\tau = m\,B \sin 90$$

$$\tau = m\,B$$

The maximum torque exerts on the loop.

Example 8.3. A rectangular coil of length 10 cm and width 6 cm is placed in a uniform magnetic field of 0.5 Wbm^{-2}. The coil has 10 turns and carries a current of $2A$. Find the torque on the coil when plane of coil is parallel to B.

Solution: Given,

$$i = 2A$$

$$n = 10$$

$$a = 10 \times 10^{-2}\text{ m}$$

$$b = 6 \times 10^{-2}\text{ m}$$

$$B = 0.5\ T$$

Then, torque when $\vec{m}$ is perpendicular to B,

$$\tau = m\,B \sin 90$$

$$= m\,B = niAB$$

$$= 10 \times 2 \times 10 \times 10^{-2} \times 6 \times 10^{-2} \times 0.5\text{ Nm.}$$

$$\tau = 6 \times 10^{-2}\text{ Nm.}$$

8.8. POTENTIAL ENERGY OF A DIPOLE IN A MAGNETIC FIELD

When a magnetic dipole is placed in a uniform magnetic field, then, a torque exerts on the magnetic dipole due to the equal and opposite forces acting on the dipole as shown in fig. 8.13.

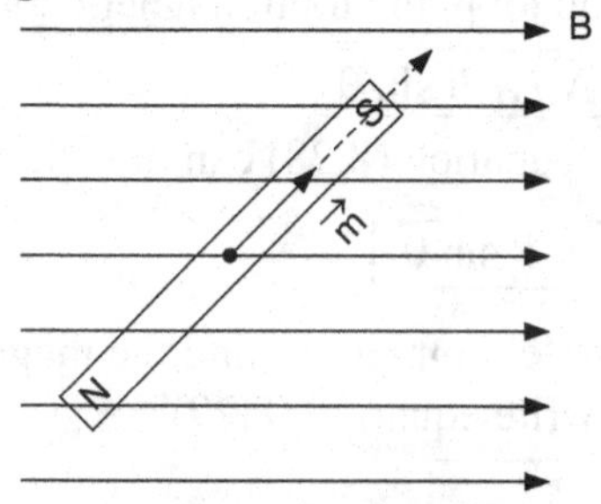

Fig. 8.13. Dipole in a magnetic field.

this torque tends to rotate the dipole. If dW is the work done in rotating the dipole through an angle $d\theta$, then,

$$dW = -\tau \, d\theta \qquad ...(8.31)$$

$$\therefore \qquad \tau = mB \sin\theta$$

The equation (8.31) becomes

$$dW = -mB \sin\theta \, d\theta \qquad ...(8.32)$$

where θ is the angle between magnetic moment $\vec{m}$ and the magnetic field $\vec{B}$. The work done by the torque is equal to the decrease in the potential energy. Thus,

$$dU = -dW$$

$$dU = m\,B \sin\theta \, d\theta \qquad ...(8.33)$$

The potential energy of the system may be obtained by integrating the equation (8.33) in the limits $\theta = 90°$ to $\theta = 0°$.

$$U = \int_{\pi/2}^{0} m\,B \sin\theta \, d\theta \qquad ...(8.34)$$

Thus, we get

$$U = -m\,B\cos\theta \qquad ...(8.35)$$

$$\Rightarrow \qquad \boxed{U = -\vec{m}\cdot\vec{B}} \qquad ...(8.36)$$

when the dipole is in stable equilibrium, $\theta = 0°$

$$U = -mB,$$

the potential energy is minimum.

When the magnetic dipole is in unstable equilibrium, $\theta = \pi$, then potential energy ($U = mB$) is maximum.

8.9. THE BIOT-SAVART LAW

A current carrying conductor produces a magnetic field in the space around it. Consider a wire carrying a current i as shown in fig. 8.14. Biot-Savart's law states that the differential magnetic field dB at any point is proportional to current element idl and the sine of angle between element dl and the line connecting to point P. The magnitude dB of the magnetic field is inversely proportion to the square of the distance from the element dl to the point P.

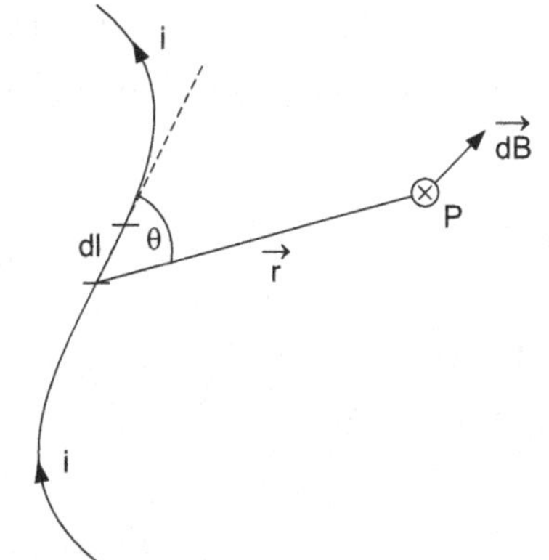

Fig. 8.14. Current carrying wire element giving rise to a magnetic field $\vec{dB}$ at a point P.

According to Biot-Savart law, the magnetic field at point P is

$$dB \propto \frac{idl \sin\theta}{r^2} \qquad ...(8.37)$$

$$\Rightarrow \qquad \boxed{dB = \frac{\mu_0}{4\pi}\frac{idl \sin\theta}{r^2}} \qquad ...(8.38)$$

In vector notation,

$$\boxed{\vec{dB} = \frac{\mu_0}{4\pi}\frac{i\,dl \times \hat{r}}{r^2}} \qquad ...(8.39)$$

where μ_0 is a constant called the permeability of free space. The vector distance of the point P from the element dl is $\vec{r} = r\hat{r}$. The total magnetic field at P due to entire current carrying wire can be computed by summing the magnetic field due to all such segments of the wire.

That is,

$$\vec{B} = \vec{dB_1} + \vec{dB_2} + \vec{dB_3} + ... + \vec{dB_i}$$

or

$$\vec{B} = \sum_i \vec{dB_i} \qquad ...(8.40)$$

The direction of the magnetic field $\vec{dB}$ is given by the right-hand rule. Here, the direction of $\vec{dB}$ is perpendicular to the plane containing dl and r and is into the paper. The Biot-Savert law is also known as Ampere's law of the current element (Do not confuse with Ampere's circuital law).

From the equation (8.40), it may be written as

$$\boxed{\vec{B} = \oint \vec{dB} = \frac{\mu_0 i}{4\pi}\int \frac{\vec{dl} \times \hat{r}}{r^2}} \qquad ...(8.41)$$

The equation (8.41) is integrated over a closed circuit. The equation (8.42) in term of current density is

$$\boxed{\vec{B} = \frac{\mu_0}{4\pi}\int \frac{\vec{J} \times \hat{r}}{r^2}\, dV} \qquad ...(8.42)$$

8.10. APPLICATIONS OF BIOT-SAVART LAW

We have some applications of the Biot-Savert law. in this section.

8.10.1. Magnetic Field Due to Long Straight Wire Carrying a Current

Consider along straight wire carrying a current i as shown in fig. 8.15. A point P, where magnetic field is be determined, is situated at a distance R from the wire. The distance between a current element idl and the point P is r.

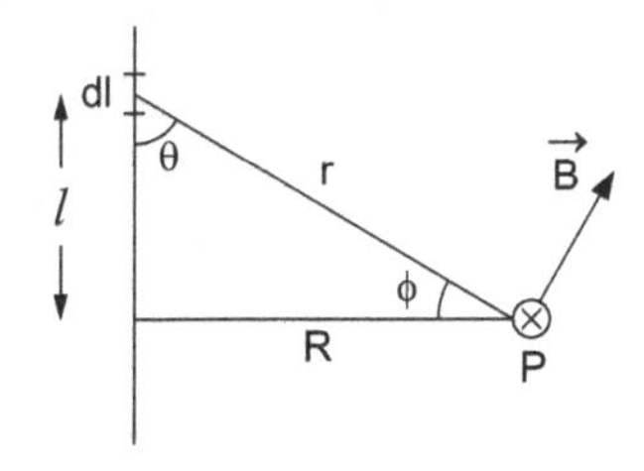

Fig. 8.15. Magnetic field due to long straight wire.

According to Biot-Savart law

$$dB = \frac{\mu_0}{4\pi}\frac{idl\sin\theta}{r^2} \quad ...(8.43)$$

Now,

$$\sin\theta = \frac{R}{r}, \cos\phi = \frac{R}{r}$$

Then,

$$dB = \frac{\mu_0 i}{4\pi}\frac{dl\cos\phi}{r^2} \quad ...(8.44)$$

Since,

$$\frac{1}{r^2} = \frac{\cos^2\phi}{R^2}, \tan\phi = \frac{l}{R}$$

$\therefore$

$$dl = \frac{R\,d\phi}{\cos^2\phi}$$

Substituting in equation (8.44), we get

$$dB = \frac{\mu_0 i}{4\pi R}\cos\phi\, d\phi \quad ...(8.45)$$

Integrating the equation (8.45) with the limits $\phi = \frac{-\pi}{2}$ to $\phi = \frac{\pi}{2}$ we have

$$B = \frac{\mu_0 i}{4\pi R}\int_{-\pi/2}^{\pi/2}\cos\phi\, d\phi \quad ...(8.46)$$

$\therefore$

$$B = \frac{\mu_0 i}{4\pi R}\cdot 2$$

$$\boxed{B = \frac{\mu_0 i}{2\pi R}} \quad ...(8.47)$$

The direction of the magnetic field B is into the plane of paper.

8.10.2. Magnetic Field at the Centre of a Current Loop

Consider a current loop carrying a current i placed in a x-y plane as shown in fig. 8.16. Since current is the cause of magnetism, we find the magnetic field at a point P situated at the centre of the loop.

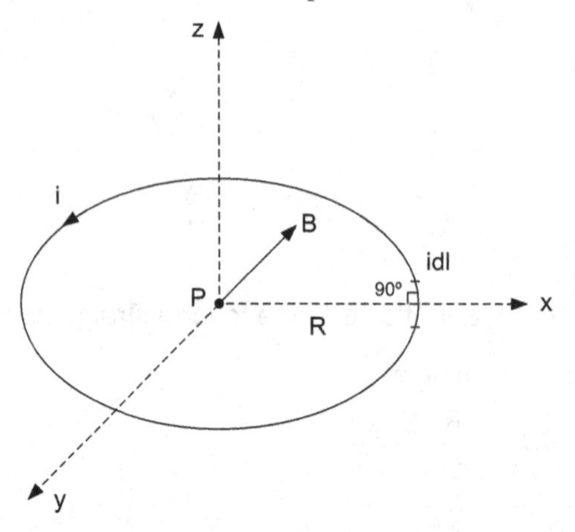

Fig. 8.16. Magnetic field at the centre of the loop.

This distance between element dl and point P is R where R is the radius of the loop. According to Biot-Savart law,

$$dB = \frac{\mu_0}{4\pi}\frac{i\,dl\sin\theta}{R^2} \qquad \text{...(8.48)}$$

The angle between dl and line joining the P is 90°,

Thus, $\sin\theta = \sin 90 = 1$, the equation (8.48) becomes

$$dB = \frac{\mu_0}{4\pi}\frac{i\,dl}{R^2} \qquad \text{...(8.49)}$$

The radius of loop is constant, thus, the magnetic field due to whole loop is given by

$$B = \frac{\mu_0 i}{4\pi R^2}\int dl \qquad \text{...(8.50)}$$

$$B = \frac{\mu_0 i}{4\pi R^2}\cdot 2\pi R$$

or

$$\boxed{B = \frac{\mu_0 i}{2R}} \qquad \text{...(8.51)}$$

The field at the centre of the loop is inversely proportional to the radius R of the loop. The direction of the magnetic field at the centre of current loop points axially outward.

8.10.3. Magnetic Field Due to Current in a Finite Straight Conductor.

Consider a conductor of finite length carrying a current i as shown in fig. 8.17.

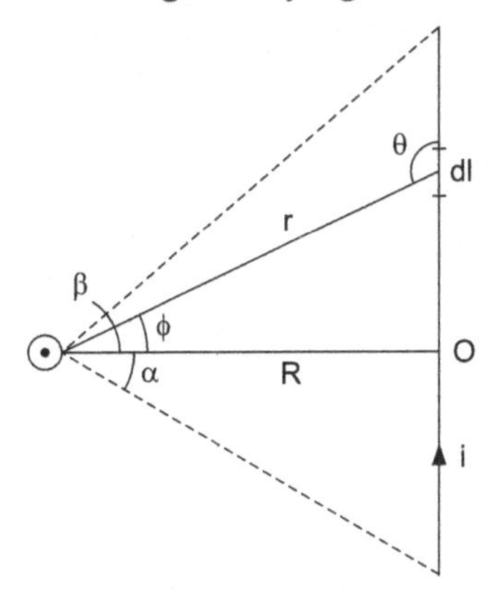

Fig. 8.17. Magnetic field due to finite length conductor

The magnetic field at point P is given by

$$dB = \frac{\mu_0}{4\pi}\frac{i\,dl\sin\theta}{r^2} \qquad ...(8.52)$$

Since, $\sin\theta = \cos\phi = \dfrac{R}{r}$

Thus, $dl\sin\theta = dl\cos\phi$

The equation (8.52) becomes

$$dB = \frac{\mu_0 i}{4\pi}\frac{dl\cos\phi}{r^2} \qquad ...(8.53)$$

Now,

$$\cos\phi = \frac{R}{r},\ l = R\tan\phi,\ dl = R\sec^2\phi\,d\phi$$

Substituting in the equation (8.53) we get

$$dB = \frac{\mu_0 i}{4\pi R}\cos\phi\,d\phi \qquad ...(8.54)$$

Integrating the equation (8.54) in the limits $\phi = -\alpha$ to $\phi = \beta$, we get

$$B = \frac{\mu_0 i}{4\pi R}\int_{-\alpha}^{\beta}\cos\phi\,d\phi$$

or

$$\boxed{B = \frac{\mu_0 i}{4\pi R}(\sin\beta + \sin\beta)} \qquad ...(8.55)$$

The direction of the magnetic field is out of the plane of paper.

8.10.4. Magnetic Field Along the Axis of a Circular Coil

But i be the current passing through a circular coil of radius R, as shown in fig. 8.18. Let P be a point along the axis of the coil at a distance x from the centre O of the coil.

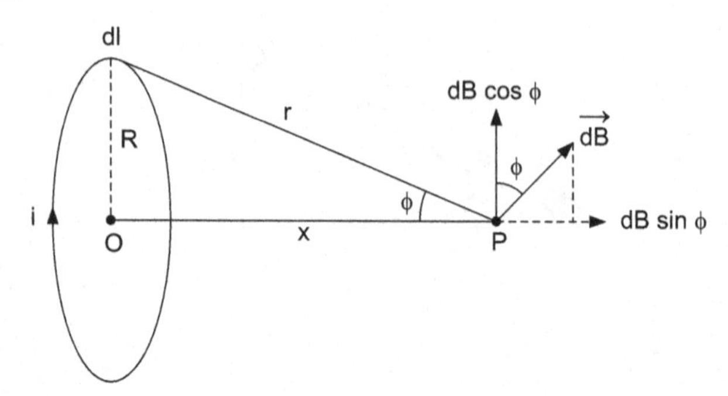

Fig. 8.18. Magnetic field along the axis of a circular coil.

According to Biot-Savart law, the magnetic field at point P is

$$dB = \frac{\mu_0}{4\pi}\frac{idl\sin\theta}{r^2} \qquad ...(8.56)$$

The all current elements dl make the angle with r, will be 90°. Thus, $\sin\theta = \sin 90 = 1$. Then

$$dB = \frac{\mu_0 i}{4\pi}\frac{idl}{r^2} \qquad ...(8.57)$$

dB lies in the plane of page, we shall take only the component of the magnetic field which lies along the axis of the coil. The component of dB not parallel to the axis of coil will cancel each other. Therefore,

$$B = \int dB\sin\phi \qquad ...(8.58)$$

or

$$B = \frac{\mu_0 i}{4\pi r^2}\int dl\sin\phi \qquad ...(8.59)$$

$$\therefore \qquad \sin\phi = \frac{R}{r}$$

Thus,

$$\therefore \qquad B = \frac{\mu_0 iR}{4\pi r^3}\int dl \qquad ...(8.60)$$

$\int dl = 2\pi R$ is the circumference of the coil, on substituting in equation (8.60), we get

$$B = \frac{\mu_0 i R^2 \cdot 2\pi}{4\pi r^3}$$

or

$$\boxed{B = \frac{\mu_0 i}{2} \frac{R^2}{(R^2 + x^2)^{3/2}}} \quad ...(8.61)$$

At the centre of the coil, $x = 0$, the expression for magnetic field is, thus, given by

$$B = \frac{\mu_0 i}{2R}$$

If the coil has number of turns n, the equation (8.61) may be written as

$$\boxed{B = \frac{\mu_0 n i}{2} \frac{R^2}{(R^2 + x^2)^{3/2}}} \quad ...(8.62)$$

8.10.5. Magnetic Field Along the Axis of a Long Solenoid

Solenoid is a helical and it is constructed from a wire tightly wound on a cylindrical surface, as shown in fig. 8.19.

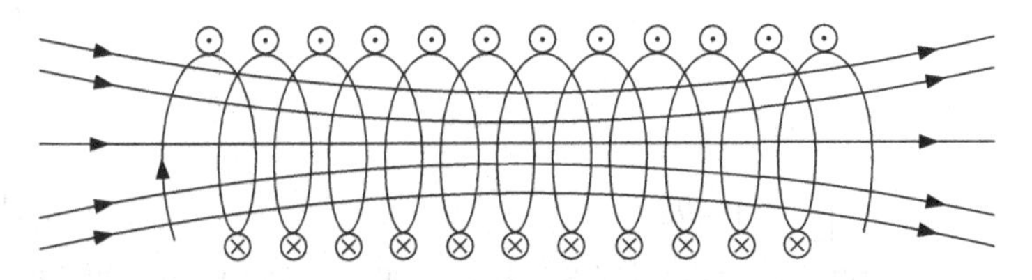

Fig. 8.19. Solenoid.

Solenoid produces a very strong uniform magnetic field along its axis. This strong magnetic field is due to the successive turns in the solenoid. The magnetic lines of force are directed along the axis of the solenoid. We calculate the magnetic field at any point lying on its axis. For the calculation of the magnetic field, consider a coil in the solenoid as shown in fig. 8.20.

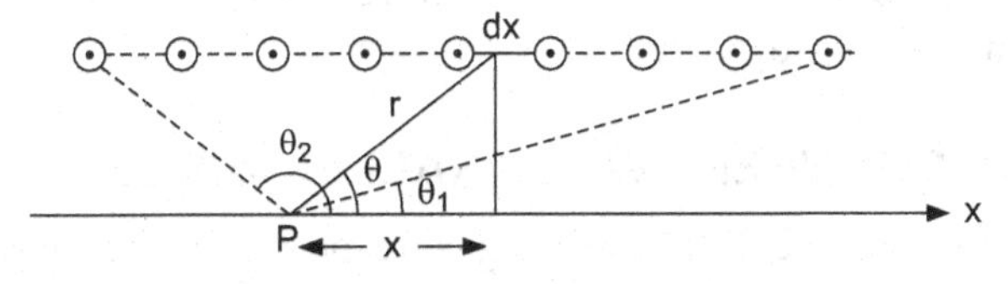

Fig. 8.20. Magnetic field due to solenoid.

If n is the number of turns per unit length of the solenoid, $dl = n\,dx$ and we have

$$dB = \frac{\mu_0 ni R^2\, dx}{2(R^2 + x^2)^{3/2}} \qquad ...(8.63)$$

where R is radius of the solenoid.

From Fig. 8.20. $x = R \cot\theta$

$$dx = -R \operatorname{cosec}^2\theta\, d\theta$$

Substituting into equation (8.63) we get

$$dB = \frac{-\mu_0 ni}{2}\frac{d\theta}{\operatorname{cosec}\theta}$$

or

$$dB = \frac{-\mu_0 ni}{2}\sin\theta\, d\theta \qquad ...(8.64)$$

Now, integrating equation (8.64). Then,

$$B = \frac{-\mu_0 ni}{2}\int_{\theta_1}^{\theta_2}\sin\theta\, d\theta \qquad ...(8.65)$$

or

$$\boxed{B = \frac{\mu_0 ni}{2}[\cos\theta_2 - \cos\theta_1]} \qquad ...(8.66)$$

If solenoid is long, $\theta_1 = \pi$ and $\theta_2 = 0°$, Thus

$$\boxed{B = \mu_0 ni} \qquad ...(8.67)$$

The magnetic field B inside the solenoid depends upon current i and the number of turns per unit length. We can calculate magnetic field at the one end of the solenoid, for this, we have $\theta_2 = 90°$ or $\pi/2$, $\theta_1 = 180°$.

Then,

$$\boxed{B = \frac{\mu_0 ni}{2}} \qquad ...(8.68)$$

The magnetic field at the end of the long solenoid is just half of the magnetic field at the centre.

8.11. FORCE BETWEEN TWO PARALLEL WIRES

We already know that a current carrying wire produces the magnetic field. Ampere showed that two parallel wires carrying current would exert a attractive force on each other. The force between two wires is purely magnetic. Consider two long parallel conducting wire separated by a distance r and carrying currents i_1 and i_2 respectively, as shown in fig. 8.21.

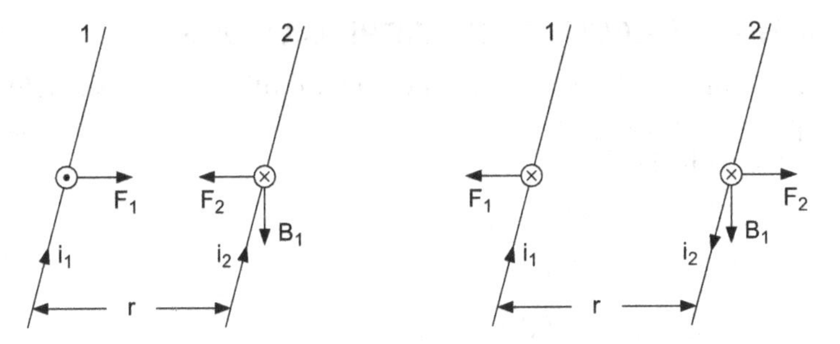

Fig. 8.21. Parallel conductors (a) Parallel currents (b) Antiparallel currents

The current i_1 in the wire 1 produces a magnetic field B_1 around it. At a distance r, the magnitude of B_1 is

$$B_1 = \frac{\mu_0 i_1}{2\pi r} \quad ...(8.69)$$

The direction of B_1 is vertically downward.

Since wire 2, carrying a current i_2 lies in magnetic field B_1. Then, the wire 2 of length l will experience a force F_2 as

$$F_2 = i_2 l B_1 \quad ...(8.70)$$

where l and B_1 are perpendicular to each other.

Substituting the value of B_1 from Equation (8.69) into Equation (8.70), we get

$$F_2 = \frac{\mu_0 i_1 i_2 l}{2\pi r} \quad ...(8.71)$$

The force per unit length l on the wire 2 is given by

$$\frac{F_2}{l} = \frac{\mu_0 i_1 i_2}{2\pi r} \quad ...(8.72)$$

If the currents are antiparallel in two conducting wires, they will repel each other with the force given by Equation (8.72). Furthermore, if $i_1 = i_2 =$ 1A and $r = 1$ m. Then

$$\frac{F}{l} = 2 \times 10^{-7} \text{ N/m}$$

Thus, one ampere is defined as the current flowing in each of two long, parallel wires situated at one meter apart, when the force per unit length on each conducting wire is 2×1^{-7} N/m.

We the ampere is defined, one coulomb is defined as 1 coulomb = (1 Ampere) (1 second).

8.12. AMPERE'S LAW AND ITS APPLICATIONS

Ampere's circuital law states that the line integral of the magnetic field over any closed path is equal to μ_0 times the net current flowing through the surface enclosed by the path. That is,

$$\boxed{\oint_C \vec{B} \cdot \vec{dl} = \mu_0 i} \qquad ...(8.73)$$

where μ_0 is the permeability of the free space.

Ampere's law is useful for calculating the magnetic fields in symmetrical situations, and in closed path around the current i. Suppose that the electric charge densities are constant and current i is not changing with time, then, Maxwell's equations can be written as

and

$$\left.\begin{aligned} \nabla \cdot \vec{B} &= 0 \\ \nabla \times \vec{B} &= \mu_0 \vec{J} \end{aligned}\right\} \qquad ...(8.74)$$

where $\vec{J}$ is called the current density. We know that the divergence of the curl of any vector is necessarily zero. That is,

$$\left.\begin{aligned} \nabla \cdot (\nabla \times \vec{B}) &= 0 \\ \therefore \quad \nabla \cdot \vec{J} &= 0 \end{aligned}\right\} \qquad ...(8.75)$$

Here, the current density is constant. Now, Stoke's theorem states that the line integral of a vector field around any closed path is equal to the surface integral of the curl of a vector field over a surface bounded by the curve.

Thus, for magnetic field, we write

$$\begin{aligned} \oint_C \vec{B} \cdot \vec{dl} &= \oint_S (\nabla \times \vec{B}) \cdot \vec{dS} \\ &= \mu_0 \int_S \vec{J} \cdot \vec{dS} \end{aligned} \qquad ...(8.76)$$

But

$$\int_S \vec{J} \cdot \vec{dS} = i$$

Then, Equation (8.76) may be written as

$$\boxed{\oint_C \vec{B} \cdot \vec{dl} = \mu_0 i} \qquad ...(8.77)$$

This is Ampere's law.

Now, we apply Ampere's law to a symmetrical case of a long straight wire. We draw an Amperian loop (analogous to Gaussian surface) to find out the magnetic field due to long straight wire carrying a current i as shown in fig. 8.22.

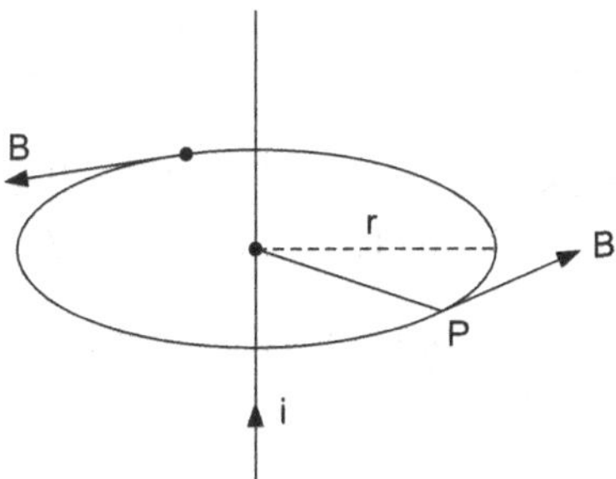

Fig. 8.22. Magnetic field of a long wire

Amperian loop passing through the point P is a circle of radius r.

Now, according to Ampere's Law

$$\oint_C \vec{B} \cdot \vec{dl} = \mu_0 i \qquad ...(8.78)$$

the magnetic field B is constant at every point on the circle. Thus,

$$B \int_C dl = \mu_0 i \qquad ...(8.79)$$

or

$$B(2\pi r) = \mu_0 i$$

$$\boxed{B = \frac{\mu_0 i}{2\pi r}} \qquad ...(8.80)$$

The direction of B is the tangent at any point on the Amperian loop (circle).

We have discussed the magnetic filed at the axial point of a solenoid. Now we calculated the magnetic field inside the solenoid using Ampere's law. Now, consider a solenoid carrying a current i as shown in Fig. 8.23. If n is the number of turns per unit length, then, applying Ampere's law to the loop $ABCDA$.

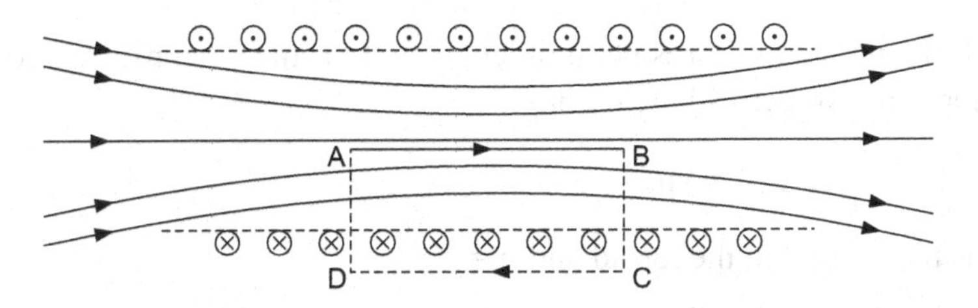

Fig. 8.23. Long Solenoid carrying a current *i*.

$$\oint_C \vec{B} \cdot \vec{dl} = N\mu_0 i \qquad ...(8.81)$$

where N is number of turns in the solenoid. Applying Eq. (8.81), we get

$$\int_A^B \vec{B} \cdot \vec{dl} + \int_B^C \vec{B} \cdot \vec{dl} + \int_C^D \vec{B} \cdot \vec{dl} + \int_D^A \vec{B} \cdot \vec{dl} = N\mu_0 i \qquad ...(8.82)$$

$$Bl + O + O + O = N\mu_0 i$$

Here, the path AB lies inside the solenoid and parallel to the magnetic field, so we get a non zero magnetic field. The path BC and DA are perpendicular to the magnetic field, thus $\vec{B} \cdot \vec{dl} = Bdl \cos 90 = 0$. The path CD lies outside the solenoid, hence $B = 0$.

Now, $$Bl = N\mu_0 i \qquad ...(8.83)$$

or $$B = \frac{N\mu_0 i}{l} \qquad ...(8.84)$$

Since $n = \frac{N}{l}$, we get

$$\boxed{B = n\mu_0 i} \qquad ...(8.85)$$

we apply Ampere's law to another symmetrical system, that is toroid. Toroid is a long solenoid bent into a circle, as shown in fig. 8.24.

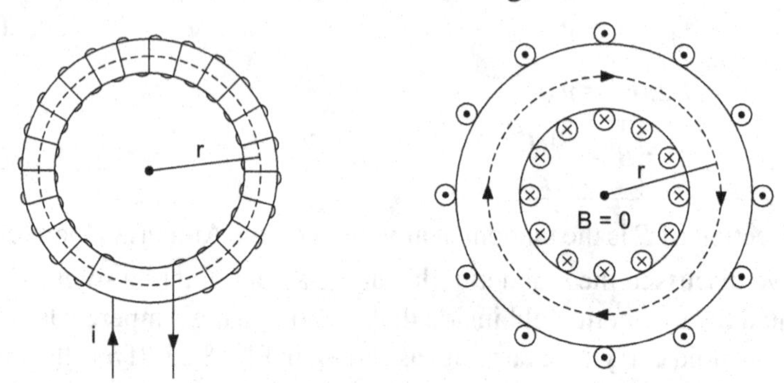

Fig. 8.24. Toroidal Coil

If n is the number of turns per unit length and l is the circumference of the toroid, applying Ampere's law, we have

$$\oint_C \vec{B} \cdot \vec{dl} = N\mu_0 i \qquad ...(8.86)$$

where i is the current in the toroid and $n = N/l$.

$\therefore$ $$B\oint dl = N\mu_0 i \qquad ...(8.87)$$

$$B(2\pi r) = N\mu_0 i$$

or
$$B = \frac{N\mu_0 i}{2\pi r} \qquad ...(8.88)$$

Since $n = \frac{N}{2\pi r} = \frac{N}{l}$, we get

$$\boxed{B = n\mu_0 i} \qquad ...(8.89)$$

In the space outside the toroidal coil, the magnetic field is zero. The magnetic field inside the toroid varies as $1/l$. On the other hand, if $l = 2\pi r$ is very small, then the variation in the magnetic field is negligible and equal to $\mu_0 ni$ as obtained in case of long solenoid.

8.13. MAGNETIC FIELD OF A MOVING POINT CHARGE

Biot-Savart law is given by

$$\vec{B} = \frac{\mu_0}{4\pi}\frac{i\,\vec{dl} \times \hat{r}}{r^2} \qquad ...(8.90)$$

If a point charge q moves with a velocity $\vec{v}$ along x-axis as shown in fig. 8.25, we replace qv in place idl. Then,

$$\vec{B} = \frac{\mu_0}{4\pi}\frac{q(\vec{v} \times \hat{r})}{r^2} \qquad ...(8.91)$$

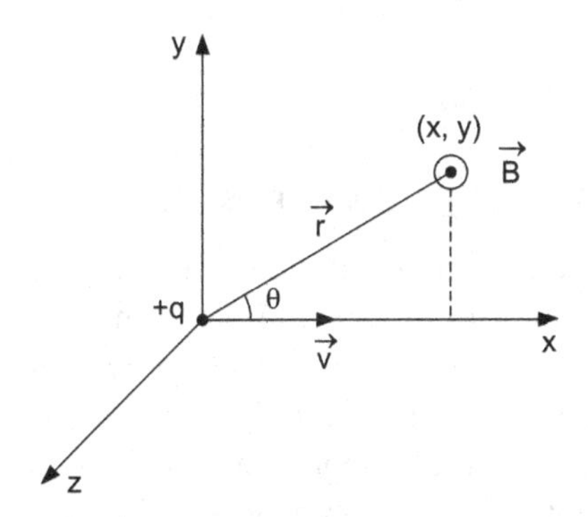

Fig. 8.25. Moving point charge

The position vector $\vec{r} = x\hat{i} + y\hat{j}$

and
$$|r| = \sqrt{x^2 + y^2}$$

$\therefore$ Unit vector along $\vec{r}$ is

$$\hat{r} = \frac{x}{r}\hat{i} + \frac{y}{r}\hat{j}$$

or

$$\hat{r} = \cos\theta\,\hat{i} + \sin\theta\,\hat{j}$$

Since, velocity of the point charge q is along x-axis, then,

$$v = v\hat{i}$$

Thus, from the Equation (8.91), we have

$$\vec{B} = \frac{\mu_0 q}{4\pi r^2}[v\hat{i} \times (\cos\theta\hat{i} + \sin\theta\hat{j})]$$

or

$$\boxed{\vec{B} = \frac{\mu_0 q}{4\pi r^2}\frac{v\sin\theta}{r^2}\hat{k}} \quad \text{...(8.92)}$$

The direction of the magnetic field is outward from the plane of paper. The magnitude of the magnetic field is given by

$$\boxed{B = \frac{\mu_0}{4\pi}\frac{qv\sin\theta}{r^2}} \quad \text{...(8.93)}$$

From the Equation (8.93), it clear that,

$$\boxed{B \propto \frac{qv\sin\theta}{r^2}} \quad \text{...(8.94)}$$

where θ is angle between $\vec{v}$ and $\vec{r}$.

8.14. MAGNETIC FIELD IN MATERIALS

The different materials behave in the different way with respect to the magnetic field. According to the atomic model, the electrons are orbiting round the positively charged nucleus. Moreover, the electrons are spinning also. Thus, an atom has magnetic dipole moments due to the orbiting and spinning electrons. In this section, we investigate the change in magnetic field due to the presence of magnetic materials. We discuss three types of magnetic behavior of the materials viz, paramagnetism, diamagnetism and ferromagnetism.

8.14.1. Magnetic Moment of an Electron

Consider an electron of charge e and mass m circulating in a Bohr orbit as shown in fig. 8.26.

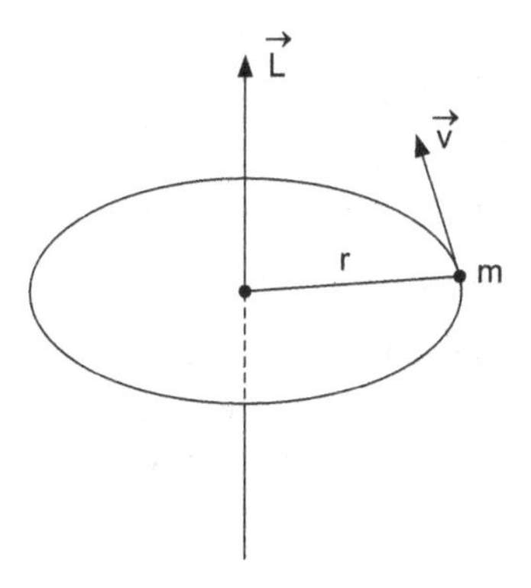

Fig. 8.26. A orbiting electron

the motion of the electron in the orbit constitutes a current loop. The current through this loop is given by

$$i = \frac{e}{T} \quad \text{...(8.95)}$$

where T is the time period of the electron in the orbit. According to Ampere's theorem, the current in the loop gives rise to a magnetic moment μ_l (we represent the magnetic moment of an electron by μ_l in Atomic spectra, a usual notation), then,

$$\mu_l = iA \quad \text{...(8.96)}$$

where $A = \pi r^2$, is area of the loop.

Thus,

$$\mu_l = \frac{e\pi r^2}{T} \quad \text{...(8.97)}$$

the time period $\quad T = \frac{\text{one revolution}}{v}$

$\therefore$ or $\quad T = \frac{2\pi r}{v}$

Substituting the value of T in Equation (8.97), we get

$$\mu_l = \frac{evr}{2} \quad \text{...(8.98)}$$

the orbital angular momentum of the electron is given by

$$L = m\,r\,v \quad \text{...(8.99)}$$

From the Equations (8.98) and (8.99) we have

$$\vec{\mu}_l = \frac{e}{2m}\vec{L} \quad \text{...(8.100)}$$

or
$$\vec{\mu}_l = \frac{-e}{2m}\vec{L} \quad ...(8.101)$$

the negative sign is due to the negative charge of the electron.

8.14.2. Magnetic Permeability

Suppose that a magnetic substance is placed in a magnetic field, as shown in fig. 8.27. The large number of magnetic lines of force will pass through the magnetic substance than in air. A degree to which the magnetic lines of force can permeate the substance is called the magnetic permeability of that substance.

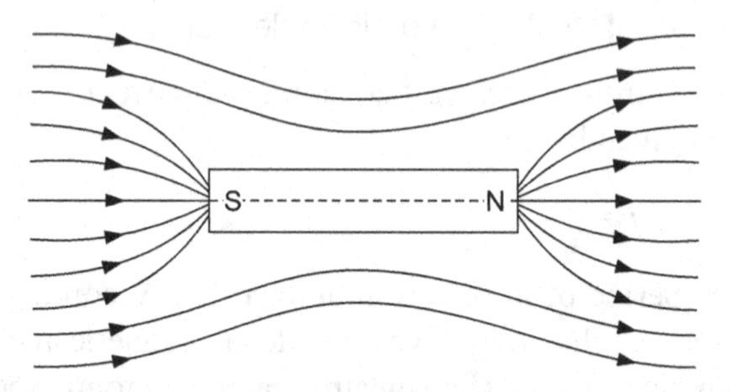

Fig. 8.27. Magnetic substance in a magnetic field.

The permeability of the material medium is defined as the ratio of the magnitude of the magnetic induction B to the magnetic field intensity H. Thus,

$$\text{Permeability } \mu = \frac{B}{H} \quad ...(8.102)$$

Further more, the relative permeability of the material medium is given as

$$\mu = \mu_r \mu_0 \quad ...(8.103)$$

where μ_0 is the permeability of free space and μ_r is the relative permeability. Then, relative permeability is

$$\mu_r = \frac{B}{B_0} = \frac{\mu}{\mu_0} \quad ...(8.104)$$

$$\mu_0 = 4\pi \times 10^{-7} \text{ Tm/A}.$$

8.14.3. Magnetization

It is a measure of magnetization of a magnetized substance. The magnetization M is defined as the magnetic dipole moment per unit volume. It is a vector quantity. If these are n magnetic dipoles having same orientation, there exists n magnetic dipole moments. Thus, Magnetization vector in is given by

$$\vec{M} = \frac{\sum_{i=1}^{n} \vec{m}_i}{V}$$

or $$\boxed{\vec{M} = \frac{1}{V}\sum_{i=1}^{n} \vec{m}_i} \qquad ...(8.105)$$

where V is the volume of the magnetic substance.

8.14.4. Magnetic Susceptibility

We know that the magnetization of a magnetic substance is proportional to the applied field. Thus, the magnetic susceptibility (χ) is defined as the ratio of the intensity of magnetization to the magnetic field intensity, i.e.

$$\boxed{\chi = \frac{M}{H}} \qquad ...(8.106)$$

It is a dimensionless quantity.

When a magnetic substance is placed in an external magnetic field B, the magnetic dipoles are oriented more or less in the direction of the magnetic field. Thus, a non-zero magnetic moment is exhibited by the magnetic substance. Therefore,

$$B = \mu_0 H + \mu_0 M \qquad ...(8.107)$$

$$\boxed{B = \mu_0 (H + M)} \qquad ...(8.108)$$

Moreover,

$$B = \mu_0 (\chi H + H) \qquad ...(8.109)$$

Here, we have used the equation (8.106),

$$B = \mu_0 H (1 + \chi) = \mu H \qquad ...(8.110)$$

or $$B = \mu_0 \mu_r H \qquad ...(8.111)$$

where $$\boxed{\mu_r = (1 + \chi) = \frac{\mu}{\mu_0}} \qquad ...(8.112)$$

8.14.5. Diamagnetism

In a diamagnetic substance, the net magnetic moment due to the atomic dipoles is zero. In diamagnetic materials the outer electronic shell is closed and the electrons are paired. The diamagnetic property in the materials occurs when the magnetic fields due to orbital and spin motion cancel each other. Therefore, each atom has zero magnetic moment. When a diamagnetic substance is placed in an external magnetic field, the small induced magnetic dipoles within the atoms oppose the applied magnetic field. Some diamagnetic materials are Cu, Ag, Zn, Bi, Au, Sb etc. The diamagnetic materials possess a negative susceptibility of the order of $\sim 10^{-7}$. Thus, we may write

$$\boxed{\mu_r < 1}$$

and $$\boxed{\chi < 0}$$

8.14.6. Paramagnetism

In paramagnetic materials, atoms possess permanent magnetic dipole moment but these are aligned spontaneously as shown in Fig. 8.28. These magnetic dipoles are randomly distributed. When an external magnetic field is applied, they tend to orient parallel to the direction of the magnetic field as shown in fig. 8.29.

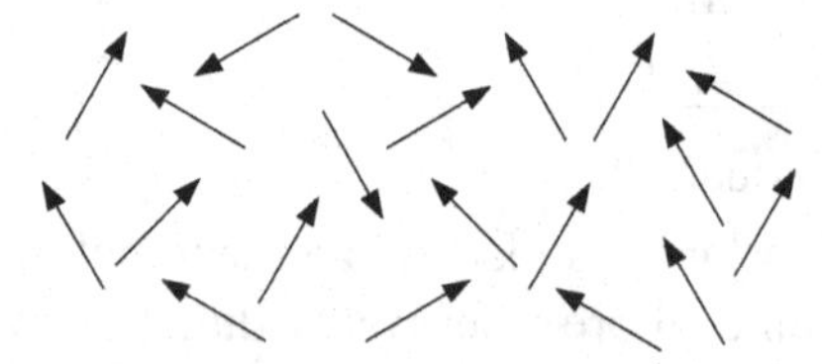

Fig. 8.28. Magnetic dipoles in paramagnetic materials.

As a result, the magnetic field or induction is increased, and magnetic substance acquires a magnetization.

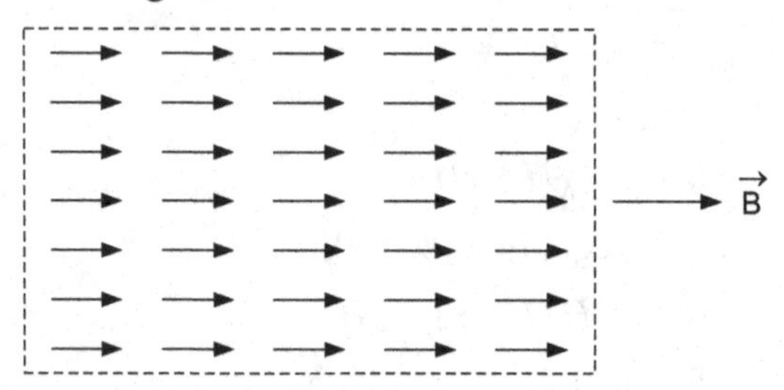

Fig. 8.29. Specimen in external magnetic field.

In the case of paramagnetic materials, the magnetization is weak and magnetic susceptibility is small at room temperature. The magnetic susceptibility is given by

$$\boxed{\chi = \frac{C}{T}} \qquad ...(8.113)$$

where C = curie constant.

For paramagnetic materials.

$$\boxed{\mu_r \geq 1}$$

and $$\boxed{\chi = \text{small and positive}}$$

From the Equation (8.113), it is shown that the magnetic susceptibility is in-versely proportional to absolute temperature. The variation of χ with T is shown in fig. 8.30.

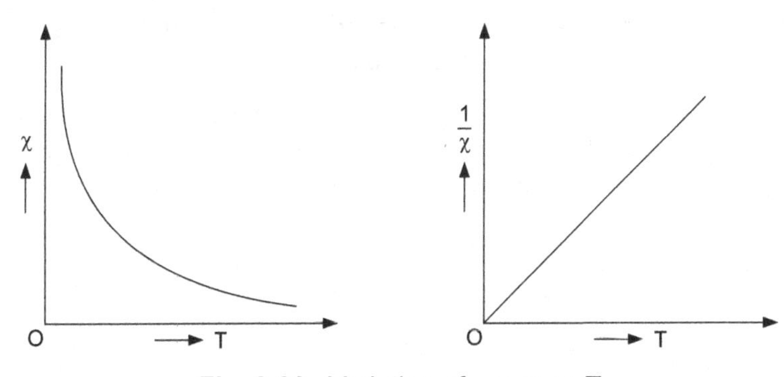

Fig. 8.30. Variation of χ versus T

8.14.7. Ferromagnetism

In ferromagnetic materials, the atoms have large permanent magnetic dipole moment. Ferromagnetism is due to interation between domains of the magnetic moments. This interaction is very strong to cause the neighboring atoms in other domains to align with their magnetic dipole moments in the same direction. The ferromagnetic materials have small regions and each region has magnetic dipole moments aligned in same direction, these regions are called domains as shown in fig. 8.31.

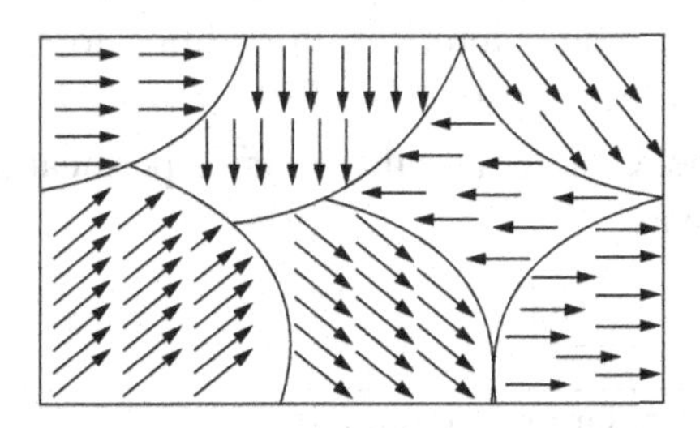

Fig. 8.31. Magnetic domains in ferromagnetic materials.

when a ferromagnetic substance is placed in an external magnetic field, a large alignment of the magnetic moments in the direction of the applied magnetic field has been found as shown in fig. 8.32.

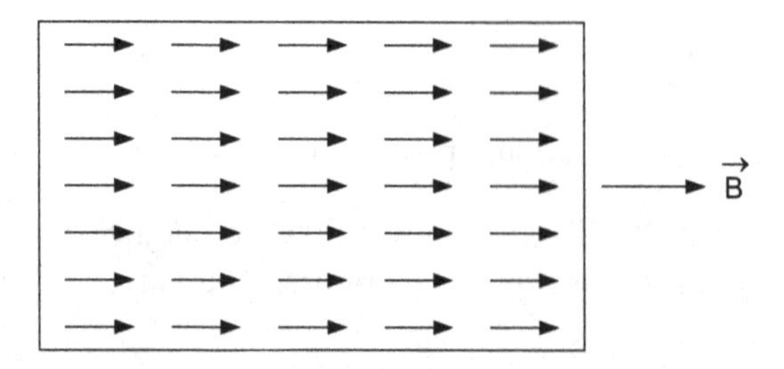

Fig. 8.32. Completely alignment of magnetic moments in the external magnetic field.

As a result of external magnetic field, the magnetization M increases abnormally and it is not linear. A plot of the magnetic field intensity H versus magnetization M is called a magnetising curve, as shown in fig. 8.33.

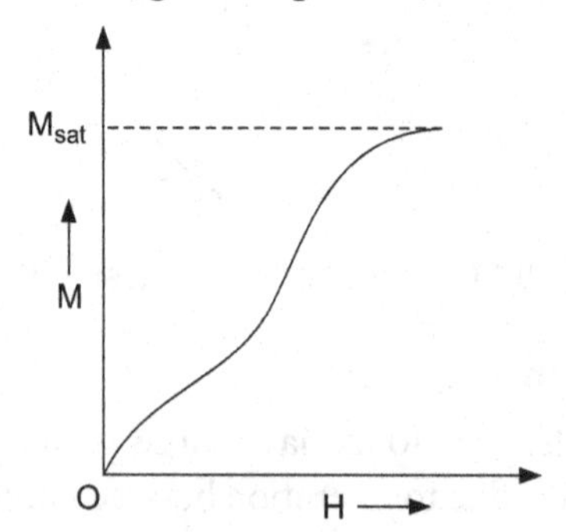

Fig. 8.33. Magnetization curve.

In this case, $\vec{M} = \chi \vec{H}$ and $\vec{B} = \mu \vec{H}$ are not applicable because χ and μ are treated as constants. So to retain these forms of the equations, we must have χ and μ as a function of H. In the magnetization curve, the M is increased enough to reach in saturation. Moreover, in addition to saturation effect, the ferromagnetic materials show hysteresis effect.

For the ferromagnetic materials, the relative permeability and magnetic susceptibility are given as

$$\boxed{\mu_r >>> 1}$$

and

$$\boxed{\chi = \text{large} + \text{ve number}}$$

Furthermore,

$$\boxed{\chi = \frac{C}{T - T_C}} \qquad ...(8.114)$$

When the temperature of the substance is raised, it increases thermal agitation and this thermal agitation breaks the alignment of all magnetic dipoles. As a result, the substance is now demagnetized and behaves like a paramagnetic substance. The temperature at which all the magnetic dipoles lose their alignment is called the Curie temperature (T_C). The ferromagnetic materials are Fe, Ni, Co (transition elements), Gd etc. The properties of paramagnetic, ferromagnetic and diamagnetic materials are given in the table 8.1.

Table 8.1. Properties of different magnetic materials.

S.No	Property	Paramagneti materials	Ferromagnetic materials	Dia magnetic materials
1.	Cause	Spin motion of electrons	Ferromagneitc domains	Orbital motion of electrons
2.	In external magnetic field	Less attracted	Strongly attracted	expels magnetic lines of force
3.	Magnetic moment	Randomly oriented	Have some magnetic moment due to alignment	zero
4.	μ_r and χ	$\mu_r \geq 1$ χ is small and +ve	$\mu_r >> 1$ χ is large and +ve	$\mu_r < 1$ χ is –ve

8.14.8. Hysteresis

When a ferromagnetic substance (with $M = 0$) is placed in an external magnetic field, it is magnetized. The magnetization in the substance changes with the strength of the magnetising field H. A non-linear relationship between magnetization M and field strength H is shown by a curve known as B–H curve. (Magnetic induction is taken for M). As in fig. 8.34, the magnetization increases and attains a constant value at A. Now, M does not increase with H. This is called magnetic saturation. Suppose that H is reduced to zero slowly, then, M does not reduced to zero. This is because the magnetic domains have been aligned by the magnetization M and remained aligned, even though H is reduced to zero. A residual intensity of magnetization (OB) left in the magnetic substance is called the retentivity or remanence of the material. Furthermore, to reduce M to zero, H is increased in reverse (negative) direction, and part BC is obtained. Here, $H = -H_C$ at point C. At, $H = -H_C$, $M = 0$ then H_C is called coercive force. Hence, the coercivity of a magnetic substance is the strength of the reverse magnetic field for which the substance is demagnetise completely. Increasing H further in negative direction until it reaches at point D, then, substance is magnetized in reverse sense and M versus H curve traces the path $DEFA$. In the plot of whole curve, M lags behind H. This lagging of M is called as hysteresis and the curve M vs H in which a core is magnetized in one direction and then in opposite direction is called hysteresis curve.

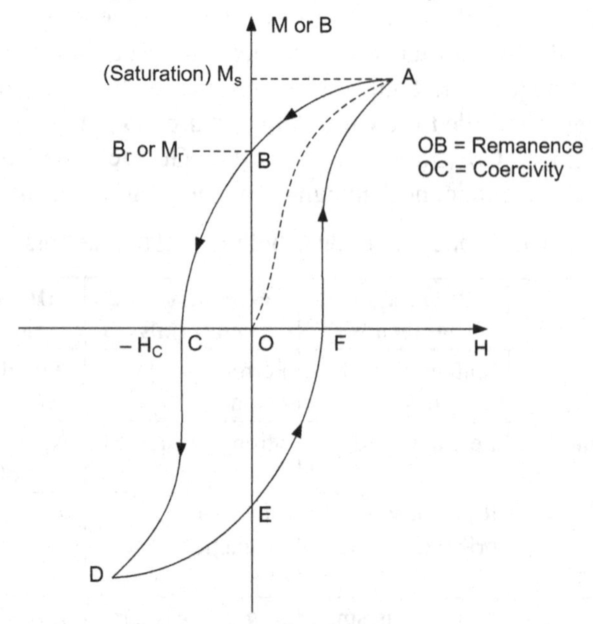

Fig. 8.34. Hysteresis loop of a ferromagnetic substance.

8.14.9. Hysteresis Loss

When an un-magnetized ferromagnetic material is placed in an external magnetic field, the atomic dipoles in the atoms are aligned in the direction of the magnetic field applied. Obviously, in the process of magnetization, the magnetising field works against mutual attraction among the elementary atomic magnets. If the magnetising field is removed, then, certain magnetization will be retained in the magnetic material. Thus, the energy supplied during the process of magnetization is not recovered completely after switching off the magnetising field. This energy is lost in form of heat during each cycle of magnetization. This is known as hysteresis loss.

Now, to calculate the energy lost per cycle, suppose that a magnetic material of unit volume, *abcd*, is placed in a magnetic field intensity *H*, as shown in fig. 8.35.

If there are *n* molecular magnets per unit volume and *m* is the magnetic dipole moment of each magnet inclined at an angle θ with magnetic field intensity *H*, then, the magnetic moment per unit volume may be given as

$$M = \sum_i m_i \cos\theta \qquad ...(8.115)$$

where *M* is magnetization field.

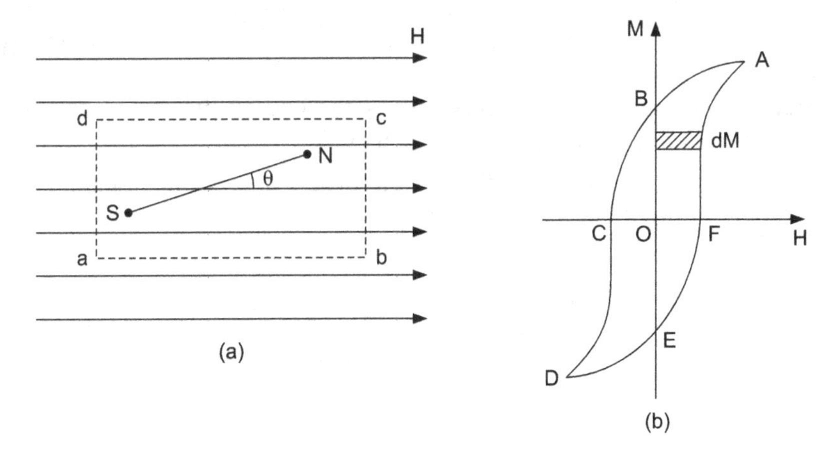

Fig. 8.35. (a) Magnetic substance in a magnetising field (b) and its hysteresis loop.

Differentiating Equation (8.115), we get

$$dM = -\sum_i m_i \sin\theta \, d\theta \qquad ...(8.116)$$

The restoring couple acting on the magnet of magnetic moment m is given by $mH \sin\theta$. Thus, workdone is rotating the magnet through $-d\theta$ angle (θ decreases with H) will be

$$dW = -mH \sin\theta \, d\theta \qquad ...(8.117)$$

Therefore, work done per unit volume is given by

$$dW = -\sum_i m_i \, H \sin\theta \, d\theta \qquad ...(8.118)$$

Here, we took dW for simplicity.

or
$$dW = -H \sum_i m_i \sin\theta \, d\theta \qquad ...(8.119)$$

$$dW = H \, dM \qquad ...(8.120)$$

we have used the Equation (8.116), again the work done for a cycle of the hysteresis loop will be

$$\boxed{W = \oint H \, dM} \qquad ...(8.121)$$

The integration $\oint$ is over a complete cycle of M-H (B-H) curve, which is equal to the area of the M-H curve ($A\,B\,C\,D\,E\,F\,A$). Moreover, consider a strip of length H and thickness dM as shown in fig. 8.35(b).

$$\text{Area of the strip} = H \, dM \qquad ...(8.122)$$

Now, whole area of the hysteresis curve is given by

$$A = \oint H\, dM \qquad ...(8.123)$$

Hence, the area of the *M-H* curve gives the energy dissipated per unit volume of the magnetic material during each cycle.

Example 8.4. Find the magnetic field at the centre of a circular segment of radius *r* shown in fig. 8.36.

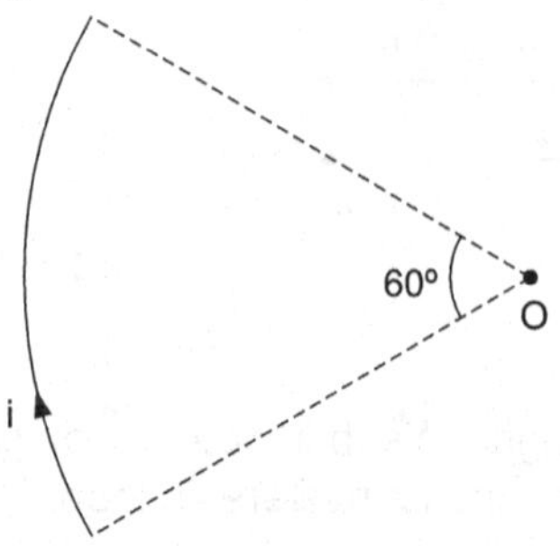

Fig. 8.36. Circular segment carrying a current *i*.

Solution: According to Biot-Savart law,

$$dB = \frac{\mu_0}{4\pi} \frac{i\, dl \sin\theta}{r^2}$$

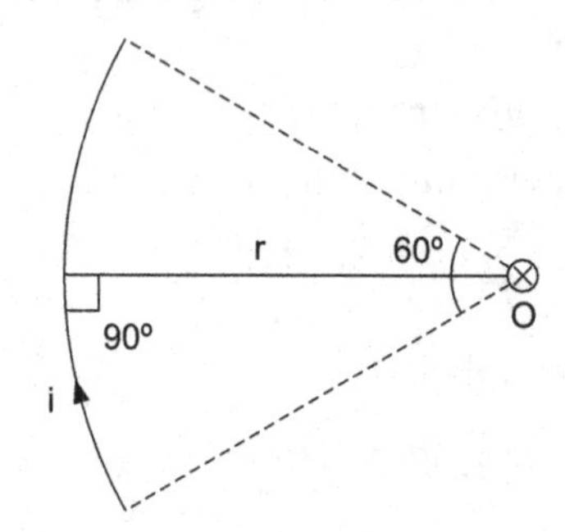

Fig. 8.37. Circular arc

Here, $\sin\theta = \sin 90 = 1$, then

$$dB = \frac{\mu_0}{4\pi} \frac{i\, dl}{r^2}$$

Net magnetic field at centre *O* is given by

$$B = \int dB = \frac{\mu_0 i}{4\pi r^2} \int dl$$

But Arc = Angle × Radius

$$= \frac{\pi}{3} \times r$$

Thus, $$B = \frac{\mu_0 i}{12r}$$

The direction of magnetic field B is into the plane of page.

Example 8.5. Find the magnetic field at the centre of a square loop of side a.

Solution: Consider a square loop carrying a current i is shown in fig. 8.38.

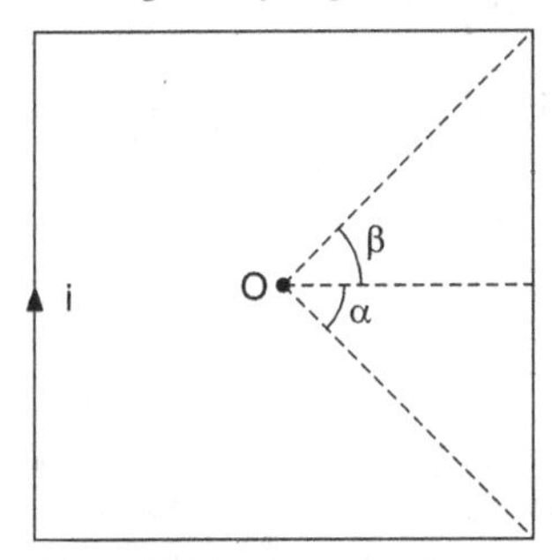

Fig. 8.38. Square loop

the magnitude of B at the centre 'O' is given by

$$B = \frac{\mu_0 i}{r\pi} \sum_{n=1}^{4} \frac{(\sin\alpha_n + \sin\beta_n)}{r_n}$$

Since square has four sides, B is given by

$$B = \frac{2\sqrt{2}\,\mu_0 i}{\pi a}$$

Note: We can find the magnetic field at the centre of a rectangle using above formula.

Example 8.6. A long straight wire carries a current $i = 100A$ along the z-axis and a constant magnetic field whose magnitude is 1×10^{-5} T is directed along the x-axis as shown in fig. 8.39. Find the magnetic field at point A and B.

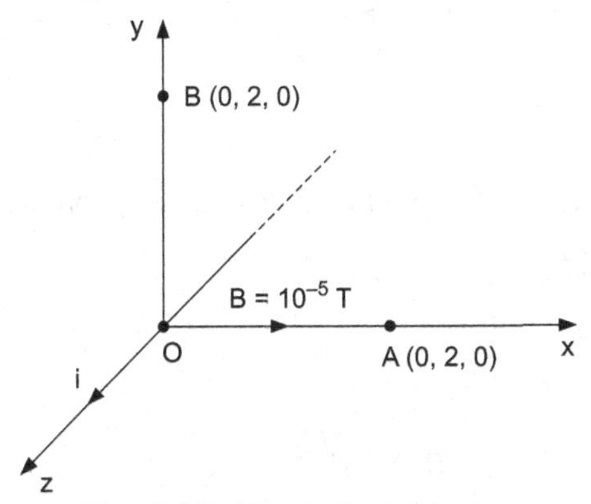

Fig. 8.39. A long straight wire.

Solution: The magnitude of the magnetic field at the point A is

$$B_A = \frac{\mu_0 i}{2\pi r}$$

$$= \frac{2 \times 100 \times 10^{-7}}{2}$$

$$B_A = 10^{-5} T$$

The direction of this field is along y-axis (right hand rule)

Thus, net magnetic field at point A is

1×10^{-5} $\quad B_A = \sqrt{B_1^2 + B_2^2}$

1×10^{-5}

$$B_A = \sqrt{2} \times 10^{-5} T$$

$$B_A = 1.414 \times 10^{-5}\ T$$

The magnetic field at the point B due to long straight wire will be $BC = 1 \times 10^{-5} T$ but it is directed along $-x$-axis.

Example 8.7. A long straight conductor carries a current i_1. Find the force experienced by a rectangular loop carrying a current i_2 shown in fig. 8.40.

Solution:

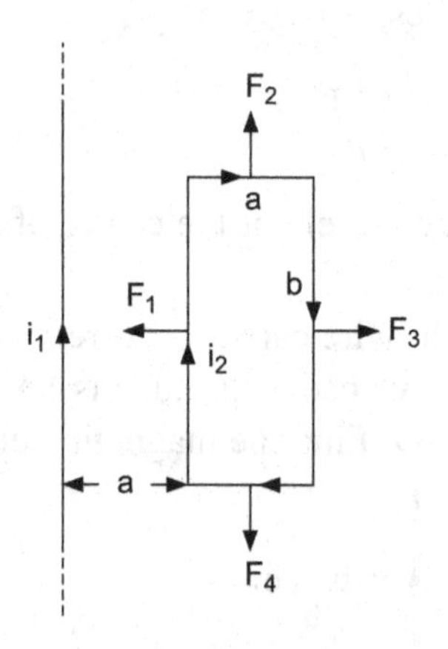

Fig. 8.40. Rectangular loop near a long conductor.

The magnetic field

$$B_1 = \frac{\mu_0 i_1}{2\pi a}$$

Then,

$$F_1 = i_2\, b\, B_1 = \frac{\mu_0 i_1 i_2 b}{2\pi a}$$

The forces F_2 and F_4 are equal and opposite and cancel each other.

the force $F_3 = i_2 \, l \, B_3$

Here, $$B_3 = \frac{\mu_0 \, i_1}{2\pi\,(2a)}$$

$$\therefore \quad F_3 = \frac{\mu_0 \, i_1 \, i_2 \, b}{2\pi\,(2a)}$$

Thus, net force on the loop, $F_1 > F_2$,

$$F = F_1 - F_2$$

$$= \frac{\mu_0 \, i_1 i_2 b}{2\pi}\left[\frac{1}{a} - \frac{1}{2a}\right]$$

$$\boxed{F = \frac{\mu_0 \, i_1 i_2 \, b}{4\pi a}}$$

Example 8.8. Two point charges q_1 and q_2 move non-relativisitically with the velocities v_1 and v_2 respectively (fig. 8.41). Compute

(a) net magnetic field at the origin.

(b) the force exerted on q_2 by q_1.

Solution: (a) From the section (8.13), a non-relativistically, the magnetic field of a moving charge q is given by

$$B = \frac{\mu_0 \, q}{4\pi} \frac{v \sin\theta}{r^2}$$

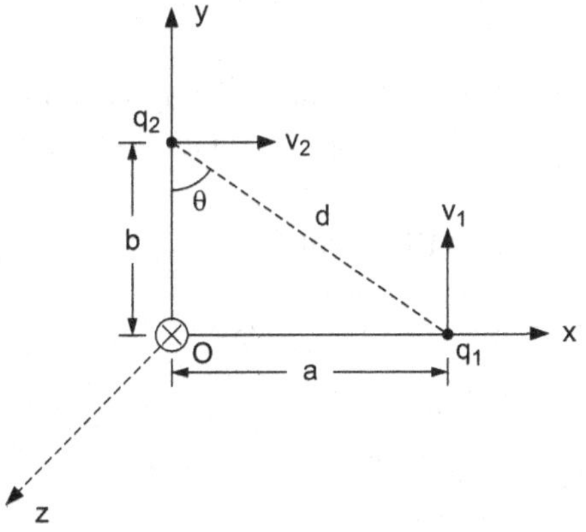

Fig. 8.41. Motion of two point charges.

The magnetic field due to charge q_1 is B_1 and given by

$$B_1 = \frac{\mu_0}{4\pi} \frac{q_1 \, v_1 \sin 90}{a^2}$$

or $$B_1 = \frac{\mu_0}{4\pi} \frac{q_1 v_1}{a^2}$$

Similarly, for q_2, the magnetic field B_2 is

$$B_2 = \frac{\mu_0}{4\pi} \frac{q_2 v_2}{b^2}$$

Since B_1 and B_2 both are directed into plane of page, then, net magnetic field at O will be

$$B = B_1 + B_2$$

or $$B = \frac{\mu_0}{4\pi}\left(\frac{q_1 v_1}{a^2} + \frac{q_2 v_2}{b^2}\right)$$

(b) The magnetic force on a moving charge is given by

$$\vec{F} = q(\vec{v} \times \vec{B})$$

Here $$B = \frac{\mu_0}{4\pi} \frac{q_1 v_1 \sin\theta}{d^2}$$

$$F = q_2 v_2 B$$

$$= \frac{\mu_0}{4\pi} \frac{q_1 q_2 v_1 v_2 \sin\theta}{d^2}$$

Since, $d = \sqrt{a^2 + b^2}$ and $\sin\theta = \frac{a}{d}$, hence

$$F = \frac{\mu_0 a}{4\pi} \frac{q_1 q_2 v_1 v_2}{d^3}$$

this is the required force.

Example 8.9. A magnet weighs 50 gm and its magnetic moment is 1000 units. If the density of the material of the magnet is 5 gm/cm^3. Compute the intensity of magnetization.

Solution: magnetic moment = 1000 units

$$\text{Volume} = \frac{\text{Mass}}{\text{Density}} = \frac{50}{5} = 10 \text{ cm}^3$$

Now, Intensity of magnetization

$$M = \frac{m}{v} = \frac{1000}{10} = 200 \text{ units.}$$

EXERCISES

8.1. Discuss a method to identify the north and south poles of a bar magnet.

8.2. What do you mean by magnetic flux. Obtain an expression of the magnetic flux originating from a current carrying wire shown in fig. 8.42, and passing through a square sheet of side 1 m.

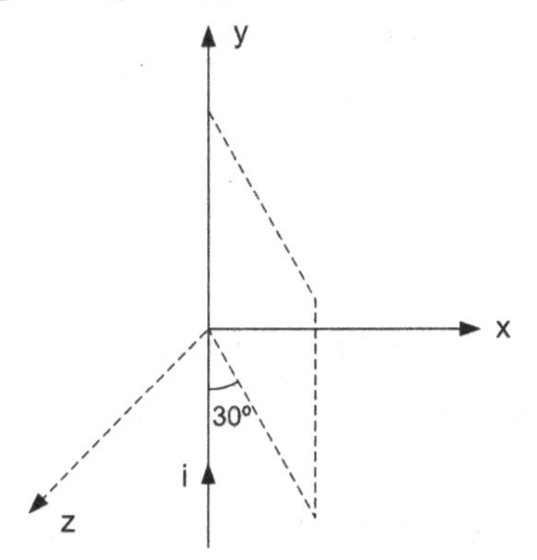

Fig. 8.42. A square sheet.

8.3. A long straight wire of length 2 m carries a current of 1.0 A lies along y-axis. Find the force on this wire, where magentic field is given by $\vec{B} = (2\hat{i} + 4\hat{j})$ Wb/m^2.

8.4. Find an expression for the flux density at an axial point of a circular coil carrying a current iA.

8.5. Show that the area of *M-H* (*B-H*) curve represents the energy dissipated per unit volume of a magnetic material during each cycle.

8.6. Explain residual magnetization, retentivity and coercivity using hysteresis curve of a magnetic material.

8.7. A point charge q moves in a uniform magnetic field with speed v. Show that the work done by the magnetic force acting on it is zero.

8.8. A long straight conductor curries a current $i_1 = 1.0$A. Compute the force experienced by a square loop of side 1.0 m and carrying a current $i_2 =$ 0.5 A, Fig. 8.43.

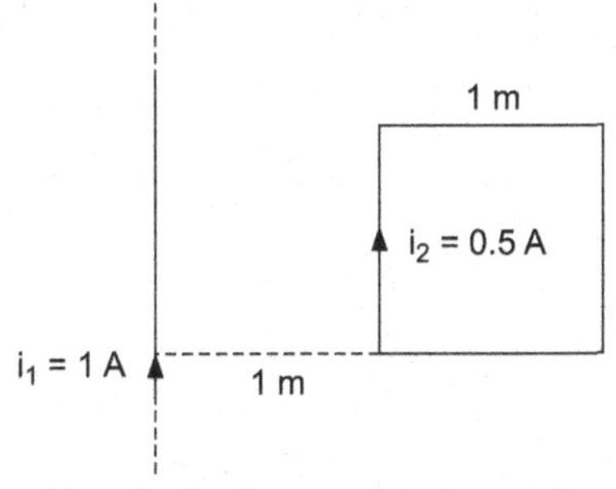

Fig. 8.43. A current element.

8.9. Determine the magnetic flux density of a toroid with radius R and number of turns n. Toroid carries a current i.

8.10. Find the magnetic flux density at the centre of a semicircle, Fig. 8.44, and carries a current of 10A.

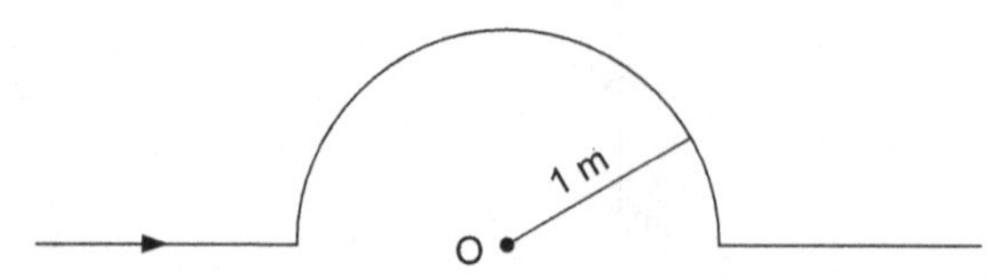

Fig. 8.44. Semi circle.

8.11. State and prove Ampere's circuital law and compute the magnetic field at an axial point of a solenoid.

8.12. Find the magnetic flux density at the centre of a square loop of side 1.0 m and carrying a current of 1.0 A.

8.13. Compute the magnetic forces that act on two parallel wires when they carry unequal currents.

8.14. A segment of wire 10 cm long carries a current of 1.0 A. Calculate the magnetic field at a point Q a distance of 0.5 m at angle of 30°.

8.15. A long solenoid of radius R and number of turns n has a length L. If a current i flows through its coil, show that magnetic flux density at its axial point is given by

$$B = \frac{\mu_0 ni}{2l}(\cos\theta_2 - \cos\theta_1).$$

8.16. A straight wire of length l carries a current i. Show that the magnetic field at a distance R is given by

$$B = \frac{\mu_0 i}{4\pi R}\frac{l}{(l^2 + R^2)^{1/2}}$$

8.17. A cube of side 1.0 m is placed in a constant magnetic field of 0.1 Wb/m^2 and directing along x-axis. Compute the magnetic flux through each face of the cube.

■ ■ ■

CHAPTER

Alternating Currents

In the previous chapters, we mostly have emphasis on source of direct current (dc). This current remains constant with time as shown in Fig. 9.1. Direct current contains zero frequency.

Fig. 9.1. A direct current source (a) and current versus time (b).

An electric current with periodically varying intensity is called an alternating current. This passes through a complete cycle of changes at regular intervals. Each cycle consists of two half cycles : positive and negative. The voltage and current are given by equations as

$$v = v_o \sin \omega t = v_o \sin 2\pi ft \quad ...(9.1)$$

$$v = i_o \sin \omega t = i_o \sin 2\pi ft \quad ...(9.2)$$

The source of alternating voltages and currents are sinusoidal as shown in fig. 9.2.

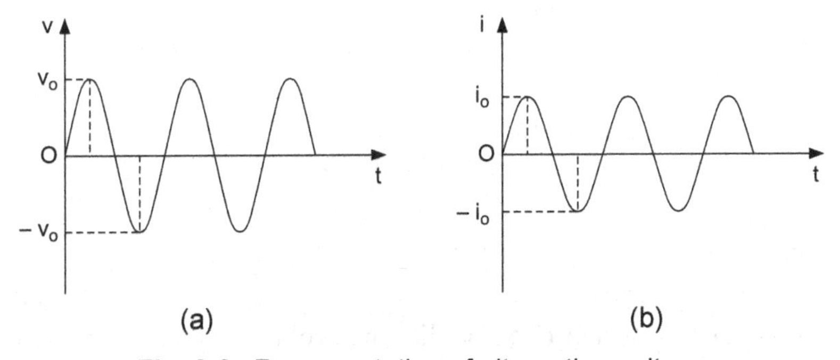

Fig. 9.2. Representation of alternating voltage (a) and alternating current (b).

where v and i are the instantaneous values of voltage and current. v_o is called maximum voltage or voltage amplitude and i_o the maximum current or current amplitude. ω is known as angular frequency ($\omega = 2\pi f$). Alternating currents in houses range from 50 cycles/s for lights, clocks and other appliances to 108 cycles/s radar use and microwave communication. The symbol of ac source is shown in fig. 9.3.

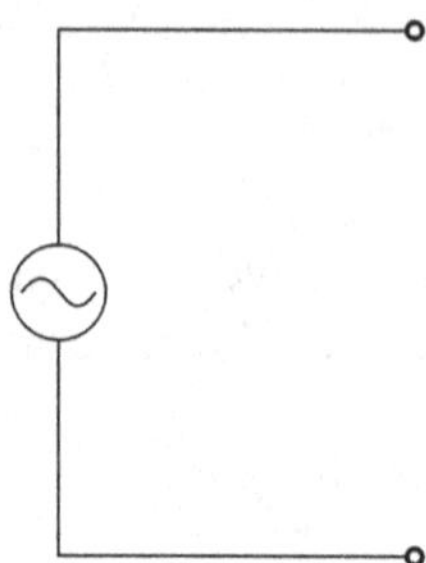

Fig. 9.3. Symbolic representation of ac source.

9.1 AVERAGE AND ROOT-MEAN-SQUARE VALUES OF VOLTAGE AND CURRENT

The alternating current and voltage are continuously varying in positive and negative directions, hence, the average values of v and i over a cycle will be zero.

The average value of ac voltage over a period:

$$\langle v \rangle = \frac{1}{T}\int_0^T v\,dt \qquad \text{...(9.3)}$$

$$= \frac{v_o}{T}\int_0^T \sin \omega t\,dt$$

$$= \frac{v_o}{2\pi}\int_0^{2\pi/\omega} \sin \omega t\,d(\omega t)$$

$$= \frac{v_o}{2\pi}\int_0^{2\pi} \sin \phi\,d\phi = 0 = 0 \qquad \text{...(9.4)}$$

Similarly, we obtain for ac current.

$$\langle i \rangle = 0 \qquad \text{...(9.5)}$$

The equations (9.4) and (9.5) predict that voltage and current oscillate symmetrically about zero, as shown in fig. 9.4.

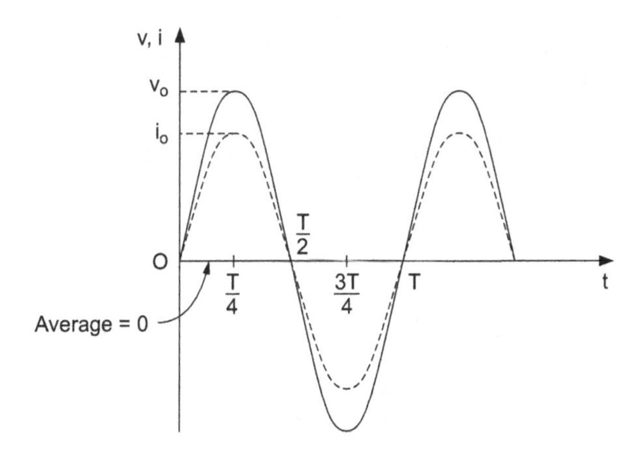

Fig. 9.4. Plots of *v* and *i*, showing average values of voltage and current.

Now, we define the root-mean-square current i_{rms} as

$$i_{rms} = \sqrt{<i^2>} \qquad ...(9.6)$$

This is also known as virtual or effective value of current. The average value of square of current for a cycle may be computed as :

$$<i^2> = \frac{1}{T}\int_0^T i_0^2 \sin^2 \omega t \, dt \qquad ...(9.7)$$

$$= \frac{i_0^2}{2\pi}\int_0^T i_0^2 \sin^2 \omega t \, dt$$

$$= \frac{i_0^2}{2\pi}\int_0^{2\pi} \sin^2 \phi \, d\phi$$

$$= \frac{i_0^2}{2} \qquad ...(9.8)$$

the square root of $<i^2>$

$$i_{rms} = \sqrt{<i^2>} = \frac{i_0}{\sqrt{2}} = 0.707 i_0 \qquad ...(9.9)$$

Similarly, root-mean-square voltage v_{rms} may be shown to be

$$v_{rms} = \sqrt{<v^2>} = \frac{v_o}{\sqrt{2}} = 0.707 \, v_0 \qquad ...(9.10)$$

The instruments read the root-mean-square values of current and voltage, and they are calibrated to read such values. Instead of using average values of *v* and *i*, we use the average values of i^2 and v^2. The graphical representation is shown in fig. 9.5.

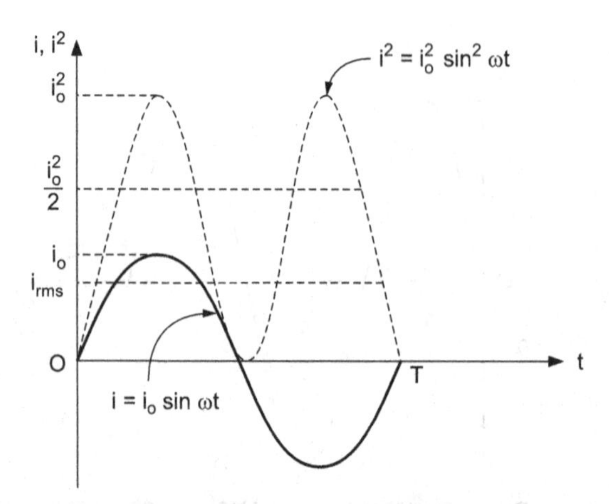

Fig. 9.5. A representation of i^2 and i versus t.

The plot of $i^2 = i_0^2 \sin^2 \omega t$ is always positive and average value of i^2 is $i_0^2/2$, which is definite. Because the average value of $\sin^2 \omega t$ is $\frac{1}{2}$.

We say that the ac voltage in houses is 200V, so that rms value of voltage is 200V. Therefore, the peak value of voltage is $v_0 = \sqrt{2}\, v_{rms} = \sqrt{2} \times 200V = 282.84$ V. This means that the ac voltage oscillates between +282.84V and –282.84V.

The alternating current depends on time, the simple procedures of solving problems cannot be applied directly. We take rms values of v and i (instantaneous values) for solving an ac circuit.

Example 9.1. The rms voltage of an ac generator is 100V and it produces a current of rms value 5A when connected in circuit. What are maximum affordable values of voltage and current?

Solution:
$$v_0 = \sqrt{2}\, v_{rms} = 141.42V$$
$$i_0 = \sqrt{2}\, i_{rms} = 7.07A$$

Thus current i oscillates between +7.07 and –7.07A and voltage oscillates between 141.42 and –141.42 V.

9.2 PHASOR DIAGRAMS

A phasor diagram represents a projection of a physical quantity which varies sinusoidally with time. It is an intuitive visualisation of phase angle of current or voltage. The projection of a physical quantity must be a uniform circular motiion. Such quantities are known as phasors. Since ac voltage and current vary sinusoidally, these may be represented by phasor diagrams. The phasor diagram of ac voltage $v = v_0 \sin \omega t$ is shown in fig. 9.6. Phasor diagrams may be plotted for resistive, capacitive and inductive circuits.

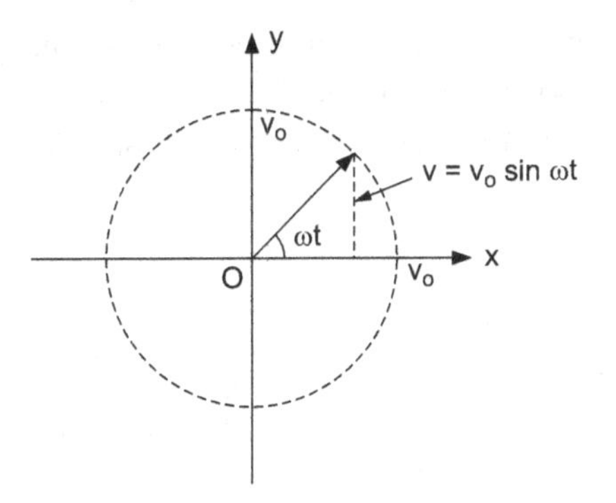

Fig. 9.6. Phasor diagram for $v = v_0 \sin \omega t$

9.3 AC CIRCUIT WITH RESISTANCE

Consider an ac circuit containing a resistance R and a source of EMF as shown in fig. 9.7.

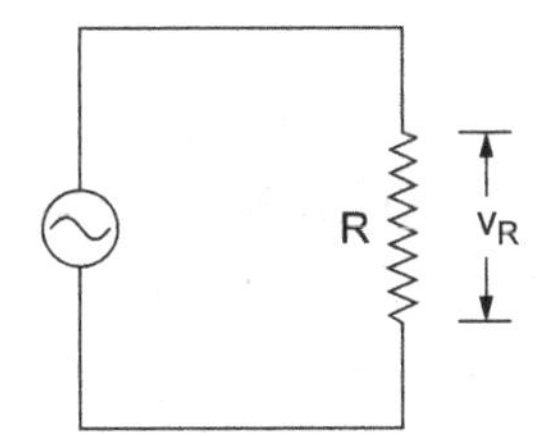

Fig. 9.7. AC circuit with a resistor and source.

This ac source supplying the current through resistor R is

$$i = i_0 \sin \omega t \qquad ...(9.11)$$

Therefore, the instantaneous potential difference across resistor R will be given by

$$v_R = Ri \qquad ...(9.12)$$

$$= R\, i_0 \sin \omega t = v_{0R} \sin \omega t \qquad ...(9.13)$$

if we write $v_{0R} = i_0 R$, thus, applied voltage and current are in phase at all times and potential difference across the resistor varies sinusoidally as shown in fig. 9.8.

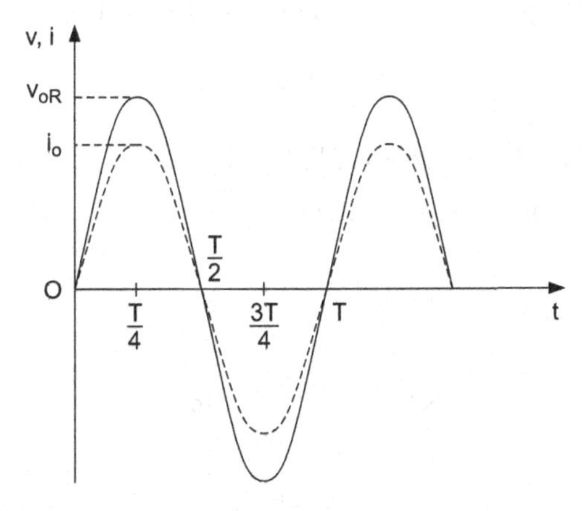

Fig. 9.8. Current through a resistor in ac circuit.

The two curves attain their maximum values and minimum values simultaneously. Thus, two waves are said to be in same phase. The phasor diagram for ac circuit containing pure resistance is shown in fig. 9.9.

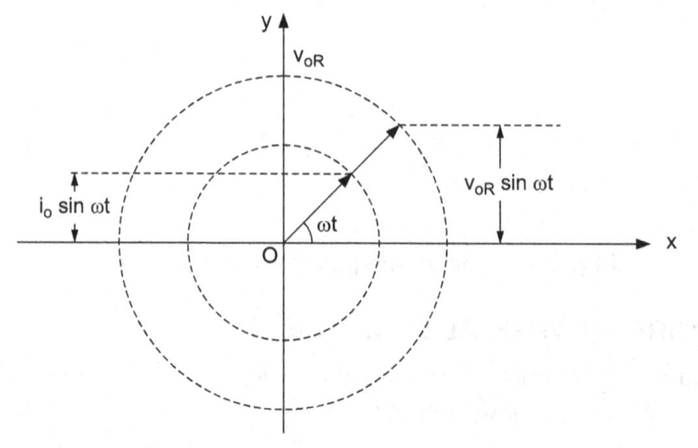

Fig. 9.9. Phasor diagram for current i_0 and voltage v_{0R}.

The value of i_0 and v_{0R} are shown as vectors rotating counterclockwise. The projections of i_0 and v_{0R} on *y*-axis give the instantaneous values of *i* and v_R. The two rotating vectors which represent the current and voltage coincide each other as shown in fig. 9.9.

Example 9.2. AC current passes through a resistance of 20Ω from a ac supply of maximum value of voltage 280V and frequency 50 Hz. Obtain instantaneous value of current.

Solution: Here $f = 50$ Hz

$$\therefore \qquad \omega = 2\pi f = 2 \times 3.14 \times 50 = 314 \text{s}^{-1}.$$

$$\text{a.c. voltage } v = v_{0R} \sin \omega t = 280 \sin 314t \text{ volt}$$

Therefore,
$$i = \frac{v}{R} = 14 \sin 314t \text{ Amp.}$$

9.4 AC CIRCUIT WITH CAPACITANCE

We apply an alternating current to a capacitor of capacitance *C* as shown in fig. 9.10.

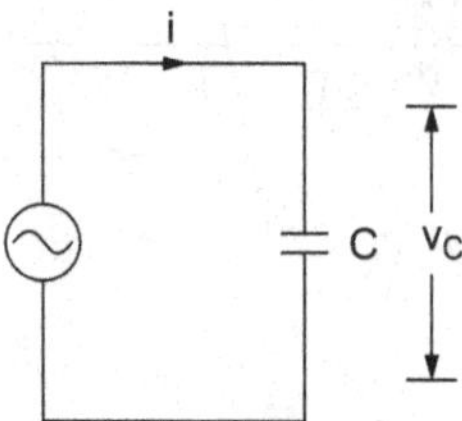

Fig. 9.10. An ac through capacitor.

The current through the circuit at any instant is given by

$$i = i_0 \sin \omega t \quad ...(9.14)$$

The instantaneous voltage v_C across the capacitor is

$$v_C = \frac{q}{C} = \frac{1}{C}\int i\, dt \quad ...(9.15)$$

$$v_C = \frac{1}{C}\int i_0 \sin \omega t\, dt$$

$$= \frac{-i_0}{C\omega} \cos \omega t$$

$$v_C = \frac{i_0}{\omega C} \sin (\omega t - 90)$$

$$v_C = v_{0C} \sin(\omega t - 90) \quad ...(9.16)$$

where $v_{0C} = i_0\left(\frac{1}{\omega C}\right)$ and the quantity $\left(\frac{1}{\omega C}\right)$ is known as capacitive reactance of the condenser. It is measured in ohms. Fig. 9.11 shows a plot of i and v_C.

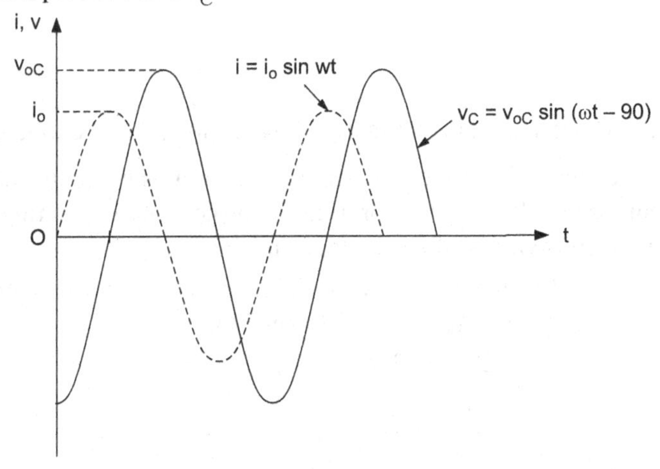

Fig. 9.11. Plots of current and voltage through capitor.

From Fig. 9.11, it is clear that the voltage v_C across the capacitor is not in phase with the current. The current leads the voltage by 90°, or current varies a quarter cycle. This implies that the energy is not dissipated. For positive half cycle voltage and current have same sign while for negative half cycle of ac, they have opposite sign, so that average power is zero. The ideal capacitor does not contain resistance, no electrical energy is dissipated in terms of heat.

The quantity $\left(\frac{1}{\omega C}\right)$ behaves like a resistance and it is represented by X_C,

$$X_C = \frac{1}{\omega C} = \frac{1}{2\pi f C} \qquad ...(9.17)$$

Here, X_C is not constant, it decreases with increasing f and C. Therefore, the alternating current passes easily through the capacitor. But for *dc*, $f = 0$, it offers infinite resistance, hence, *dc* does not pass through the capacitor.

The phasor diagram of capacitor is shown in fig. 9.12.

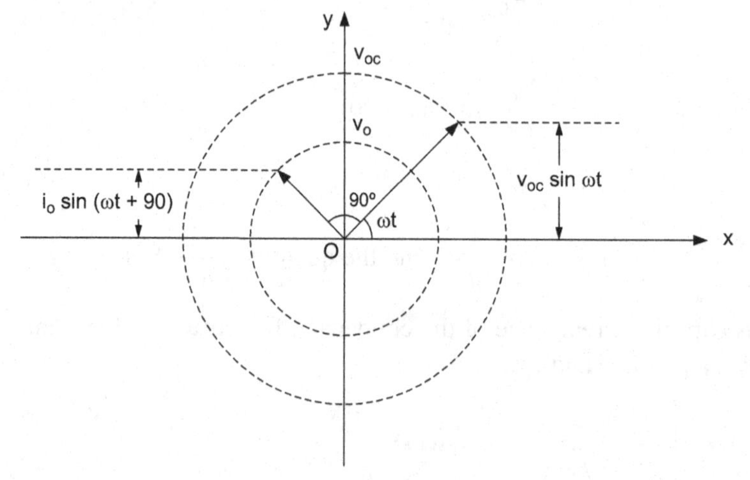

Fig. 9.12. Current leads voltage by phase angle π/2 in a capacitor.

The phase relationship between voltage and current is given in such a way that the current leads voltage by 90° or π/2. The projections of rotating vectors give the instantaneous values of the voltage and current.

Example 9.3. Calculate the reactance and the current when a voltage across the capacitor of capitance 10 μ*f* is $v_C = 280 \sin 314t$.

Solution: $v_C = 280 \sin 314t$

$$\therefore \qquad X_C = \frac{1}{\omega C} = \frac{1 \times 10^6}{314 \times 10} = 318.5\Omega$$

Now,

$$i = \frac{280}{318.5} \sin(314t + 90°) = 0.88 \sin(314t + \pi/2).$$

9.5 AC CIRCUIT WITH INDUSTANCE

An ac circuit with source and an inductor is shown in fig. 9.13.

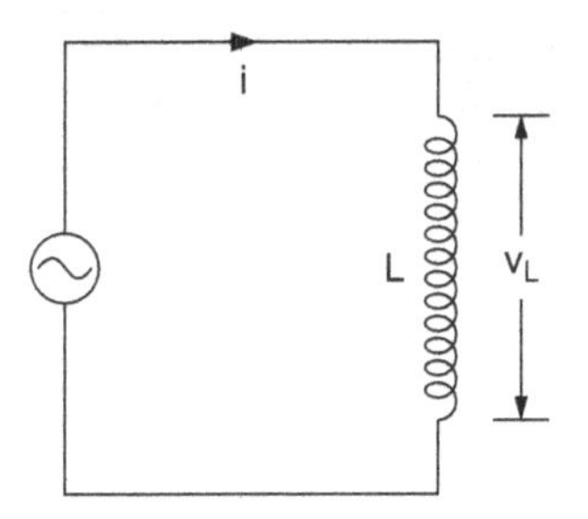

Fig. 9.13. An ac flowing through an inductor.

If the inductor consists of self-inductance, it opposes the alternating current to flow, because of nature of ac, which is continuously changing.

Let the current through the circuit be

$$i = i_0 \sin \omega t \qquad ...(9.18)$$

The voltage drop across the coil is

$$v_L = L\frac{di}{dt} \qquad ...(9.19)$$

Here, L is the inductance of the inductor.

Now, $$v_L = \frac{d}{dt}(i_0 \sin \omega t)$$

$$v_L = Li_0\, \omega \cos \omega t \qquad ...(9.20)$$

we may write the equation (9.20) as

$$v_L = v_{0L} \sin (\omega t + \pi/2) \qquad ...(9.21)$$

where, $$v_{0L} = i_0\,(\omega L) \qquad ...(9.22)$$

The resultant voltage across the inductor is always zero, thus, energy is not dissipated therein.

The plots of current and voltage for a coil is shown in fig. 9.14. This indicates that the current and voltage are not in phase, but voltage drop v_L across the inductor leads the current by 90° or $\pi/2$.

Again, Fig. 9.14 shows that voltage leads the current by a quarter cycle or a phase angle 90°. The quantity $\omega L = 2\pi fL$ behaves like a resistance and it is known as inductive reactance, X_L.

$$X_L = \omega L = 2\pi fL \qquad ...(9.23)$$

Here, the units of X_L are ohms and its magnitude increase with increasing frequency f of *ac*.

For dc, the frequency $f = 0$, it offers zero reactance and dc may pass easily through inductor. Since $X_L \propto fL$, it offers reactance of ac, and greater the frequency of ac, the magnetic flux changes more rapidly in the inductor. Moreover,

$$v_L = X_L \, i_{rms} \qquad ...(9.24)$$

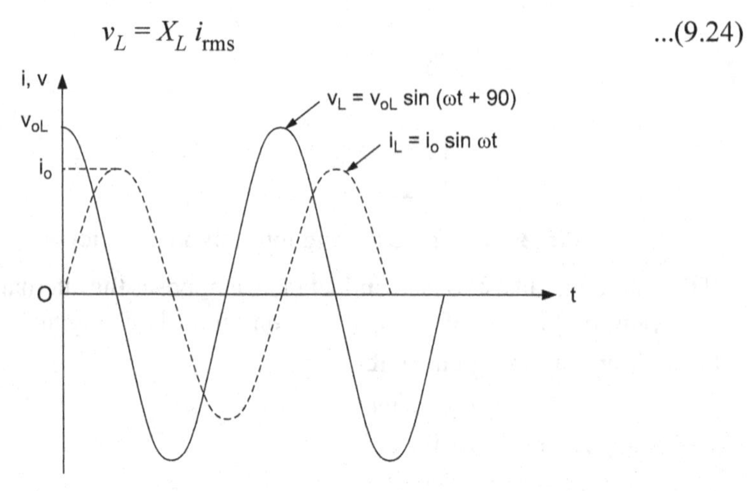

Fig. 9.14. Plots of current and voltage through an inductor.

The phasor diagram for an inductor in ac circuit is shown in fig 9.15, which indicates that current lags voltage by a quarter cycle or 90°.

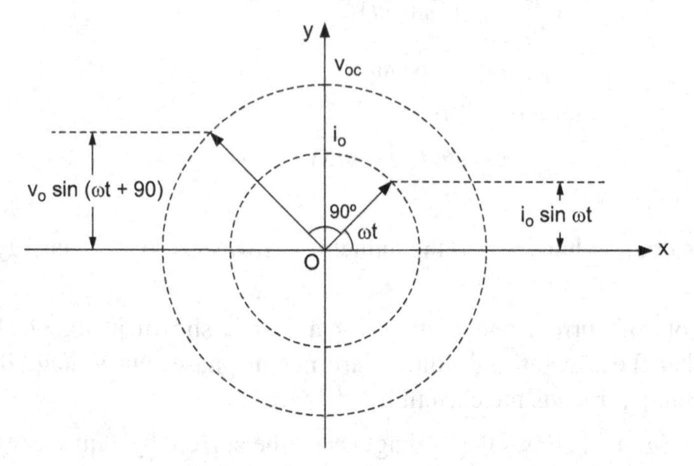

Fig. 9.15. Phasor diagram for current and voltage in inductor.

Example 9.4. An ac, $i = 5 \sin(314\, t)$ is passing through an inductor of inductance 0.3 H. Determine inductive reactance and v_L.

Solution: $X_L = \omega_L = 314 \times 0.3$ ohms

$= 94.2$ ohms.

$v_L = X_L \, i_{rms} = 94.2 \times 3.54$

$= 333$ volts.

9.6 AC CIRCUIT WITH R AND C IN SERIES

A series connection of a resistor and a capacitor in ac circuit is shown in fig. 9.16.

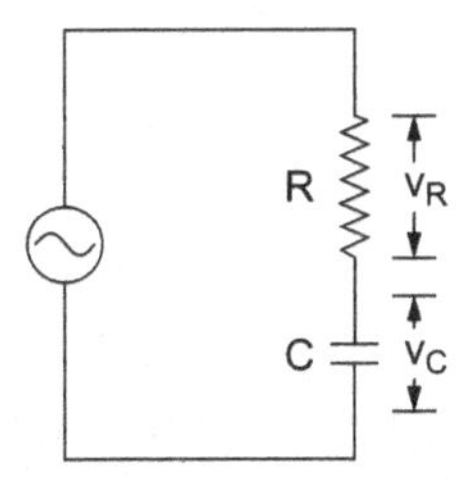

Fig. 9.16. R and C in series.

If $i = i_0 \sin \omega t$ is the instantaneous value of current through the circuit containing R and C, then voltage drops across R and C will be

$$v_R = Ri = Ri_0 \sin \omega t \qquad ...(9.25)$$

and
$$v_C = \frac{i_0}{\omega C} \sin(\omega t - \pi/2) \qquad ...(9.26)$$

Thus,

$$v = v_R + v_C$$

$$v = R\, i_0 \sin \omega t + \frac{i_0}{\omega C} \sin(\omega t - \pi/2) \qquad ...(9.27)$$

If we substitute,

$$\left.\begin{aligned} R &= Z \cos\phi \\ \frac{1}{\omega C} &= Z \sin\phi \end{aligned}\right\} \qquad ...(9.28)$$

Then, equation (9.27) comes out to be

$$v = Zi_0 \sin(\omega t - \phi) \qquad ...(9.29)$$

where, $Z = \sqrt{R^2 + \frac{1}{\omega^2 C^2}}$ is known as impedance of the combination, and

$$\tan \phi = \frac{1}{RC\omega} \qquad ...(9.30)$$

From, equation (9.29), it is clear that the voltage v lags behind the current by an angle ϕ, as given in fig. 9.17. (phase diagram).

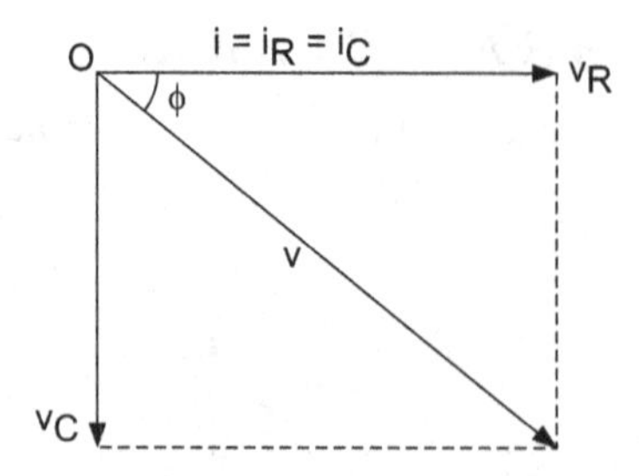

Fig. 9.17. Phasor diagram of *v* and *i*.

Phasor diagram shows that the phase shift is between 0° and –90°. It tends to 0° for high frequencies and –90° for low frequencies.

If we take a parallel combination of *R* and *C*, then

$$\text{admittance} = \sqrt{\frac{1}{R^2} + \omega^2 C^2} \qquad \text{...(9.31)}$$

9.7 AC CIRCUIT WITH *R* AND *L* IN SERIES

Consider a resistor *R* in series with an inductor *L* with an alternating current source $i = i_0 \sin \omega t$, as shown in fig. 9.18.

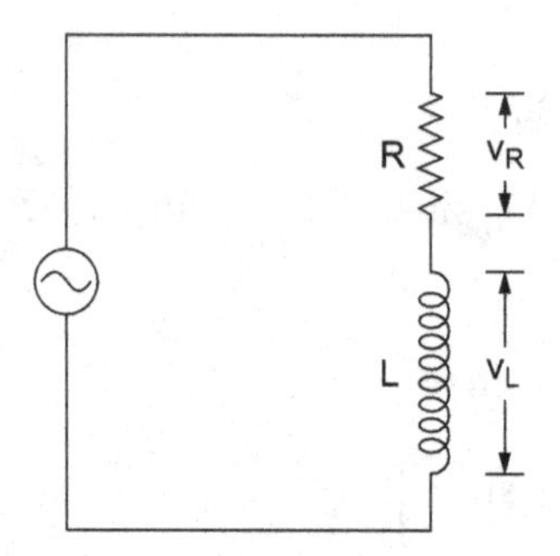

Fig. 9.18. R and L in series.

An inductor coil is one which passes both resistance and self inductance. If v_R and v_L are voltage drops across resistor *R* and inductor *L*. Then,

$$v_R = Ri = Ri_0 \sin \omega t \qquad \text{...(9.32)}$$

and

$$v_L = X_L i = \omega L \sin (\omega t + \pi/2) \qquad \text{...(9.33)}$$

Now, the instantaneous value of the voltage across the combination of *R* and *L*, is given by

$$\begin{aligned} v &= v_R + v_L \\ &= Ri_0 \sin \omega t + X_L\, i_0 \sin (\omega t + \pi/2) \qquad \text{...(9.34)} \\ &= Ri_0 \sin \omega t + X_L\, i_0 [\sin \omega t \cos \pi/2 + \cos \omega t \sin \pi/2] \end{aligned}$$

$$v = Ri_0 \sin \omega t + X_L\, i_0 \cos \omega t \qquad ...(9.35)$$

If we substitute,

$$\left.\begin{aligned} R &= Z \cos\phi \\ X_L &= Z \sin\phi \end{aligned}\right\} \qquad ...(9.36)$$

we get,

$$v = z\, i_0 \sin(\omega t + \phi) \qquad ...(9.37)$$

where,

$$Z = \sqrt{R^2 + X_L^2} \qquad ...(9.38)$$

$$Z = \sqrt{R^2 + \omega^2 L^2} \qquad ...(9.39)$$

and

$$\tan\phi = \frac{X_L}{R} = \frac{\omega L}{R} \qquad ...(9.40)$$

Here, $X_L = \omega L$ is called the inductive reactance of the coil and Z is the impedance.

The applied voltage leads the current by an angle ϕ and this is shown in phase diagram, fig. 9.19. It is clear that the phase shift is between 0° and 90°. It approaches 0° for law frequencies and 90° for high frequencies.

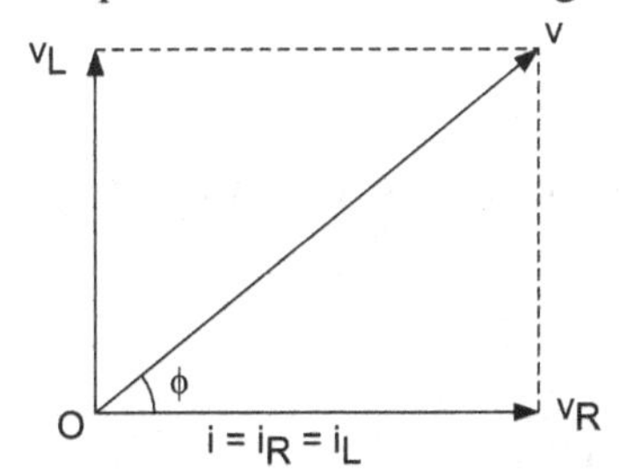

Fig. 9.19. Phasor diagram of *v* and *i*

If we take a parallel combination of R and L, then

$$\text{admittance} = \sqrt{\frac{1}{R^2} + \frac{1}{\omega^2 L^2}} \qquad ...(9.41)$$

9.8 AC CIRCUIT WITH L AND C IN SERIES

An alternating circuit containing an inductor and a capacitor in series is shown in fig. 9.20.

Suppose the current through the circuit at any instant is given by

$$i = i_0 \sin \omega t \qquad ...(9.42)$$

If v_L and v_C are the voltage drops across the inductor and the capacitor, then,

$$v = v_L + v_C \qquad ...(9.43)$$

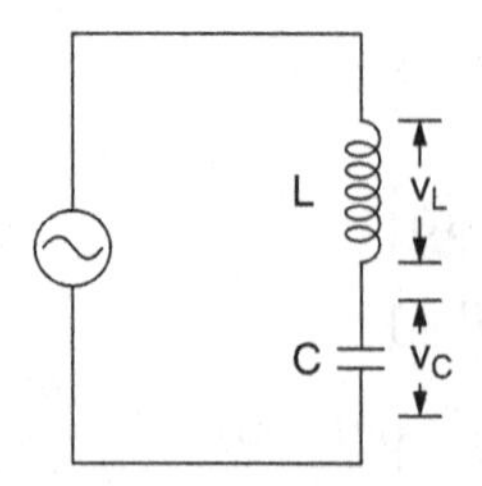

Fig. 9.20. L and C in series

Here,
$$v_L = x_L\, i$$
$$v_C = X_C\, i$$

then,
$$v = X_L i + X_C i \qquad ...(9.44)$$
$$= \left(\omega L - \frac{1}{\omega C}\right) i \qquad ...(9.45)$$

$\therefore$
$$v = \left(\omega L - \frac{1}{\omega C}\right) i_0 \sin \omega t \qquad ...(9.46)$$

Now
$$v = Z i_0 \sin \omega t \qquad ...(9.47)$$

where
$$Z = \left(\omega L - \frac{1}{\omega C}\right) \qquad ...(9.48)$$

The specific driving frequency at which the amplitude of the current is maximum

$$X_L = X_C \qquad ...(9.49)$$
$$\omega_L = \frac{1}{\omega C} \qquad ...(9.50)$$

which gives
$$f = \frac{1}{2\pi\sqrt{LC}} \qquad ...(9.51)$$

This is known as natural frequency of the LC circuit.

Phasor diagram of *LC* circuit is shown in fig. 9.21.

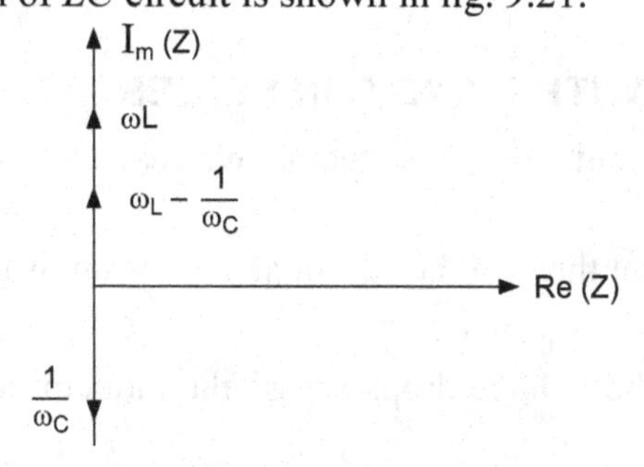

Fig. 9.21. Phasor diagram of LC circuit.

9.9 AC CIRCUIT WITH LCR IN SERIES

Now we do an ac circuit that contains, a resistor R, a capacitor C and an inductor L as shown in fig. 9.22.

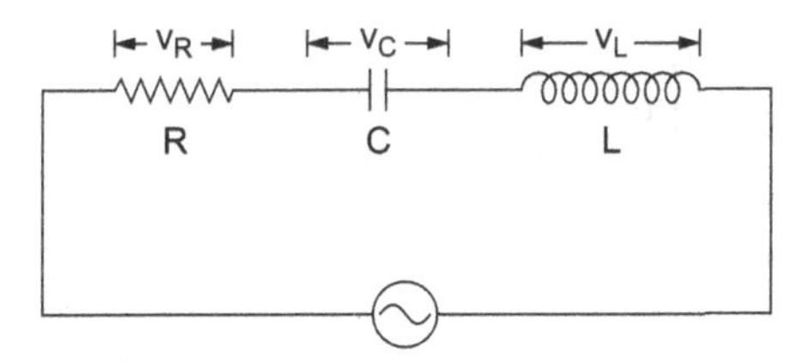

Fig. 9.22. An ac circuit with R, C, L in series.

Let the instantaneous current i in the circuit is given by

$$i = i_0 \cos \omega t \qquad ...(9.52)$$

Suppose that v_R, v_L and v_C are the instantaneous value of the voltages across R, L and C respectively. Then, we may write

$$v = v_R + v_C + v_L \qquad ...(9.53)$$

Since we known that:

(a) In resistive circuit, the current and the voltage are in same phase.

(b) In capacitive circuit, current leads the voltage by 90°.

(c) While in inductive circuit, current lags the voltage by 90°.

Then, we have,

$$v_R = v_{0R} \cos \omega t = Ri_0 \cos \omega t,$$

$$v_C = v_{0C} \cos (\omega t - \pi/2) = \frac{i_0}{\omega C} \cos (\omega t - \pi/2),$$

and $$v_L = v_{0L} \cos (\omega t + \pi/2) = i_0\, \omega L \cos (\omega t + \pi/2)$$

where $v_{0R} = Ri_0$, $v_{0C} = \dfrac{i_0}{\omega C}$ and $v_{0L} = i_0 \omega L$ are the peak values of the voltages that drop across R, C and L respectively. Substituting the values of v_R, v_C and v_L in equation (9.53), we get,

$$v = Ri_0 \cos \omega t + \frac{i_0}{\omega C} \cos \left(\omega t - \frac{\pi}{2} \right) + i_0 \omega L \cos \left(\omega t + \frac{\pi}{2} \right) \qquad ...(9.54)$$

$$= i_0 \left[R \cos \omega t - \left(\omega L - \frac{1}{\omega C} \right) \sin \omega t \right] \qquad ...(9.55)$$

Let $$\left.\begin{aligned} R &= Z \cos \phi \\ \left(\omega L - \frac{1}{\omega C} \right) &= Z \sin \phi \end{aligned}\right\} \qquad ...(9.56)$$

and

then $$v = Zi_0 \cos(\omega t + \phi) \quad ....(9.57)$$

where, $$Z = \sqrt{R^2 + \left(\omega L - \frac{1}{\omega C}\right)^2} \quad ...(9.58)$$

and $$\tan\phi = \frac{\left(\omega L - \frac{1}{\omega C}\right)}{R} \quad ...(9.59)$$

Again, $$X_C = \frac{1}{\omega C}$$

and $$X_L = \omega L$$

Equation (9.58) and (9.59) may re-write as

$$\boxed{Z = \sqrt{R^2 + (X_L - X_C)^2}} \quad ...(9.60)$$

and $$\boxed{\tan\phi = \frac{(X_L - X_C)}{R}} \quad ...(9.61)$$

Also, equation (8.57) may be re-written as

$$\boxed{v = v_0 \cos(\omega t + \phi)} \quad ...(9.62)$$

where, $$v_0 = i_0 Z \quad ...(9.63)$$

$Z = \sqrt{R^2 + (X_L - X_C)^2}$ is known as the impadance of the circuit.

The equation (9.60) is represented by the diagram shown in fig. 9.23.

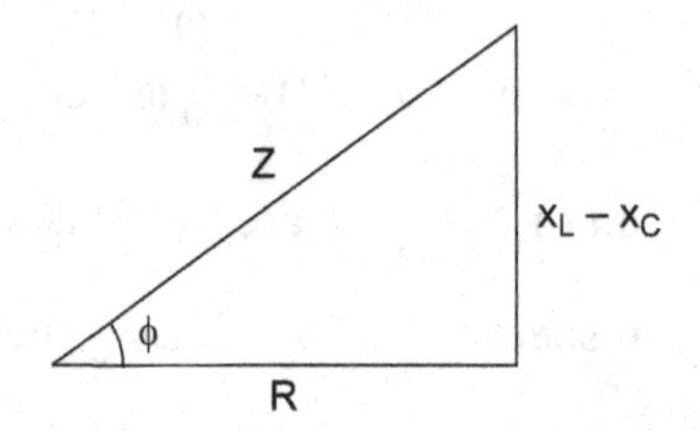

Fig. 9.23. Relation between *R*, X_L, X_C and *Z*.

The equations (9.52) for current and (9.62) for voltage indicate that voltage and current are out of phase by an angle ϕ. The angle ϕ depends on the values of X_L, X_C and *R*.

There are three cases for the angle ϕ, which are

(a) If $X_L = X_C$. Then

$$\tan\phi = 0$$

$$\therefore \quad \phi = 0$$

then impedance $Z = R$, and the voltage and the current are in same phase.

(b) If $X_L > X_C$. Then, tan ϕ will be positive. Hence ϕ will be positive. In this situation, voltage leads the current be angle ϕ. In a purely inductive circuit, $R \to 0$ and $C \to \infty$ that gives $X_C = 0$ and $R = 0$. Therefore, $\phi = 90°$.

(c) If $X_L < X_C$. Then ϕ will be negative, hence voltage lags the current by angle ϕ. In a purely capacitance circuit, $L \to 0$ and $R \to 0$. Thus, $Z = \dfrac{1}{\omega C}$ and $\phi = -\pi/2$.

9.10 SERIES RESONANCE

In a series resonance circuit, R, C and L are connected in a series with ac source. The rms value of current is given by

$$i_{rms} = \frac{v_{rms}}{Z} = \frac{v_{rms}}{\sqrt{R^2 + (X_L - X_C)^2}} \qquad ...(9.64)$$

when $X_L = X_C$, the voltage and current are in same phase.

$$\therefore \qquad \omega_0 L = \frac{1}{\omega C} \qquad ...(9.65)$$

$$\therefore \qquad \omega_0^2 = \frac{1}{LC} \qquad ...(9.66)$$

This condition determines the specific frequency for which inductive reactance cancels the capacitive reactance.

From equation (9.66),

$$\boxed{f_0 = \frac{1}{2\pi\sqrt{LC}}} \qquad ...(9.67)$$

The frequency f_0 is known as resonant frequency. At $X_L = X_C$, $Z = R$(minimum). Thus, current becomes maximum. The amplitude of the current

$$i_0 = \frac{v_0}{Z} = \frac{v_0}{R} \qquad ...(9.68)$$

$$\therefore \qquad \left.\begin{aligned} \tan\phi = 0 \\ \phi = 0 \end{aligned}\right\} \qquad ...(9.69)$$

At resonance frequency f_0 the voltages across capacitor C and inductor L are equal and opposite. The variation of current as a function of frequency is shown in fig. 9.24. The curves obtained by plotting current against the frequency ω are known as resonance curves. Here three plots of current i versus ω for different values of resistance R are shown in fig. 9.24.

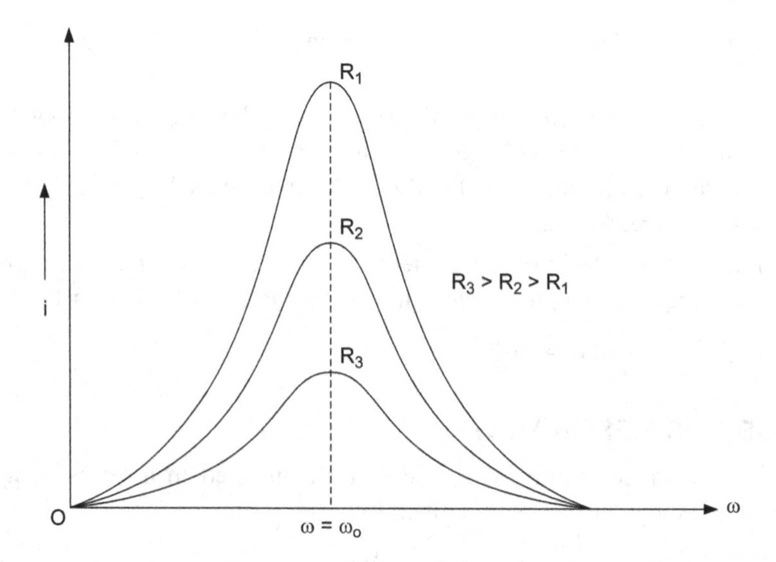

Fig. 9.24. The plots of current versus ω for different values of R.

From fig. 9.24, it is observed that

(a) the maximum current occurs at $\omega = \omega_0$.

(b) the resistance of the circuit is real and has least value at $\omega = \omega_0$.

(c) the peak of any curve depicts the selectivity or sharpness of the circuit.

(d) for $\omega < \omega_0$, the voltage lags the current and for $\omega > \omega_0$, the voltage leads the current.

(e) at $\omega = \omega_0$, the voltage and current are in same phase.

9.11 POWER IN AC CIRCUITS

The power is dissipated for resistors while capacitors and inductors store and release energy to the circuit, there is no dissipation of energy due to capacitors and inductors. When current flows through the circuit, the inductor stores magnetic energy and this energy is fed from the inductor back into circuit when current tends to zero. Similarly, capacitor stores electric energy during charging. When capacitor is discharged, the electric energy is fed back to the circuit.

Now power dissipation for a resistor is given by

$$P = vi \quad \text{...(9.70)}$$

where v and i are real function of time.

For voltage and current, we may write,

$$v = v_0\, e^{i\omega t} = v_0 \cos \omega t \quad \text{...(9.71)}$$

and $$i = i_0\, e^{i(\omega t + \phi)} = e^{i\omega t}\, e^{i\phi} \qquad ...(9.72)$$
$$= i_0\, (\cos \omega t + i \sin \omega t)\, (\cos \phi + i \sin \phi)$$

The real value of i is given by

$$i = i_0\, (\cos \omega t \cos \phi - \sin \omega t \sin \phi) \qquad ...(9.73)$$

From equation (9.70), the power is

$$P = v_0\, i_0\, (\cos^2 \omega t \cos \phi - \cos \omega t \sin \omega t \sin \phi) \qquad ...(9.74)$$

The average power over a complete cycle is given by

$$P_{av} = \frac{\int_0^{2\pi/\omega} P\, dt}{\int_0^{2\pi/\omega} dt} = \frac{1}{2} v_0 i_0 \cos\phi \qquad ...(9.75)$$

Since $\langle \sin^2 \omega t) = \dfrac{1}{2}$ and $\langle \sin \omega t \cos \omega t \rangle = 0$

$$P_{av} = \frac{v_0}{\sqrt{2}} \cdot \frac{i_0}{\sqrt{2}} \cos\phi = v_{\text{rms}}\, i_{\text{rms}} \cos\phi$$

$$\therefore \qquad \boxed{P_{av} = v_{\text{rms}}\, i_{\text{rms}}\, \text{cas}\, \phi} \qquad ...(9.76)$$

The factor $\cos \phi$ is known as power factor of the circuit

Again, $$\tan \phi = \frac{\left(\omega L - \dfrac{1}{\omega C}\right)}{R}$$

Then, $$\cos \phi = \frac{R}{\sqrt{R^2 + \left(\omega L - \dfrac{1}{\omega L}\right)^2}}$$

$$\cos \phi = \frac{R}{Z} \qquad ...(9.77)$$

For a complete cycle of an ac current the power consumed by L and C is zero. The equation (9.76) represents the power consumed by the resistor only.

$\cos \phi$ is always positive when $\dfrac{-\pi}{2} < \phi < \dfrac{\pi}{2}$, this means that the energy per unit time is always expanded by the source.

$$\therefore \qquad P_{av} = v_{\text{rms}}\, i_{\text{rms}} \cos \phi$$
$$= Z\, i_{\text{rms}}^2 \cdot \frac{R}{Z} = i_{\text{rms}}^2 R \qquad ...(9.78)$$

Thus, all the energy supplied to the ac circuit by the source is dissipated as the heat in the resistor R at the average rate of $i_{rms}^2 R$.

For resistor R only, $Z = R$, hence $\phi = 0$

$$\therefore \quad P_{av} = v_{rms}\, i_{rms}$$

If $R = 0$ and ac circuit must contain L and C, $\phi = 90°$ or $\frac{\pi}{2}$,

$$\cos\phi = 0$$

hence,
$$P_{av} = 0$$

Then, current flowing through the circuit is known as wattless current and the average power consumed in the circuit is zero.

9.12 AC CIRCUIT WITH LCR IN PARALLEL

Consider an ac circuit that contains R, L and C in parallel as shown in fig. 9.25.

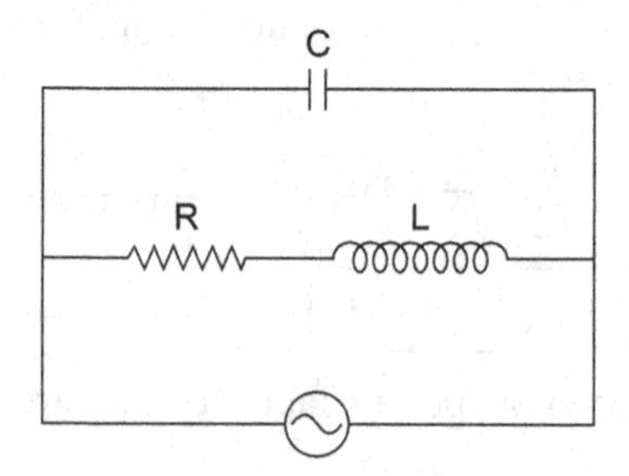

Fig. 9.25. An ac circuit with R, L and C in parallel connection.

The impedance of the circuit may be calculated as

$$\frac{1}{Z} = \frac{1}{R + j\omega L} + \frac{j\omega C}{1}$$

$$= \frac{1}{R + j\omega L} + j\omega C \qquad ...(9.79)$$

$$\frac{1}{Z} = \frac{(R - j\omega L)}{(R + j\omega L)(R - j\omega L)} + j\omega C \qquad ...(9.80)$$

we have multiplied by $(R - j\omega L)$ in numerator and denominator.

Now,

$$\frac{1}{Z} = \frac{R}{(R^2 + \omega^2 L^2)} - \frac{j\omega L}{(R^2 + \omega^2 L^2)} + j\omega C$$

$$= \frac{R}{(R^2 + \omega^2 L^2)} + j\left(\omega C - \frac{\omega L}{(R^2 + \omega^2 L^2)}\right) \qquad ...(9.81)$$

For resonant frequency of the circuit, the imaginary term of the equation (9.81) should be equal to zero.

$$\omega_0 C - \frac{\omega_0 L}{(R^2 + \omega_0^2 L^2)} = 0 \qquad ...(9.82)$$

$$\Rightarrow \qquad \omega_0^2 = \left(\frac{1}{LC} - \frac{R^2}{L^2}\right)$$

$$\boxed{\omega_0 = \sqrt{\frac{1}{LC} - \frac{R^2}{L^2}}} \qquad ...(9.83)$$

This is the resonant frequency for which the current and voltage are in same phase. For resonance to occur, ω_0 should be real, i.e.

$$\frac{1}{LC} > \frac{R^2}{L^2} \qquad ...(9.84)$$

$$\Rightarrow \qquad \frac{L}{C} > R^2$$

$$\therefore \qquad \sqrt{\frac{L}{C}} > R \qquad ...(9.85)$$

Now, current may be found as

$$i_0 = \frac{v_0}{Z} \qquad ...(9.86)$$

Since, $$\frac{1}{Z} = \frac{1}{(R + j\omega L)} + j\omega C$$

and its complex conjugate is given by

$$\therefore \qquad \frac{1}{Z^*} = \frac{1}{(R - j\omega L)} - j\omega C$$

on multiplying we get

$$\frac{1}{Z^2} = \left(\frac{1}{R + j\omega L} + j\omega C\right)\left(\frac{1}{R - j\omega L} - j\omega C\right) \qquad ...(9.87)$$

$$= \left(\frac{1}{R^2 + \omega^2 L^2} + \omega^2 C^2\right) + j\omega C\left(\frac{1}{R - j\omega L} - \frac{1}{R + j\omega L}\right)$$

$$\frac{1}{Z^2} = \frac{1}{(R^2 + \omega^2 L^2)} + \omega^2 C^2 - \frac{2\omega^2 LC}{R^2 + \omega^2 L^2}$$

$$\Rightarrow \quad \frac{1}{Z^2} = \frac{1}{R^2+\omega^2L^2}\left[\omega^2R^2C^2+(1-\omega^2LC)^2\right] \quad ...(9.88)$$

Then,

$$\frac{1}{Z} = \sqrt{\frac{\omega^2R^2C^2+(1-\omega^2LC)^2}{R^2+\omega^2L^2}} \quad ...(9.89)$$

Substituting in equation (9.86) we get

$$\boxed{i_0 = v_0\sqrt{\frac{\omega^2R^2C^2+(1-\omega^2LC)^2}{R^2+\omega^2L^2}}} \quad ...(9.90)$$

The current i_0 is minimum at $\omega^2LC = 1$. From above expression of the current, it appears that the current is minimum at a slightly lower frequency, and it rejects the current to pass through the circuit. Hence, parallel resonance circuit is known as a rejector circuit. The parallel *LC* circuit offers maximum *Z* for resonance condition.

9.13 QUALITY FACTOR Q OF SERIES RESONANCE

The quality factor *Q* of series resonant circuit is defined as the ratio of the voltage across inductor *L* to the voltage across resistor *R*. That is,

$$Q = \frac{v_L}{v_R} = \frac{\omega_0 L}{R} \quad ...(9.91)$$

where $$\omega_0 = \frac{1}{\sqrt{LC}}$$

$$\therefore \quad \boxed{Q = \frac{1}{R}\sqrt{\frac{L}{C}}} \quad ...(9.92)$$

Now, the average power is

$$P_{av} = v_{\text{rms}}\, i_{\text{rms}} \cos\phi \quad ...(9.93)$$

and $$\cos\phi = \frac{R}{Z},\ i_{\text{rms}} = \frac{v_{\text{rms}}}{Z}$$

Then,

$$P_{av} = \frac{v_{\text{rms}}^2 R}{Z^2}$$

$$\Rightarrow \quad P_{av} = \frac{v_{\text{rms}}^2 R}{R^2+\left(\omega L - \frac{1}{\omega C}\right)^2}$$

$$= \frac{v_{\text{rms}}^2 R}{R^2 + \frac{L^2}{\omega^2}\left(\omega^2 - \frac{1}{LC}\right)^2} \qquad \text{...(9.94)}$$

Since, resonance frequency,

$$\omega_0^2 = \frac{1}{LC}$$

substituting it in equation (9.94), we get

$$P_{av} = \frac{v_{\text{rms}}^2 R}{R^2 + \frac{L^2}{\omega^2}(\omega^2 - \omega_0^2)^2}$$

$$\Rightarrow \qquad \boxed{P_{av} = \frac{v_{\text{rms}}^2 R\omega^2}{\omega^2 R^2 + L^2(\omega^2 - \omega_0^2)^2}} \qquad \text{...(9.95)}$$

The variation of the average power as a function of the frequency is shown in fig. 9.26.

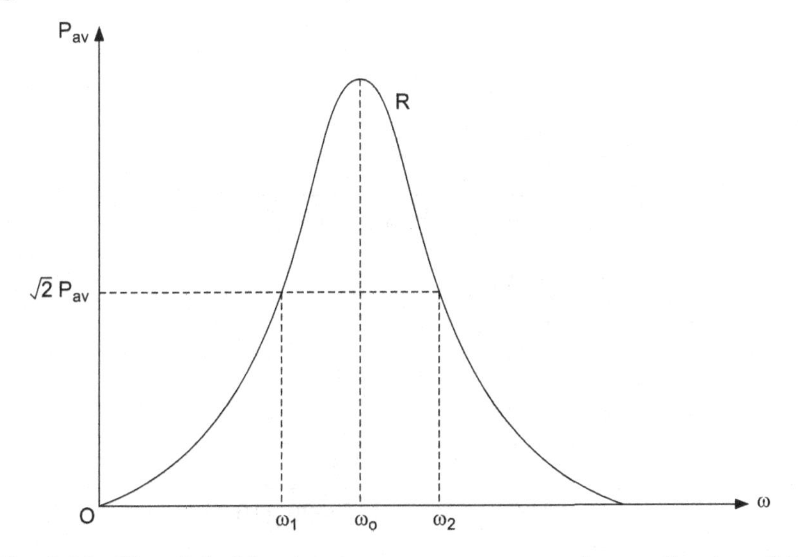

Fig. 9.26. The plot of the average power versus ω for small value of *R*.

when *R* is small, the quality factor *Q* is large. The full width at half maximum determines the *Q*.

$$Q = \frac{\omega_0}{\omega_2 - \omega_1} = \frac{\omega_0}{\Delta\omega} \qquad \text{...(9.96)}$$

where $\Delta\omega$ is known as resonance width. The full width at half maximum determines the sharpness of the resonance (at $\sqrt{2}$ times of the peak), which is the half power. Hence, a large value of Q gives narrow and sharp resonance curve.

Example 9.5. An ac circuit with *L, C* and *R* is shown in fig. 9.27.

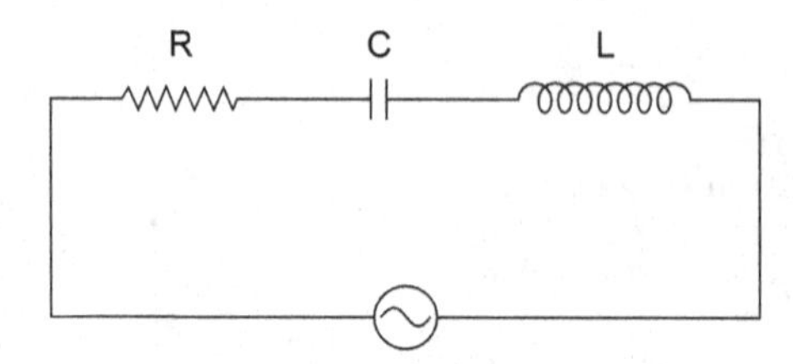

Fig. 9.27. LCR in series.

Here, $v = 120 \cos 314t$; $R = 10\Omega$, $L = 40$ mH and $C = 20\mu$F. Then Find

(a) Impedance of the circuit.

(b) Phase angle ϕ.

(c) Resonant frequency ω_0.

Solution:

$$X_L = \omega L = 314 \times 40 \times 10^{-3}\Omega = 12.56\Omega$$

$$X_C = \frac{1}{\omega C} = \frac{1}{314 \times 20 \times 10^{-6}}\Omega = 159.23\Omega$$

(a) $\therefore$

$$Z = \sqrt{R^2 + (X_L - X_C)^2}$$

$$= \sqrt{(10)^2 + (12.56 - 159.23)^2}$$

$$Z = 147\ \Omega$$

(b)

$$\tan\phi = \frac{X_L - X_C}{R}$$

$$\phi = -1°\ 39'$$

(c)

$$\omega_0 = \frac{1}{\sqrt{LC}} = \frac{1}{\sqrt{40 \times 10^{-3} \times 20 \times 10^{-6}}}$$

$$= \frac{1}{\sqrt{8 \times 10^{-7}}}$$

$$\omega_0 = 1118\ \text{sec}^{-1}.$$

9.14 TRANSFORMERS

We have many situations, where low voltage or high voltage is needed. For example home appliances need a voltage of 200V. In electronic equipments, we need either 18 V or 12V. A transformer is a device that transform an ac at high voltage into low voltage and vice versa with very small loss of power. Ideal transformer decreases or increases an ac voltage without loss of power. If the output voltage is smaller than the input voltage, it is called step down transformer, moreover, if the output voltage of a transformer is higher than the input voltage, it is called step up transformer.

A schematic diagram of transformer is shown in fig. 9.28. It consists of two coils known as primary and secondary coils. These coils are electrically insulated from each other and wound on a same soft iron core which is laminated to prevent the eddy-current losses.

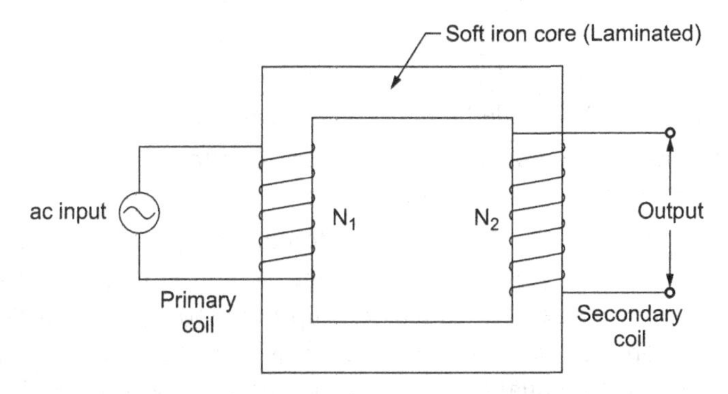

Fig. 9.28. A typical view of a transformer.

The symbolic view of a transformer is shown in fig. 9.29.

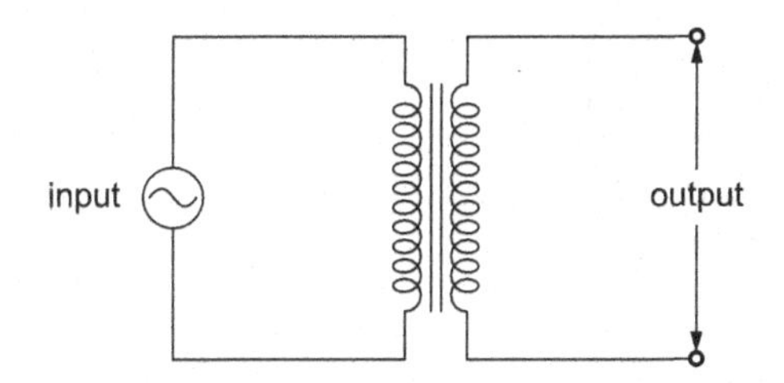

Fig. 9.29. Conventional view of transformer.

The working of transformer is based on the principle of mutual induction. That is, the magnetic flux produced by an ac in primary coil passes through secondary coil. When an ac current flows through the primary coil produces an induced emf (voltage) in the secondary coil. The frequency of the voltage in the secondary coil will be same as that in primary coil.

From Faraday's law, the rate of change of magnetic flux in primary coil is equal to emf (voltage) produced, then,

$$V_1 = -N_1 \frac{d\phi}{dt} \qquad ...(9.97)$$

where N_1 is the number of turns in the primary coil. This flux is linked to the secondary coil, so that, the emf or voltage produced in the secondary coil will be

$$V_2 = -N_2 \frac{d\phi}{dt} \qquad ...(9.98)$$

The magnetic flux in two coils will be same. From the equations (9.97) and (9.98), we have

$$\boxed{\frac{V_1}{V_2} = \frac{N_1}{N_2}} \qquad ...(9.99)$$

The equation (3.99) is called transformer equation.

Hence, the ratio of voltages in the primary coil to the secondary coil is equal to the ratio of their corresponding terms.

If $N_1 > N_2$, then $V_1 > V_2$. Thus, it is a step down transformer. Moreover, if $N_1 < N_2$, $V_1 < V_2$, the voltage in the secondary coil is larger than voltage in the primary coil, hence it is called step up transformer. A well designed transformer is more efficient. In the otherwords, we may say that the input power in primary coil is equal to the output power in the secondary. Thus,

$$I_1 V_1 = I_2 V_2$$

$$\boxed{\frac{V_1}{V_2} = \frac{I_2}{I_1}} \qquad ...(9.100)$$

From the equations (9.99) and (9.100) we get,

$$\boxed{\frac{N_1}{N_2} = \frac{I_2}{I_1}} \qquad ...(9.101)$$

Here, I_1 and I_2 are rms values of the currents.

The equation (9.101) reveals that the currents in the primary coil and secondary coil are inversely proportional to the number of terms.

Transformers donot operate on direct current (dc), because dc current does not produce a charge in flux in the primary coil, hence no emf is produced in the secondary coil.

EXERCISES

9.1. A resistor R and an inductor L are connected in series with an ac source of 100V. An ac voltmeter is connected across R and then across L and giving same readings. What does it predict. **Ans.** 70.7

9.2. $L = 40$ mH, $C = 20$ μF and $R = 10\Omega$

$v = 100 \sin 314t.$

Find

(a) voltage across each element.

(b) instantaneous value of current.

(c) total impedance of the circuit.

9.3. For the preceding problem. Find the average power of the circuit.

9.4. In a parallel LCR circuit, the current is not minimum when $X_L = X_C$ $\left(\omega_0 = \frac{1}{\sqrt{LC}}\right)$. Find the frequency for which current is minimum.

9.5. Find the expression for the current in the parallel LCR circuit.

9.6. In a LCR circuit, $L = 40$ mH, $C = 20$ μF and $R = 100\Omega$. Find the frequencies for which power factor is equal to 0.2.

9.7. An alternating voltage of amplitude 100V and angular frequency of 100π rad s^{-1} is connected in series circuit with $R = 10\Omega$, $L = 2$ mH and $C = 20$ μF. Determine

(a) The amplitude and phase of current

(b) Potential difference across R, L and C

9.8. Why the voltage and current in L and C of ac circuits have a phase difference of 90°, explain.

9.9. A step up transformer is designed to change a voltage 100V into 440V. The primary coil contains 1000 terns. If transformer has 100% efficiency, how many terns (N_2) in the secondary coil.

9.10. A resistance R and a capacitance C are connected in series across a source of ac voltage of 100V. A voltmeter across the capacitance reads 75V. Find the voltmeter reading when it is connected across the resistance R.

Ans. 66.14V

9.11. A resistor R and a capacitor C are connected in series as shown in fig. 9.30.

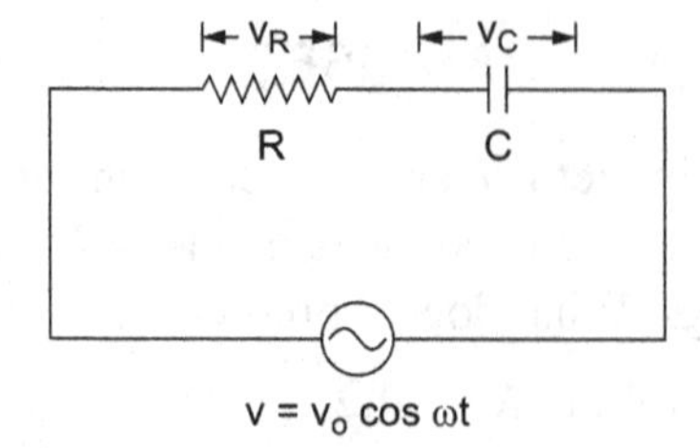

Fig. 9.30. CR in series.

Explain, why $|V_0| \neq |V_R| + |V_C|$

9.12. A 10μF capacitor is connected across an ac source that has rms value of 100V and frequency $f = 50$ Hz. What is the amplitude of current.

■ ■ ■

10 CHAPTER Time-Varying Fields

A static charge produces an electric field and a moving charge produces static magnetic field. In the previous chapter, we have seen how the electricity produces magnetism. If we think about the converse of this and a possibility arises that the magnetic effect produces electric current. After Hans-Christian Oersted, Michael Faraday started the experiments and showed that the magnetic field may produce the electric current. In 1931, Faraday observed that if a magnet is moved in the vicinity of a coil, a current is induced in the coil and is indicated by a galvenometer, as shown in fig. 10.1. He concluded that the induced current was dependent not only on the magnetic flux itself, but on its time rate of change also. The changing magnetic field produces an induced emf. Faraday gave a relation between time rate of change of magnetic flux and the induced emf. This relation is called the Faraday's law of electromagentic induction.

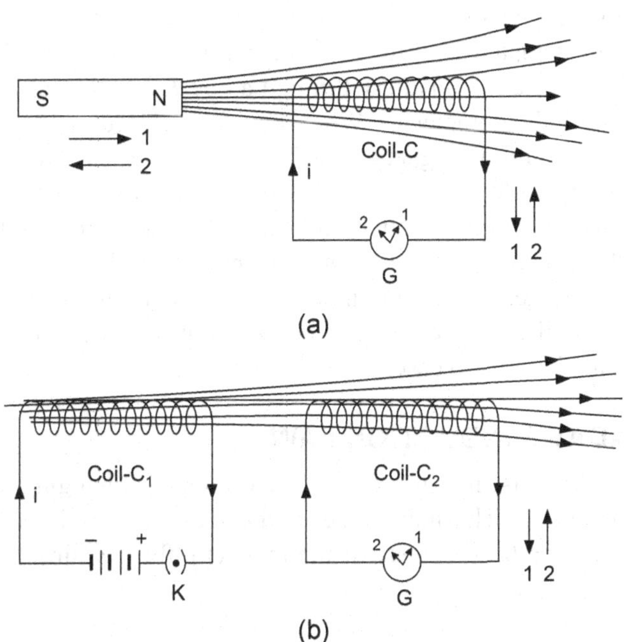

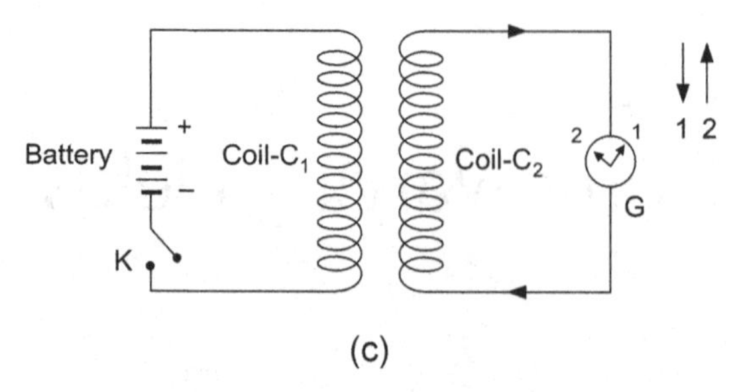

(c)

Fig. 10.1. (a) The induced emf when a magnet moves towards and away from a coil (b) moving a current carrying coil and (c) switching on and off a current in a coil.

Fig. 10.1 (a) depicts an experiment, in which a magnet moves toward a coil C. As a result of this, an emf is induced in the coil C due to the change in magnetic flux passing through the coil. A current flows in the galvenometer *G*. On the other hand, if the magnet moves away from the coil C, the magnetic flux linked with the coil decreases and a current is observed in the galvenometer but in opposite direction.

In the fig. 10.1 (b), the magnet is now replaced with a coil C_1 connected to a battery. When a current flows in the coil C_1, it produces a magnetic field. Thus, a magnetic lines of force (magnetic flux) is linked to the coil C_1. When coil C_1 moves toward or away from the coil C_2 rapidly, the change in flux causes an emf is induced in the coil C_2, and a current flows in the coil C_2 as indicated by galvenometer *G*.

In the circuit depicted by Fig. 10.1 (c), we have two coils C_1 and C_2. The coil C_1 is connected to a battery through a key *K* and the second coil C_2 is connected to a galvenometer *G*. When key *K* is closed, a current passes through the coil C_1 and a magnetic field is produced. It shows that an emf is induced in the coil C_2. A deflection in the galvenometer is produced. The direction of deflection of galvenometer indicates that the current in the coil C_2 is opposite to the direction of the current in the coil C_1. If the key *K* is opened, the current becomes zero very soon in the coil C_1, and a momentary current is induced in the coil C_2 in the same direction as in C_1. The direction of the current is opposite in the two cases.

10.1. FARADAY'S INDUCTION LAW

If the magnetic flux passing through a closed circuit is changing, an emf is induced in that circuit. The magnitude of the induced emf is equal to the negative of the time rate of change of the magnetic flux passing through the circuit.

Changing magnetic field produces an induced emf

Mathematically, Faraday's law can be written as

$$\varepsilon = \frac{-Nd\phi_B}{dt} = -\frac{d(N\phi_B)}{dt} \qquad ...(10.1)$$

where N is the number of turns in the coil and induced emf ε is in volts. The quantity $N\phi_B$ is the total flux linked to the coil.

Since, the induced emf is due to the motion of the electrons, a force is exerted on the electrons associated with the induced emf. Thus, we write

$$\varepsilon = \oint_l \vec{E} \cdot \vec{dl} \qquad ...(10.2)$$

From the equations (10.1) and (10.2) we have

$$\oint_l \vec{E} \cdot \vec{dl} = -\frac{d\,(N\phi_B)}{dt} \qquad ...(10.3)$$

But we know,

$$N\phi_B = \int_S \vec{B} \cdot \vec{dS} \qquad ...(10.4)$$

$N\phi_B$ is the total flux linked to the surface S. Thus we may write.

$$\oint_l \vec{E} \cdot \vec{dl} = \frac{-d}{dt}\int_S \vec{B} \cdot \vec{dS} \qquad ...(10.5)$$

or

$$\oint_l \vec{E} \cdot \vec{dl} = -\int_S \frac{d\vec{B}}{dt} \cdot \vec{dS} \qquad ...(10.6)$$

Now, applying stoke's theorem, we obtain

$$\int_S (\nabla \times \vec{E}) \cdot \vec{dS} = -\int_S \frac{d\vec{B}}{dt} \cdot \vec{dS} \qquad ...(10.7)$$

For any arbitrary surface, we can write the equation (10.7) as

$$\nabla \times \vec{E} = -\frac{\partial B}{\partial t} \qquad ...(10.8)$$

In the region of time varying magnetic field the electric field is not conservative and cannot be expressed as the negative of a gradient of a scalar potential. The Equation (10.8) is known as Faraday's law. The Equation (10.1) holds for a stationary circuit (dc networks).

10.2. LENZ'S LAW

Since Faraday's law does not give the direction of the induced emf, the direction of the induced emf is given by Lenz's law. This law is based on the conservation of energy principle. According to Lenz's law, the induced emf produces a current and the direction of the induced current is always such as to oppose the cause that produces it.

To apply the Lenz's law, consider a single loop as shown in fig. 10.2.

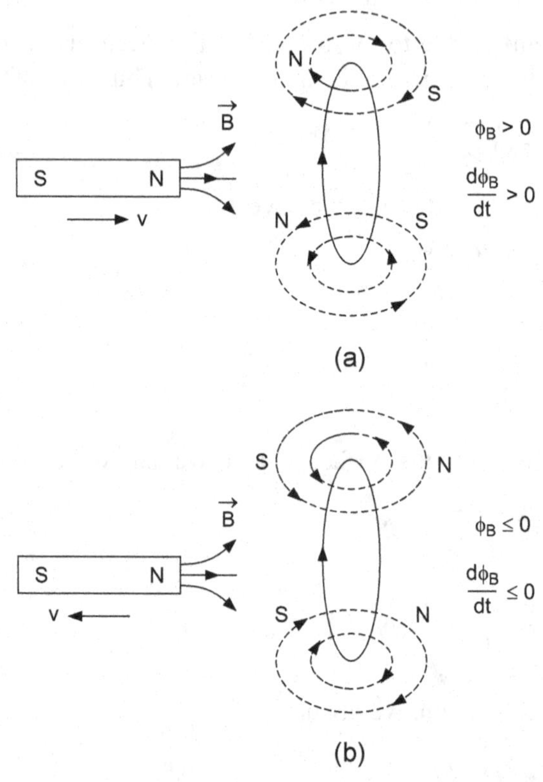

Fig. 10.2. Direction of the induced current.

When a magnet moves towards the loop, the magnetic flux through the loop will increases, $\phi_B > 0$ and the magnetic field B will increase, thus $\frac{d\phi_B}{dt} > 0$. As a result of this, a current is induced, the direction of the B is pointing to the right. The induced current will be in a direction such that the induced field B_{ind} is in left direction to reduce the effect. As in Fig. 10.2 (b), when the magnet moves away from the loop, the magnetic flux linked to the loop decreases and $\phi_B \leq 0$, $\frac{d\phi_B}{dt} \leq 0$. As a result, the magnetic across the loop will decrease, and B_{ind} should point to the right. From the Fleming's right hand rule this is possible only if the current in the loop is in counter-clock-wise direction. In both the cases direction of the induced current opposes the change in magnetic flux linked to the loop.

Example 10.1. A square loop of 50 turns and side 10 cm is placed with its plane perpendicular to the magnetic field. Calculate the induced emf if the magnetic field changes from $0.5T$ to $1.0T$ in $0.5s$.

Solution: $N = 50$ turns

Area of the loop $S = 0.1 \times 0.1 = 0.001\ \text{m}^2$

The plane of loop is normal to B, $\cos\theta = \cos 0 = 1$

$$\text{Induced emf } \varepsilon = \frac{dB}{dt} = \frac{1.0 - 0.5}{0.5} = 1\ \text{volts/m}^2$$

$$\varepsilon = \frac{-Nd\phi_B}{dt} = -\frac{N\,d(BS)}{dt}$$

or

$$\varepsilon = -NS\frac{dB}{dt}$$

$$\varepsilon = -50 \times 0.001 \times 1\ \text{volts}$$

$$\varepsilon = 0.05\ \text{volts}$$

10.3. FLEMING'S RIGHT HAND RULE

This is also called a generator rule because it is useful in generators and motors. It is applicable when a current is induced in a circuit. When a conductor moves through a magnetic field, a current is induced in the conductor the direction of the current induced is given by Fleming's right hand rule. When right hand is stretched in such a way that the thumb, forefinger and the middle finger are mutually perpendicular to each other, then

(a) the thumb represents the direction of motion of the conductor.
(b) forefinger is pointed in the direction of the magnetic field.
(c) and the middle finger points in the direction of induced current as shown in fig. 10.3.

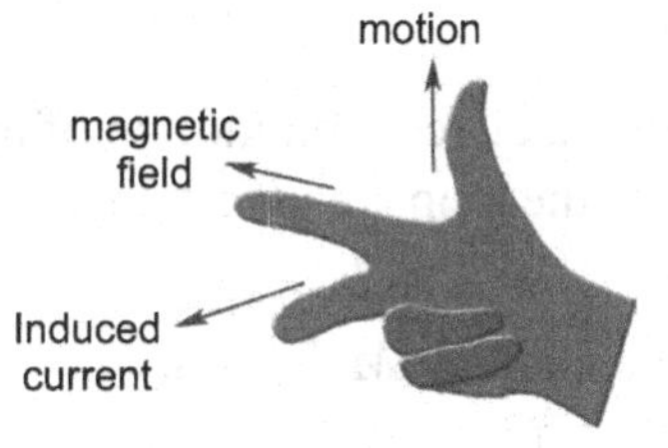

Fig. 10.3. Fleming's right hand rule

10.4. EMF INDUCED IN A MOVING CONDUCTOR

According to Faraday's law of magnetic induction, an induced emf always produces a current which gives rise to the electric field associated with it.

When a conductor moves through a uniform magnetic field, the emf is induced in the conductor also.

When a conductor MN moves through a static field by sliding on a U-shaped conductor as shown in fig. 10.4.

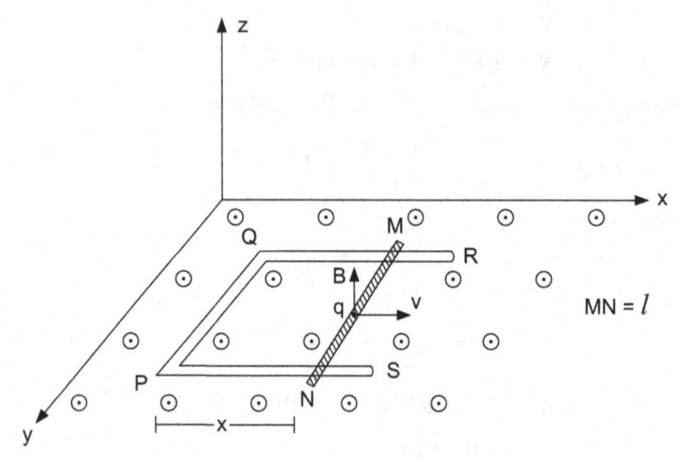

Fig. 10.4. EMF is produced when a conductor moves in a uniform magnetic field.

The area of the loop *MNPQ* is changing with time as the conductor *MN* moves. A uniform magnetic field is perpendicular to the area bounded by the loop and is parallel to the *z*-axis. Now, the force on a charge *q* moving with a uniform velocity *v* is given by

$$\vec{F} = q(\vec{v} \times \vec{B}) \quad ...(10.9)$$

Since $\vec{v}$ and $\vec{B}$ are perpendicular, then

$$F = q\,v\,B \quad ...(10.10)$$

or

$$E = \frac{F}{q} = vB \quad ...(10.11)$$

where *E* is the electric field associated with motion of the charge. Since *MN* is moving and the other parts of the loop are in rest, the emf induced in the loop is

$$\varepsilon = V_M - V_N = El \quad ...(10.12)$$

where *l* is the length of the conductor *MN*. Thus,

$$\boxed{\varepsilon = vBl} \quad ...(10.13)$$

If $v = 0$, $\varepsilon = 0$

The Equation (9.13) can be obtained as follows. The magnetic flux through the loop *MNPQ* is

$$\phi_B = \oint_S \vec{B} \cdot d\vec{S} = BS \qquad ...(10.14)$$

or

$$\phi_B = Blx \qquad ...(10.15)$$

since x is changing with time, then

$$\frac{d\phi_B}{dt} = Bl\frac{dx}{dt} = Blv \qquad ...(10.16)$$

the magnitude of the induced emf is given by

$$\varepsilon = \frac{d\phi_B}{dt}$$

or

$$\boxed{\varepsilon = vBl} \qquad ...(10.17)$$

the induced current i is, then, given by

$$i = \frac{\varepsilon}{R}$$

or

$$\boxed{i = \frac{vBl}{R}} \qquad ...(10.18)$$

where R is the resistance of the loop *MNPQ*, and its direction is clockwise

10.5. EMF INDUCED IN A ROTATING COIL : PRINCIPLE OF ELECTRIC GENERATOR

Consider a rectangular coil of turns N moving in a static field B as shown in fig. 10.5. When this coil rotates in a static field, the magnetic flux ϕ_B passing through the coil changes with time.

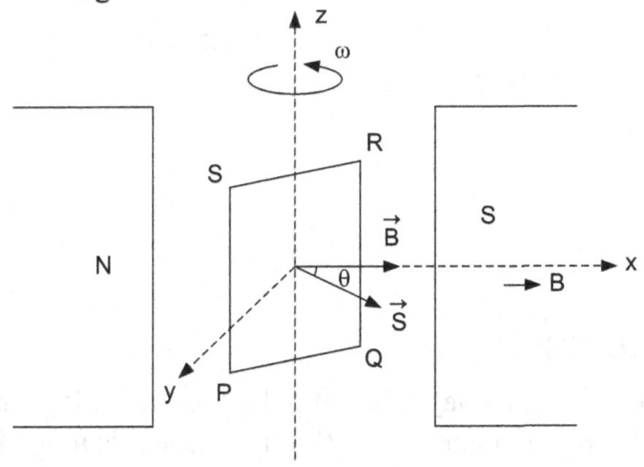

Fig. 10.5. Rotating coil in a static field.

Since ϕ_B is changing with time, an emf is induced in the coil. In fig. 10.5, the normal surface vector $\vec{S}$ makes an angle θ with the magnetic field B. Therefore, Magnetic flux through the coil *PQRS* is

$$\phi_B = BS\cos\theta \quad ...(10.19)$$

where S is the area of the rectangular coil. The coil is rotating about z-axis with the angular velocity $\omega = \frac{\theta}{t}$. Thus,

$$\phi_B = BS\cos\omega t \quad ...(10.20)$$

The emf induced in the coil

$$\varepsilon = -N\frac{d\phi_B}{dt} = NBS\frac{d}{dt}(\cos\omega t) \quad ...(10.21)$$

$$\varepsilon = \omega NBS\sin\omega t \quad ...(10.22)$$

or

$$\boxed{\varepsilon = \varepsilon_0 \sin\omega t} \quad ...(10.23)$$

where $\varepsilon_0 = \omega NBS$ is maximum induced emf.

If R is the resistance of the coil, the induced current is given by

$$i = \frac{\varepsilon}{R} = \frac{\varepsilon_0}{R}\sin\omega t \quad ...(10.24)$$

or

$$\boxed{i = i_0 \sin\omega t} \quad ...(10.25)$$

where $i_0 = \frac{\varepsilon_0}{R}$ is amplitude of the current.

Example 10.2. A conductor of length 0.5 m is sliding along two conducting rails with the speed 0.5 m/s. If the resistance of the system is 1 Ω, calculate the induced current, given $B = 1.0T$.

Solution: According to Faraday's law, the induced emf is

$$\varepsilon = vBl$$

$$= (0.5 \text{ m/s}) \times (1.0T) \times (0.5 \text{ m})$$

$$\varepsilon = 0.25 \text{ volts}$$

the induced current

$$i = \frac{\varepsilon}{R} = \frac{0.25}{1}$$

or

$$i = 0.25\,A.$$

10.6. EDDY CURRENTS

Eddy currents are due to Faraday's law of induction. The eddy currents occur when an ac flows through a conductor. If a metal sheet is moving across the magnetic field, the magnetic flux ϕ_B is changing with time and an emf is produced, this emf will produce eddy currents also, that are circulating on the surface of the metal as shown in fig. 10.6.

These eddy currents decrease the change in magnetic flux $\left(\frac{d\phi_B}{dt}\right)$,

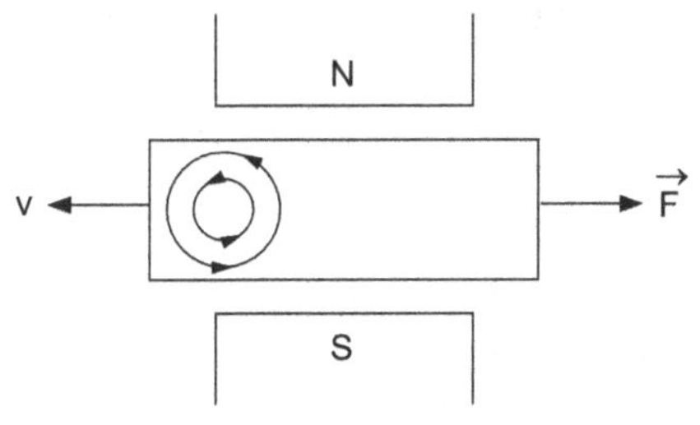

Fig. 10.6. Eddy currents in a metal.

and produces a force due to which the motion of the conductor occurs. The detail of eddy currents and skin effect may be found in "a plane EM wave in a conducting media".

10.7. SELF-INDUCTANCE

A choke or an inductor is a coil of the wire. If a current flowing in a coil is changed, an emf is produced. This is due to the change in flux linked with the coil. This inductive effect is called self-induction.

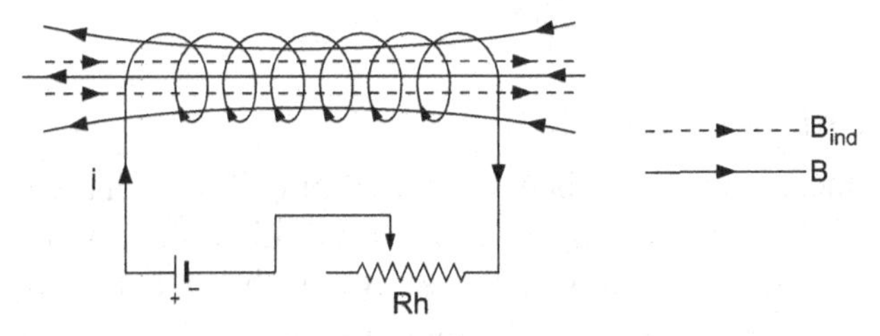

Fig. 10.7. Self induction

Suppose that the current i is flowing through a coil and as a result, a magnetic field setup across the coil as shown in fig. 10.7. The current i is increased or decreased by a rheostat Rh and it changes the magnetic field. This will result in changing the magnetic flux ϕ_B. If there are N turns in the coil, the magnetic flux $N\phi_B$ is proportional to the current i. Thus,

$$N\phi_B = Li \qquad ...(10.26)$$

or

$$\boxed{L = \frac{N\phi_B}{i}} \qquad ...(10.27)$$

where the proportionality constant L is called the coefficient of self inductance or simply self inductance.

According to Faraday's law of induction, the induced emf is

$$\varepsilon = \frac{-N d\phi_B}{dt} = -\frac{d\,(N\phi_B)}{dt} \qquad ...(10.28)$$

or

$$\varepsilon = -\frac{d\,(Li)}{dt}$$

or

$$\boxed{\varepsilon = -L\frac{di}{dt}} \qquad ...(10.29)$$

From the Equation (10.29), the coefficient of self induction is given by

$$\boxed{L = -\frac{\varepsilon}{(di/dt)}} \qquad ...(10.30)$$

The Equation (10.29) shows a back emf and it is due to Lenz's law.

The SI unit of self inductance L is Wb/A or Volt per ampere per second $\left(\frac{\text{Volt}}{\text{Ampere}\cdot s}\right)$ or henry, in the honor of Joseph Henry.

$$1 \text{ henry} = \frac{1 \text{ volt}}{1 \text{ Ampere/Second}} \qquad ...(10.31)$$

or

$$1 \text{ henry} = \frac{1 \text{ Wb.}}{1 \text{ Ampere}} \qquad ...(10.32)$$

10.8. MUTUAL INDUCTANCE

The mutual inductance is arises between a pair of coils which are linked by a magnetic flux. The change in current in one coil will produce an induced emf in the other coil, and this phenomenon is known as the mutual induction.

Consider two coils c_1 and c_2 consisting of N_1 and N_2 turns respectively as shown in fig. 10.8.

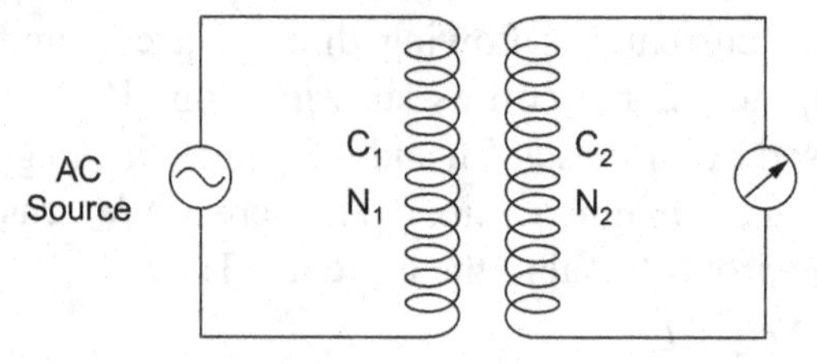

Fig. 10.8. A current in coil c_1 produces a current in the coil c_2.

The current i_1 in the coil c_1 will produce a magnetic flux $(N_2\phi_B)$ linking two circuits. That is, the magnetic flux $(N_2\phi_B)$ is proportional to the current i_1. Thus,

$$N_2\phi_B \propto i_1 \quad ...(10.33)$$

or

$$N_2\phi_B = Mi_1 \quad ...(10.34)$$

where the proportionality constant M is called the coefficient of mutual inductance. M is purely a geometric factor and contains permeability, and it depends on shapes and size of coil and turns N etc.

Now, according to Faraday's law, the induced emf in the coil c_2 is given by

$$\varepsilon_2 = \frac{-d\,(N_2\phi_B)}{dt} \quad ...(10.35)$$

or

$$\boxed{\varepsilon_2 = -M\frac{di_1}{dt}} \quad ...(10.36)$$

Here, we have substituted the value of $(N_2\phi_B)$ from the Equation (10.34) into equation (10.35). Therefore,

$$M = -\frac{\varepsilon_2}{(dt_1/dt)} \quad ...(10.37)$$

Now, we will calculate the effect on the current i_1 due the time rate of change of current in the coil c_2.

The emf induced in the coil c_1 is given as

$$\boxed{\varepsilon_1 = -M\frac{di_2}{dt}} \quad ...(10.38)$$

From the Equations (10.36) and (10.38), we have

$$\frac{di_1}{dt} = \frac{di_2}{dt} \quad ...(10.39)$$

Thus, we write

$$\boxed{M = \frac{N_1\phi_B}{i_2} = \frac{N_2\phi_B}{i_1}} \quad ...(10.40)$$

The SI unit of M is same as that of L.

Example 10.3. Compute the self inductance per unit length of a long solenoid.

Solution: Consider a solenoid of length l consisting of number of turns per unit length n. The magnetic field at axial point is given by

$$B = \frac{\mu_0 ni}{l}$$

The magnetic flux through the solenoid is

$$\phi_B = BA$$

where A is the cross-sectional area of the solenoid.

$$\therefore \qquad \phi_B = \frac{\mu_0 n i A}{l}$$

the self inductance L is given as

$$L = \frac{n\phi_B}{i} = \frac{\mu_0 n^2 A}{l}.$$

10.9. ENERGY STORED IN A MAGNETIC FIELD

Since the capacitor stores an electric energy which is equal to $\frac{1}{2}\epsilon_0 E^2$, an inductor stores magnetic energy when it is connected to a battery. Consider an inductor in which a current i is flowing. Then induced emf is given by

$$\varepsilon = -L\frac{di}{dt} \qquad ...(10.41)$$

If dW represents the amount of work done in moving the charge dq against emf ε, then

$$dW = -\varepsilon dq \qquad ...(10.42)$$

$$= \frac{L\,di\,dq}{dt}$$

or $$dW = \frac{L\,dq\,di}{dt}$$

$$\therefore \qquad dW = Lidi \qquad ...(10.43)$$

Integrating the Equation (10.43) to obtain total work done which is equal to the energy stored in the inductor.

$$U_B = \int dW = \int_0^i L\,i\,di \qquad ...(10.44)$$

or $$\boxed{U_B = \frac{1}{2}Li^2} \qquad ...(10.45)$$

For a particular case, we may calculate the magnetic energy per unit volume, that is, the magnetic energy density.

Example 10.4. Obtain an expression for the magnetic energy density of a long solenoid.

Solution: The magnetic field of a long solenoid of length l having number of turns n is given by

$$B = \frac{\mu_0 n i}{l}$$

The magnetic flux through the solenoid is

$$\phi_B = \frac{\mu_0 niA}{l}$$

where A is area of cross-section of the solenoid. Thus, inductance of solenoid is then, give as

$$L = \frac{\mu_0 n^2 A}{l}$$

$\therefore$ $$U_B = \frac{1}{2}Li^2$$

$$= \frac{1}{2}\left(\frac{\mu_0 n^2 A}{l}\right)\left(\frac{Bl}{\mu_0 n}\right)$$

or $$U_B = \frac{1}{2}\frac{B^2}{\mu_0}Al$$

Here Al is the volume of the solenoid, the energy density

$$u_B = \frac{U}{Al} = \frac{1}{2}\frac{B^2}{\mu_0}$$

or $$\boxed{u_B = \frac{B^2}{2\mu_0}}$$

10.10. RL CIRCUIT

Consider an inductor L and a resistor R is series with a battery of V volts as shown in fig. 10.9.

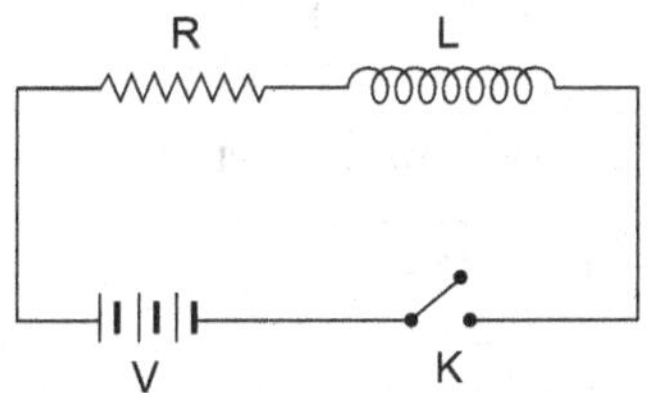

Fig. 19.9. RL circuit with battery.

when the switch K is closed, a current i is flowing through the circuit. For Fig. 10.9, applying Kirchhoff's loop rule, we have

$$V = V_L + V_R \quad \text{...(10.46)}$$

where V_R is the voltage drop across the resistance R and V_R is the induced emf which is equal to $L\,\frac{di}{dt}$.

$$V_L = \frac{Ldi}{dt} \qquad ...(10.47)$$

$$V = \frac{Ldi}{dt} + Ri \qquad ...(10.48)$$

or
$$\frac{Ldi}{V - Ri} = dt \qquad ...(10.49)$$

on integration of the Equation (10.49), we get

$$\frac{-L}{R}\log_e(V - Ri) + C = t \qquad ...(10.50)$$

The initial conditions are given as
when $t = 0$, $i = 0$

Therefore,

$$C = \frac{L}{R}\log_e V$$

Substituting the value of C into Equation (10.50), we have

$$\text{Log}_e(V - Ri) - \log_e V = \frac{-R}{L}t \qquad ...(10.51)$$

or
$$\frac{\log_e(V - Ri)}{V} = \frac{-R}{L}t \qquad ...(10.52)$$

$\because$
$$1 - \frac{Ri}{V} = -e^{(R/L)t} \qquad ...(10.53)$$

or
$$i = \frac{V}{R}(1 - e^{-(R/L)t}) \qquad ...(10.54)$$

Thus, the Equation (10.54) may be written as

$$\boxed{i = i_0(1 - e^{-(R/L)t})} \qquad ...(10.55)$$

where $i_0 = \dfrac{V}{R}$ is the maximum value of current when $t \to \infty$. The time constant τ is given by

$$\boxed{\tau = \frac{L}{R}} \qquad ...(10.56)$$

The plot of i versus t is shown in fig. 10.10.

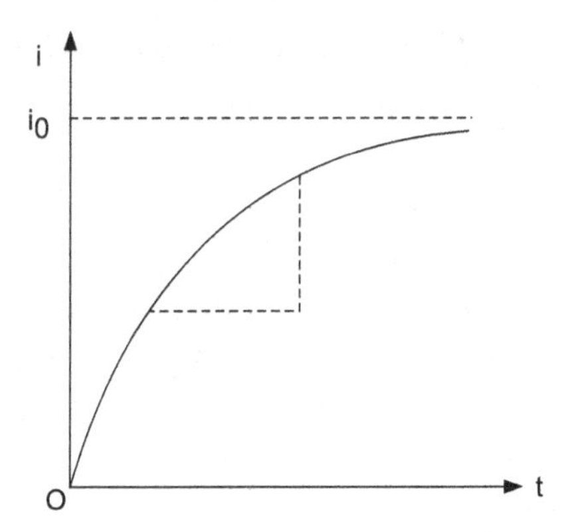

Fig. 10.10. Growth of current in RL circuit

Now, differentiating Equation (10.55) with respect to t, we have,

$$\frac{di}{dt} = \frac{Ri_0}{L} e^{-(R/L)t} \quad ...(10.57)$$

or $$\frac{di}{dt} = \frac{i_0}{\tau} e^{-(R/L)t} = \frac{(i_0 - i)}{\tau} \quad ...(10.58)$$

The Equation (10.58) gives the slop of the exponential curve.

After reaching the current to its maximum value i_0, the switch K is now opened. It means the applied voltage $V = 0$. Then, the Equation (10.48) takes the form

$$iR + \frac{Ldi}{dt} = 0 \quad ...(10.58(1))$$

The current starts falling from its maximum value i_0.

Now,

$$\frac{-Ldi}{dt} = Ri \quad ...(10.59)$$

or $$\frac{-L}{R}\frac{di}{i} = dt \quad ...(10.60)$$

on integration, we get

$$\frac{-L}{R}\log_e i + C = t \quad ...(10.61)$$

Initially at $t = 0$, $i = i_0$. Thus,

$$C = \frac{L}{R}\log_e i_0$$

Substituting the value of C into Equation (10.61), we get

$$\text{Log}_e \frac{i}{i_0} = e^{-(R/L)t} \quad ...(10.62)$$

or

$$\boxed{i = i_0\, e^{-(R/L)t}} \quad ...(10.63)$$

The Equation (10.63) shows that the current in the RL circuit is falling exponentially. The slop of the Equation (10.63) gives the rate at which current is falling. The plot of i versus t is shown in fig. 10.11.

Differentiating (10.63) w.r. to t, we get

$$\frac{di}{dt} = -\frac{i_0 R}{L} e^{-(R/L)t}$$

The negative sign shows that the current is decreasing exponentially.

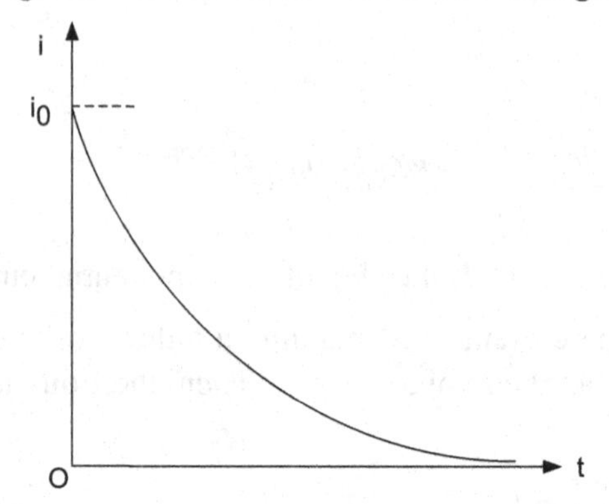

Fig. 10.11. Decay of current in *RL* circuit.

The energy is conserved in the process of rise and decay of current in *RL* circuit. When the battery is removed or switch K is opened, the energy is stored in the inductor, as

$$U = \frac{1}{2} L i_0^2$$

This energy maintains the current in the circuit.

Example 10.5. A conductor of length l rotates in a static magnetic field with angular speed ω. Obtain an expression for the emf induced in the conductor.

Solution: Consider an element dy of the conductor moving with a velocity v as shown in fig. 10.12.

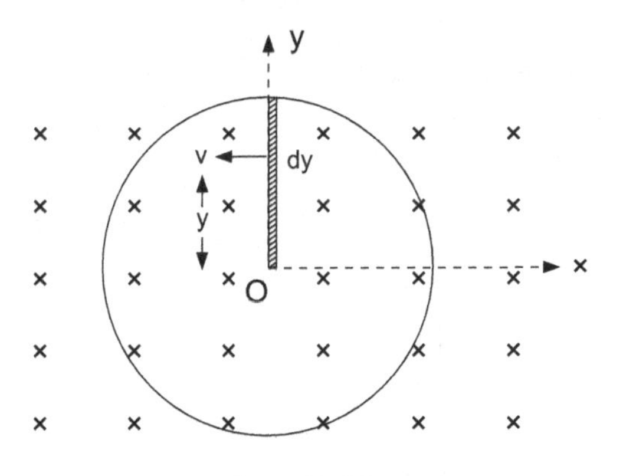

Fig. 10.12.

The conductor is moving in a magnetic field which is directed into the page. Since velocity $\vec{v}$ is perpendicular to the magnetic field B. Then, the emf induced in the conductor is

$$\varepsilon = \int_0^l v B \, dy$$

or

$$\varepsilon = \int_0^l B \omega y \, dy$$

or

$$\boxed{\varepsilon = \frac{1}{2} B \omega l^2}$$

Example 10.6. A stationary square loop of 50 turns and side 1 m lies in x-y plane, as shown in fig. 9.13. If the magnetic field $B = B_0 \sin \omega t$ points in z-axis, compute the induced emf in the loop.

Solution:

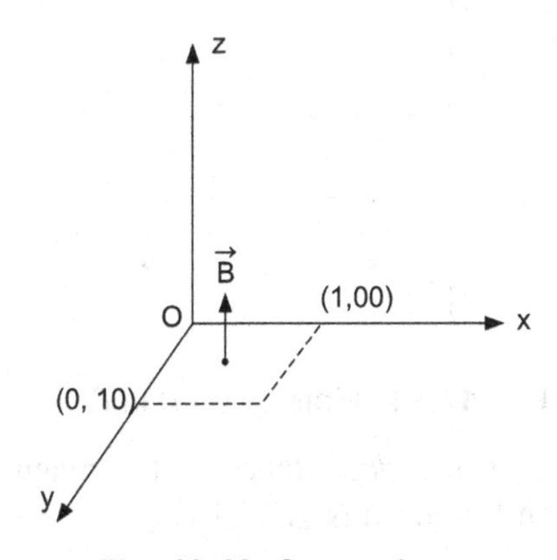

Fig. 10.13. Square loop.

$$\therefore \qquad \phi_B = \int_S \vec{B} \cdot \vec{dS}$$

Thus, induced emf is

$$\varepsilon = -N \frac{d}{dt} \phi_B$$

or

$$\varepsilon = -N \int \frac{\partial \vec{B}}{\partial t} \cdot \vec{dS}$$

Now

$$\frac{\partial B}{\partial t} = \frac{d}{dt}(B_0 \sin \omega t)$$
$$= \omega B_0 \cos \omega t$$

and surface vector

$$\vec{dS} = dx\, dy\, \hat{j}$$

Thus,

$$\varepsilon = -50\, \omega B_0 \cos \omega t \int_0^1 dx \int_0^1 dy$$

or

$$\varepsilon = -50\, \omega B_0 \cos \omega t$$

Example 10.7. A rod of length *L*, in fig. 10.14, is moving with a uniform velocity along the direction of the current *i* in a long wire. Calculate the emf induced across the rod.

Solution:

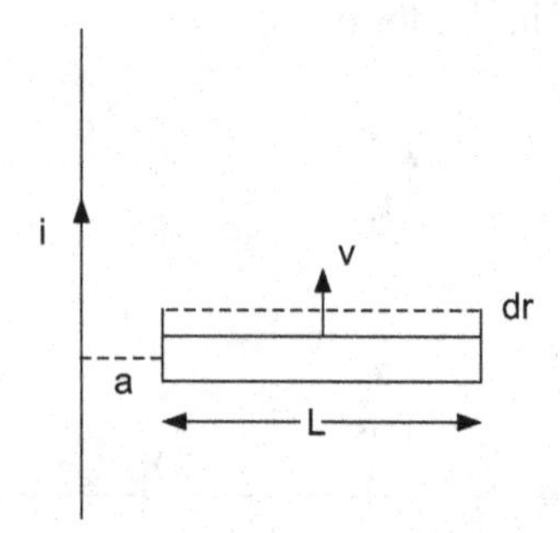

Fig. 10.14. Motion of rod in *B*

If the moving rod moves a *dr* distance in the magnetic field which is directed into page. The emf induced is given by

$$\varepsilon = \int (\vec{v} \times \vec{B}) \cdot \vec{dr}$$

But
$$B = \frac{\mu_0 i}{2\pi r}$$

$$\varepsilon = \frac{\mu_0 i v}{2\pi} \int_a^{a+L} \frac{dr}{r}$$

$$\varepsilon = \frac{\mu_0 i v}{2\pi} \log_e \frac{a+L}{a}$$

Example 10.8. Obtain an expression for mutual inductance between a long wire carrying a current i_1 and a rectangle with current i_2 as shown in fig. 10.15.

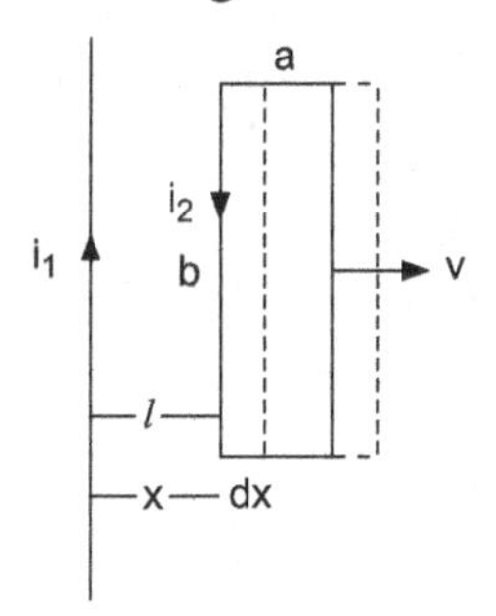

Fig. 10.15. A current loop.

Solution: The magnetic flux across the rectangle is

$$\phi_B = \int B\, bdx$$

$$= \frac{\mu_0 i_1 v}{2\pi} \int_l^{l+a} \frac{dx}{x}$$

or
$$\phi_B = \frac{\mu_0 b}{2\pi} \log_e \left(\frac{l+a}{l}\right)$$

the mutual inductance is given as

$$M = \frac{\phi_B}{i_1}$$

or
$$M = \frac{\mu_0 b}{2\pi} \log_e \left(\frac{l+a}{l}\right)$$

Example 10.9. Find the emf induced in the rectangle, in fig. 10.16.

Solution: We take a strip of width dx at a distance x from the wire carrying a current i. The magnetic flux through the strip is

$$d\phi_B = BdS$$

Here, $$B = \frac{\mu_0 i}{2\pi x}$$

and dS = area of the strip

$= bdx$

$$\therefore \quad d\phi_B = \frac{\mu_0 ib}{2\pi}\frac{dx}{x}$$

$\therefore$ total flux is

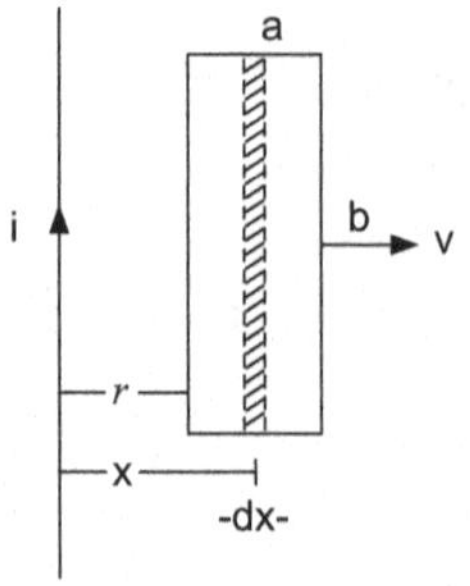

$$\phi_B = \int d\phi_B = \int_x^{x+a} \frac{\mu_0 ib}{2\pi}\frac{dx}{x}$$

Fig. 10.16. Current loop.

$$\phi_B = \frac{\mu_0 ib}{2\pi}[\log_e(x+a) - \log_e x]$$

Thus, emf induced in rectangle is

$$\varepsilon = \frac{-\partial\phi}{\partial t} = \frac{-\mu_0 ib}{2\pi}\frac{\partial}{\partial t}[\log_e(x+a) - \log_e x]$$

$$= \frac{-\mu_0 ib}{2\pi}\left[\frac{1}{x+a}\cdot\frac{dx}{dt} - \frac{1}{x}\cdot\frac{dx}{dt}\right]$$

Since $\frac{dx}{dt} = v$

$$\therefore \quad \varepsilon = \frac{\mu_0 ibv}{2\pi}\left[\frac{1}{x+a} - \frac{1}{x}\right]$$

put $x = r$, we get

$$\varepsilon = \frac{\mu_0 iabv}{2\pi}$$

EXERCISE

10.1. Define the principle of the electromagnetic induction and deduce the Faraday's law of electromagnetic induction.

10.2. Obtain an expression for the growth of current in *RL* circuit.

10.3. Show that the magnetic energy density of a long solenoid is given by

$$u_B = \frac{B^2}{2\mu_0}$$

10.4. Define the following physical quantities.

(a) self inductance

(b) Mutual inductance

10.5. Deduce an expression for the self-inductance of a toroid.

10.6. Show that the magnetic energy stored in an inductor L is given by

$$U_B = \frac{1}{2}Li^2$$

10.7. Find the inductance per unit length of a co-axial cable of inner radius r_1 and outer radius r_2.

10.8. Two hundred turns of a copper wire are wrapped around a wooden cylinder of radius 0.1 m. If the current in the coil is 2A and a magnetic field through the coil is changed from 0.1 Wb/m^2 to 1 Wb/m^2 then, calculate (a) emf induced in the coil (b) energy stored in the coil.

10.9. A square loop of side 0.2 m lies in a plane of long straight wire at a distance 0.5 m from it. Find the mutual inductance when a current of 1A is flowing in the wire.

10.10. Obtain an expression for the magnetic flux passing through the hemisphere, fig. 10.17. **Ans.** $-\pi r^2 B$

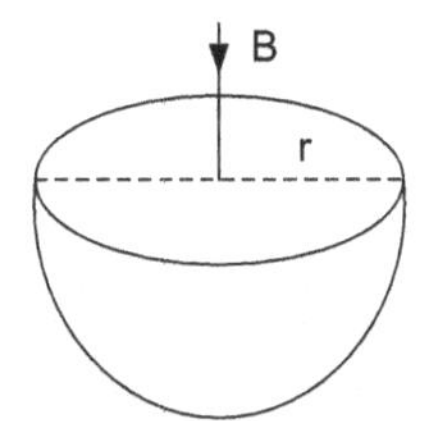

Fig. 10.17. Hemisphere.

10.11. When a current in one coil changes from 0A to 10A in 10^{-3} *s*, and induced emf is 100 V. Calculate self inductance of the coil.

10.12. Calculate the speed of the rod of length 0.5 m in a magnetic field of 0.2 Wb/m^2 and perpendicular to the motion of the rod.

10.13. A resistor of resistance 100Ω and an inductor of inductance 0.5 H are connected to a battery of 18V. Calculate steady state current and value of it for $t = 1s$. Plot i versus t.

10.14. If an inductor operates at the voltage 200V and 50 Hz frequency. The inductor draws a current of 10A, calculate the inductance of the inductor.

■ ■ ■

Maxwell's Equations and Electromagnetic Waves

James Clerk Maxwell (1832–1879) gave the relation between charges, current and electromagnetic field in the form of four mathematical equations. These equations are a type of unified theory, which describe all the phenomena of classical electromagnetism. This theory proves that the light waves are the electromagnetic waves and the speed of the electromagnetic waves is given by

$$c = \frac{1}{\sqrt{\mu_0 \epsilon_0}}$$

$$\epsilon_0 = 8.85 \times 10^{-2} \ \text{C}^2/\text{Nm}^2$$
$$\mu_0 = 1.26 \times 10^{-6} = 4\pi \times 10^{-7} \ \text{Wb/A-m}$$

Then, we get

$$c = 2.99 \times 10^8 \ \text{m/s}$$

or

$$c \approx 3.0 \times 10^8 \ \text{m/s}$$

This unified theory of Maxwell predicts that the accelerated charge produces energy in form of electromagnetic (EM) waves.

11.1 MAXWELL'S EQUATIONS IN DIFFERENTIAL FORM

There are four equations given by J.C. Maxwell. The differential form of the set of four equations is given as follows.

(i) $\boxed{\nabla \cdot \vec{D} = \rho}$ — Gauss's Law

(ii) $\boxed{\nabla \cdot \vec{B} = 0}$ — No magnetic monopole or charge

(iii) $\boxed{\nabla \times \vec{E} = \frac{\partial \vec{B}}{\partial t}}$ — Faraday's Law

(iv) $\boxed{\nabla \times \vec{H} = \vec{J} + \frac{\partial \vec{D}}{\partial t}}$ — Modified Ampere's Law where

$$\vec{E} = \text{Electric field intensity}$$

$$\vec{D} = \text{Electric flux density or electric displacement}$$

$$\vec{B} = \text{Magnetic induction or Magnetic flux density}$$

$$\vec{H} = \text{Magnetic field intensity}$$

$$\rho = \text{Electric charge density}$$

$$\vec{J} = \text{Electric current density}$$

$$\vec{D} = \epsilon_0 \vec{E} \text{ and } \vec{D} = \epsilon \vec{E}$$

$$\vec{B} = \mu_0 \vec{H} \text{ and } \vec{B} = \mu \vec{H} = \mu_0 \mu_r \vec{H}$$

The Maxwell's first equation shows that the divergence of $\vec{E}$ is equal to volume charge density bounded by a surface. This equation predicts that how much is the electric field intensity spread out from the source of charges. The value of the $\nabla \cdot \vec{D}$ will be greater if more charges are present in the volume. If charge source is not present, $\nabla \cdot \vec{D} = 0$.

If we take a straight conductor carrying a current *i*, the magnetic lines of force across the conductor is shown in fig. 11.1.

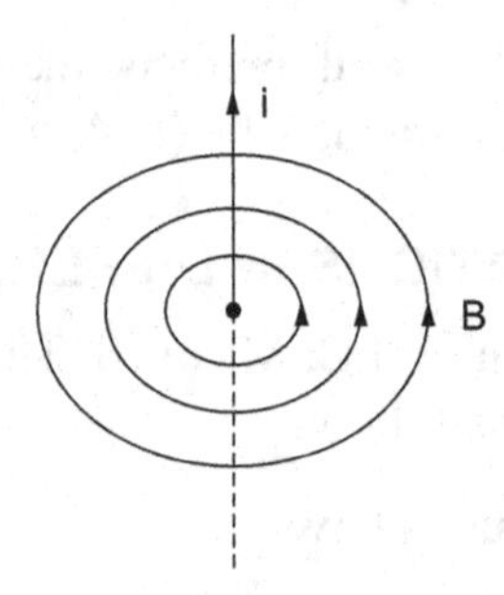

Fig. 11.1. Magnetic lines of force for a conductor

It is very clear that, fig. 11.1, the magnetic lines of force don't begin or end. Hence, the divergence of magnetic induction *B* is always zero, and shows non-existence of the monopole.

The third equation of Maxwell shows that the rot or curl of the electric field intensity is equal to negative the time rate of change of the magnetic induction *B*. The negative sign is due to Lenz's law. It predicts that the time varying magnetic field produces an electric field. If the magnetic field B is not varying with time, there is a no effect on the electric field. Thus, The electric lines of force are straight.

The fourth equation of Maxwell is a modified Ampere's law. This equation contains a term $\frac{\partial D}{\partial t}$, is called a displacement current. This is very helpful in dealing with capacitors.

Moreover, the Ampere's circuital law is given by

$$\oint \vec{B} \cdot \vec{dl} = \mu_0 i \qquad ...(11.1)$$

using stoke's theorem, we write

$$\int_S (\nabla \times \vec{B}) \cdot \vec{dS} = \mu_0 \int_S i \, dS \qquad ...(11.2)$$

Since $B = \mu_0 H$ and the current density is

$$J = \int_S i \, dS$$

Thus, Equation (11.2) may be written for any arbitrary surface as

$$\nabla \times \vec{H} = \vec{J} \qquad ...(11.3)$$

we know that the divergence of curl of any vector is equal to zero. Taking the divergence of the Equation (11.3), we get

$$\nabla \cdot (\nabla \times \vec{H}) = \nabla \cdot \vec{J} = 0 \qquad ...(11.4)$$

However, the equation of continuity predicts that $\nabla \cdot \vec{J} \neq 0$ in time varying field.

$$\nabla \cdot \vec{J} + \frac{\partial \rho}{\partial t} = 0 \qquad ...(11.5)$$

or

$$\nabla \cdot \vec{J} = -\frac{\partial \rho}{\partial t} \qquad ...(11.6)$$

Hence, $\frac{\partial \rho}{\partial t}$ must be added to the right side of the Equation (11.4), we get

$$\nabla \cdot (\nabla \times \vec{H}) = \nabla \, \vec{J} + \frac{\partial \rho}{\partial t} = 0 \qquad ...(11.7)$$

According to Maxwell's first equation, we have

$$\nabla \cdot \vec{D} = \rho$$

Substituting in the Equation (11.7), we get

$$\nabla \cdot (\nabla \times \vec{H}) = \nabla \cdot \vec{J} + \nabla \cdot \frac{\partial \vec{D}}{\partial t}$$

or $$\nabla \cdot (\nabla \times \vec{H}) = \nabla \cdot \left(\vec{J} + \frac{\partial \vec{D}}{\partial t} \right) \quad ...(11.8)$$

Hence

$$\boxed{\nabla \times \vec{H} = \vec{J} + \frac{\partial \vec{D}}{\partial t}} \quad ...(11.9)$$

The Equation (11.9) contains the curl of the magnetic field intensity H, this means that the field lines involve circulation. The term $\frac{\partial \vec{D}}{\partial t}$ is called the displacement current to differ it from the conduction current ($\vec{J}$), and the rate of change of the electric displacement gives rise to magnetic field.

11.2 DISPLACEMENT CURRENT

The second and third equation of Maxwell are homogeneous and these are source free.

and $$\left.\begin{aligned} \nabla \cdot \vec{B} &= 0 \\ \nabla \cdot \vec{E} &= -\frac{\partial \vec{B}}{\partial t} \end{aligned}\right\} \quad ...(11.10)$$

while, the first and fourth equations are inhomogeneous and depend on source, permitivity and permeability of the medium.

and $$\left.\begin{aligned} \nabla \cdot \vec{D} &= \rho \\ \nabla \times \vec{H} &= \vec{J} + \frac{\partial \vec{D}}{\partial t} \end{aligned}\right\} \quad ...(11.11)$$

These equations show that electric and magnetic fields are related to each other and symmetrical equations should describe them. This symmetry led to introduce the concept of the displacement current.

To describe the concept of the electric displacement current, suppose that an electrical circuit contains a capacitor C as shown in fig. 11.2.

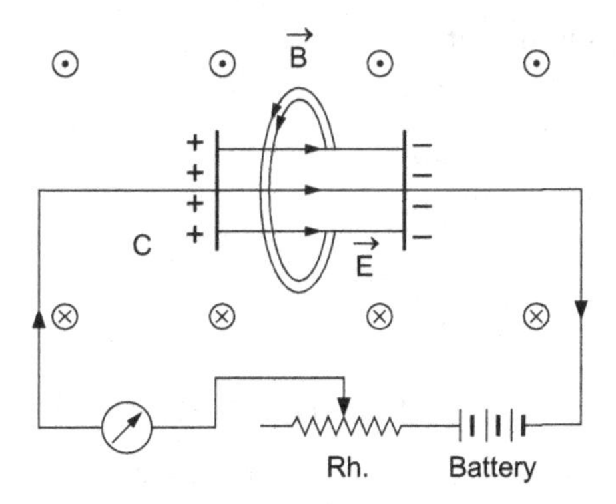

Fig. 11.2. Concept of displacement current.

When rheostat is at static position, no current flows but if we vary the position of rheostat by sliding it, current flows through the capacitor and an electric field is developed across the C. This electric field E varies with time until current reaches maximum value. So, this $\frac{\partial E}{\partial t}$ is known as displacement current.

The magnetic field will remain same. At the outside of the capacitor, the displacement current is equal to the conduction current. That is,

$$i_d = i \text{ (outside the capacitor)} \qquad ...(11.12)$$

The capacity of the capacitor C is

$$C = \frac{\epsilon_0 A}{d} \qquad ...(11.13)$$

where A is area of cross-section and d is distance between the plates of a capacitor. The electric field between the plates.

$$\therefore \qquad E = \frac{q}{\epsilon_0 A} \qquad ...(11.14)$$

differentiating w.r.to t

$$\frac{\partial E}{\partial t} = \frac{1}{\epsilon_0 A}\frac{\partial q}{\partial t} \qquad ...(11.15)$$

or

$$\frac{\partial E}{\partial t} = \frac{1}{\epsilon_0 A}\cdot i_d \qquad ...(11.16)$$

or

$$i_d = A\frac{\partial (\epsilon_0 E)}{\partial t} \qquad ...(11.17)$$

Thus,

$$i_d = A\frac{\partial \vec{D}}{\partial t} \qquad ...(11.18)$$

Displacement current density

$$J_d = \frac{i_d}{A} = \frac{\partial D}{\partial t} \quad \text{...(11.19)}$$

or

$$\boxed{\vec{J}_d = \frac{\partial \vec{D}}{\partial t}} \quad \text{...(11.20)}$$

11.3 MAXWELL'S EQUATIONS IN INTEGRAL FORM

We have described Maxwell's Equations in differential form. To describe Maxwell's Equations in the integral form, consider an arbitrary volume bounded by a surface. Thus,

$$\boxed{\int_S \vec{D} \cdot \vec{dS} = \int_V \rho \, dV} \quad \text{Gauss's Law} \quad \text{...(11.21)}$$

$$\boxed{\int_S \vec{B} \cdot \vec{dS} = 0} \quad \text{No magnetic Monopole exist} \quad \text{...(11.22)}$$

$$\boxed{\oint_l \vec{E} \cdot \vec{dl} = -\int_S \frac{\partial \vec{B}}{\partial t} \cdot \vec{dS}} \quad \text{Faraday's Law} \quad \text{...(11.23)}$$

$$\boxed{\oint \vec{H} \cdot \vec{dl} = \int_S \vec{J} \cdot \vec{dS} + \int_S \frac{\partial \vec{D}}{\partial t} \cdot \vec{dS}} \quad \text{Modified Ampere's Law} \quad \text{...(11.24)}$$

Here, we have used Stoke's theorem and Gauss's theorem, as stated below.

$$\boxed{\int_S (\nabla \times \vec{A}) \cdot \vec{dS} = \int_l \vec{A} \cdot \vec{dl}} \quad \text{Stoke's theorem} \quad \text{...(11.25)}$$

and

$$\boxed{\int_V \nabla \cdot \vec{A} \, dV = \int_S \vec{A} \cdot \vec{dS}} \quad \text{Gauss theorem} \quad \text{...(11.26)}$$

11.4 MAXWELL'S EQUATIONS FOR STATIC ELECTRIC AND MAGNETIC FIELDS

(a) The electric flux per unit volume passing through an arbitrary infinitesimally small volume is equal to the volume charge density.

Thus, we write

$$\nabla \cdot \vec{D} = \rho \quad ...(11.27)$$

or

$$\nabla \cdot \vec{E} = \frac{\rho}{\epsilon_0} \quad ...(11.28)$$

where $\rho = \int_V q\, dV$ is the volume charge density.

(b) Consider a surface enclosing a volume *V* as shown in fig. 11.3.

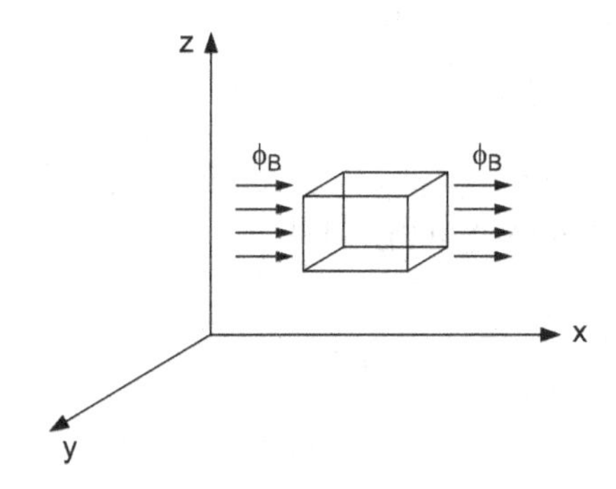

Fig. 11.3. Flux through an elementary volume

The magnetic flux entering is equal to the flux leaving the volume. Thus, net outgoing magnetic flux is zero.

$$\boxed{\nabla \cdot \vec{B} = 0} \quad ...(11.29)$$

(c) For a static charge, $\frac{\partial B}{\partial t} = 0$

Thus,

$$\nabla \times \vec{E} = 0$$

and potential is given by

$$\boxed{V = -\int_l \vec{E} \cdot \vec{dl}} \quad ...(11.30)$$

(d) For a static magnetic field,

$$\nabla \times \vec{H} = \vec{J}$$

or

$$\boxed{\oint \vec{B} \cdot \vec{dl} = \mu_0 i} \quad ...(11.31)$$

As per above discussion, one can say that the Maxwell's Equations are the general equations which describe electromagnetism.

11.5 ENERGY FLOW IN ELECTROMAGNETIC WAVES: POYNTING THEOREM

When an electromagnetic wave propagates through the space, it transports the electromagnetic energy from one point to another. According to Poynting theorem, the Poynting vector $\vec{S}$ is a measure of amount of energy crossing per unit area per unit time.

Mathematically, Poynting vector $\vec{S}$ is a cross product of the electric field intensity and the magnetic field intensity. Hence,

$$\vec{S} = \vec{E} \times \vec{H} \quad ...(11.32)$$

If $\vec{E}$ and $\vec{B}$ are in the directions of z and y axes respectively, Then, energy will flow along x-axis as shown in fig. 11.4.

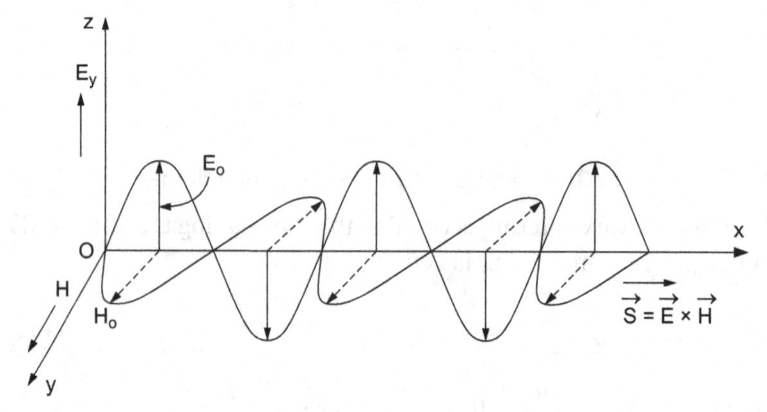

Fig. 11.4. Direction of flow of energy.

We have Maxwell's equations as

$$\nabla \times \vec{E} = -\mu_0 \frac{\partial H}{\partial t} \quad ...(11.33)$$

and

$$\nabla \times \vec{H} = \vec{J} + \epsilon_0 \frac{\partial E}{\partial t} \quad ...(11.34)$$

But $J = \sigma E$, Then we write the Equation (11.34) as

$$\nabla \times \vec{H} = \sigma E + \epsilon_0 \frac{\partial E}{\partial t} \quad ...(11.35)$$

Taking the dot product of E in Equation (11.35) we get

$$\vec{E} \cdot (\nabla \times \vec{H}) = \sigma E^2 + \epsilon_0 \vec{E} \cdot \frac{\partial \vec{E}}{\partial t} \quad ...(11.36)$$

Now, we have a vector identity, and give by

$$\nabla \cdot (\vec{F} \times \vec{G}) = \vec{G} \cdot (\vec{\nabla} \times \vec{F}) - \vec{F} \cdot (\nabla \times \vec{G})$$

using this identity in Equation (10.36), we get

$$\vec{H} \cdot (\nabla \times \vec{E}) - \nabla(\vec{E} \times \vec{H}) = \sigma E^2 + \epsilon_0 \vec{E} \cdot \frac{\partial \vec{E}}{\partial t} \quad \text{...(11.37)}$$

or

$$-\vec{H} \cdot \left(\mu_0 \frac{\partial \vec{H}}{\partial t} \right) - \nabla \cdot (\vec{E} \times \vec{H}) = \sigma E^2 + \epsilon_0 \vec{E} \cdot \frac{\partial \vec{E}}{\partial t} \quad \text{...(11.38)}$$

Here, we have substituted the value of $(\nabla \times \vec{E})$ from Equation (11.33) into equation (11.37).

Now,

$$-\nabla \cdot (\vec{E} \times \vec{H}) = \sigma E^2 + \epsilon_0 \vec{E} \cdot \frac{\partial \vec{E}}{\partial t} + \mu_0 \vec{H} \cdot \frac{\partial \vec{H}}{\partial t} \quad \text{...(11.39)}$$

or

$$-\nabla \cdot (E \times H) = \sigma E^2 + \frac{1}{2} \epsilon_0 \frac{\partial E^2}{\partial t} + \frac{1}{2} \mu_0 \frac{\partial H^2}{\partial t} \quad \text{...(11.39)}$$

$$= \sigma E^2 + \frac{\partial}{\partial t} \left[\frac{1}{2} \epsilon_0 E^2 + \frac{1}{2} \mu_0 H^2 \right] \quad \text{...(11.40)}$$

Integrating both sides of the Equation (11.40) over a volume, we get

$$-\int_V \nabla \cdot (\vec{E} \times \vec{H})\, dV = \int_V \sigma E^2 dV + \int_V \frac{\partial u}{\partial t} dV \quad \text{...(11.41)}$$

where $u = u_e + u_B = \frac{1}{2} \epsilon_0 E^2 + \frac{1}{2} \mu_0 H^2$ is the energy of the electromagnetic field.

Applying Gauss's divergence theorem to the left side, we get

$$-\int_S (\vec{E} \times \vec{H}) \cdot \vec{dS} = \int_V \sigma E^2 dV + \int_V \frac{\partial u}{\partial t} dV \quad \text{...(11.42)}$$

The physical interpretation of the terms involved in the Equation (11.42) are as follows;

(a) $\int_V \sigma E^2 dV$ represents the ohmic power dissipated in the volume V.

(b) $\int_V \frac{\partial u}{\partial t} dV$ represents the time rate at which electromagnetic energy stored in the volume V.

(c) $-\int_S (\vec{E} \times \vec{H}) \cdot \vec{dS}$ represents the amount of power crossing per unit area, and the cross product of $\vec{E}$ and $\vec{H}$ is represented by $\vec{S}$, Poynting vector. Thus,

$$\vec{S} = \vec{E} \times \vec{H} \quad ...(11.43)$$

For the plane electromagnetic waves, we have

$$\boxed{H = \sqrt{\frac{\epsilon_0}{\mu_0}}\, E} \quad ...(11.43a)$$

Now,

$$|S| = |E \times H| \quad ...(11.44)$$

$$= EH$$

$$= \frac{E \epsilon_0 E}{\sqrt{\epsilon_0 \mu_0}}$$

or

$$S = c\epsilon_0 E^2 \quad ...(11.45)$$

we can write,

$$\boxed{S = cu} \quad ...(11.46)$$

where

$$u = \frac{1}{2}\epsilon_0 E^2 + \frac{1}{2}\mu_0 H^2$$

From Equation (11.43), we may write

$$\boxed{\vec{S} = \frac{\vec{E} \times \vec{B}}{\mu_0}} \quad ...(11.47)$$

The SI unit of Poynting vector is watt/m^2 or Joules/m^2 – sec.

11.6 MAXWELL'S EQUATIONS FOR FREE SPACE AND DIELECTRIC MEDIA

For free space, we have

$$\sigma = 0, \quad \rho = 0 \quad \text{and} \quad J = \sigma E = 0$$

$$\epsilon = \epsilon_0 \text{ and } \mu = \mu_0 \text{ (For free space)}$$

Thus, Maxwell's equations may be written as

(a) $$\nabla \cdot \vec{D} = 0$$

or $$\nabla \cdot \vec{E} = 0 \quad ...(11.48)$$

(b) $$\nabla \cdot \vec{B} = 0$$

or $$\nabla \cdot \vec{H} = 0 \qquad \text{...(11.49)}$$

(c) $$\nabla \times \vec{E} = -\mu_0 \frac{\partial \vec{H}}{\partial t} \qquad \text{...(11.50)}$$

(d) $$\nabla \times \vec{H} = \epsilon_0 \frac{\partial \vec{E}}{\partial t} \qquad \text{...(11.51)}$$

11.7 MAXWELL'S EQUATIONS FOR CONDUCTING MEDIA

Suppose that the conducting medium is linear and isotropic. Then we can write,

$$\rho = 0,\ \vec{D} = \epsilon \vec{E},\ \vec{J} = \sigma \vec{E} \text{ and } \vec{B} = \mu \vec{H}$$

There is no charge in the conducting medium.

(a) $$\nabla \cdot \vec{D} = \nabla \cdot \vec{E} = 0 \qquad \text{...(11.52)}$$

(b) $$\nabla \cdot \vec{B} = \nabla \cdot \vec{H} = 0 \qquad \text{...(11.53)}$$

(c) $$\nabla \times \vec{E} = \frac{-\mu \partial \vec{H}}{\partial t} \qquad \text{...(11.54)}$$

(d) $$\nabla \times \vec{H} = \sigma \vec{E} + \epsilon \frac{\partial E}{\partial t} \qquad \text{...(11.55)}$$

11.8 ELECTROMAGNETIC WAVE EQUATION

A wave is a function of space and time co-ordinates. The most beautiful result of Maxwell's equations is the electromagnetic wave equation. We are developing a general wave equation for a material medium. For a conducting or material medium.

$$\vec{J} = \sigma \vec{E}$$

where σ is the conductivity of the material medium. Suppose that the medium consists of permitivity ϵ and permeability μ. No charge is present in the medium, Thus, $\rho = 0$.

In the absence of the external charge, Maxwell's equations are given by

$$\nabla \cdot \vec{E} = 0 \qquad \text{...(11.56)}$$

$$\nabla \cdot \vec{H} = 0 \qquad \text{...(11.57)}$$

$$\nabla \times \vec{E} = \frac{-\mu \partial \vec{H}}{\partial t} \qquad \text{...(11.58)}$$

and $$\nabla \times \vec{H} = \sigma E + \in \frac{\partial E}{\partial t} \qquad ...(11.59)$$

Taking the cross-product of del (∇) in Equation (11.58), we get

$$\nabla \times (\nabla \times \vec{E}) = -\mu \frac{\partial}{\partial t} (\nabla \times \vec{H}) \qquad ...(11.60)$$

Substituting the value of $(\nabla \times \vec{H})$ from Equation (11.59) into equation (11.60). Then,

$$\nabla \times (\nabla \times \vec{E}) = -\mu \frac{\partial}{\partial t} \left[\sigma E + \in \frac{\partial E}{\partial t} \right] \qquad ...(11.61)$$

Using the vector identify,

$$\nabla \times (\nabla \times \vec{E}) = \nabla (\nabla \cdot \vec{E}) - \nabla^2 \vec{E}$$

Thus, we get

$$\nabla (\nabla \cdot \vec{E}) - \nabla^2 \vec{E} = -\mu\sigma \frac{\partial E}{\partial t} - \mu \in \frac{\partial^2 E}{\partial t^2} \qquad ...(11.62)$$

But $\nabla \cdot E = 0$

Hence, we get

$$\boxed{\nabla^2 \vec{E} - \in \mu \frac{\partial^2 \vec{E}}{\partial t^2} - \mu\sigma \frac{\partial E}{\partial t} = 0} \qquad ...(11.63)$$

Similarly, Taking the cross product of (∇) del in Equation (11.59) and substituting the values of $\nabla \cdot \vec{H}$ and $\nabla \times \vec{E}$. From the Equations (11.57) and (11.58), we get

$$\boxed{\nabla^2 \vec{H} - \in \mu \frac{\partial^2 \vec{H}}{\partial t^2} - \mu\sigma \frac{\partial H}{\partial t} = 0} \qquad ...(11.64)$$

The Equations (11.63) and (11.64) are the differential equations of second order and involving first order term also as in differential equation of damped harmonic oscillator. These are the wave equations for conducting median.

11.9 PLANE ELECTROMAGNETIC WAVES IN FREE SPACE

We have already derived the general differential equations for the electromagnetic waves. For free space, we write, $\sigma = 0$, $\rho = 0$, $\mu = \mu_0$ and $\in = \in_0$. Then the equations (11.63) and (11.64) take form as

$$\nabla^2 \vec{E} - \epsilon_0 \mu_0 \frac{\partial^2 E}{\partial t^2} = 0 \quad ...(11.65)$$

and $$\nabla^2 \vec{H} - \epsilon_0 \mu_0 \frac{\partial^2 \vec{H}}{\partial t^2} = 0 \quad ...(11.66)$$

Here, the velocity of the wave

$$v = \frac{1}{\sqrt{\epsilon_0 \mu_0}}$$

$$= \frac{1}{\sqrt{8.85 \times 10^{-12} \times 4 \times 3.14 \times 10^{-7}}} \text{ m/s}$$

or $$v = 3 \times 10^8 \text{ m/s} \quad ...(11.67)$$

which is equal to velocity of light. It means that the electromagnetic wave propagates with the speed of light in the free space.

The Maxwell's equations for free space are

$$\nabla \cdot \vec{E} = 0 \quad ...(11.68)$$

$$\nabla \cdot \vec{H} = 0 \quad ...(11.69)$$

$$\nabla \times \vec{E} = -\mu_0 \frac{\partial \vec{H}}{\partial t} \quad ...(11.70)$$

and $$\nabla \times \vec{H} = \epsilon_0 \frac{\partial \vec{E}}{\partial t} \quad ...(11.71)$$

The plane wave solutions for the equations (10.65) and (11.66) are given by

and $$\left.\begin{aligned} \vec{E}(\vec{r}\, t) &= E_0\, e^{i(\vec{k} \cdot \vec{r} - \omega t)} \\ \vec{H}(\vec{r}, t) &= H_0\, e^{i(\vec{k} \cdot \vec{r} - \omega t)} \end{aligned}\right\} \quad ...(11.72)$$

Now, we define operators as

$$\boxed{\nabla \rightarrow i\vec{k}} \quad ...(11.73)$$

and $$\boxed{\frac{\partial}{\partial t} \rightarrow -i\omega}$$

Substituting these operators into equations (10.68), (10.69) (11.70) and (11.71) we get

$$\vec{k} \cdot \vec{E} = 0 \qquad ...(11.74)$$

$$\vec{k} \cdot \vec{H} = 0 \qquad ...(11.75)$$

$$\vec{k} \times \vec{E} = \mu_0 \omega \vec{H} \qquad ...(11.76)$$

and

$$\vec{k} \times \vec{H} = -\epsilon_0 \omega \vec{E} \qquad ...(11.77)$$

To find the propagation constant, taking the cross product of $\vec{K}$ in (11.76) we have,

$$\vec{k} \times (\vec{k} \times \vec{E}) = \mu_0 \omega (\vec{k} \times \vec{H}) \qquad ...(11.78)$$

or

$$k(\vec{k} \cdot \vec{E}) - k^2 \vec{E} = -\epsilon_0 \mu_0 \omega^2 \vec{E} \qquad ...(11.79)$$

or

$$k^2 = \epsilon_0 \mu_0 \omega^2 \qquad ...(11.80)$$

we have used Equations (11.74) and (11.77).

Thus,

$$k^2 = \frac{\omega^2}{c^2}$$

or

$$\boxed{k = \frac{\omega}{c}} \qquad ...(11.81)$$

From the Equation (11.76), we write

$$\vec{H} = \frac{1}{\mu_0 \omega} (\vec{k} \times \vec{E}) \qquad ...(11.82)$$

Poynting vector is

$$\vec{S} = \vec{E} \times \vec{H} \qquad ...(11.83)$$

$$\Rightarrow \quad \vec{S} = \frac{\vec{E} \times (\vec{k} \times \vec{E})}{\mu_0 \omega}$$

or

$$\vec{S} = \frac{\vec{k}(E^2) - \vec{E}(\vec{k} \cdot \vec{E})}{\mu_0 \omega} \qquad ...(11.84)$$

using equation (11.68) we get

$$\vec{S} = \frac{\vec{k} E^2}{\mu_0 \omega} \qquad ...(11.85)$$

$$|\vec{S}| = \left|\frac{kE^2}{\mu_0\omega}\right|$$

or
$$\boxed{S = \frac{E^2}{\mu_0 c}} \qquad ...(11.86)$$

For intrinsic impedance of the free space,

$$S = \frac{E^2\sqrt{\mu_0 \in_0}}{\mu_0}$$

or
$$\boxed{S = \sqrt{\frac{\in_0}{\mu_0}}\,E^2 = \frac{E^2}{Z}} \qquad ...(11.87)$$

where $Z = \sqrt{\frac{\in_0}{\mu_0}}$ is called intrinsic impedance of the free space.

$$\boxed{Z = \sqrt{\frac{\in_0}{\mu_0}} = 377\ \Omega} \qquad ...(11.88)$$

However, the average power per unit area is

$$\boxed{\langle S \rangle = \frac{E^2}{2Z}} \qquad ...(11.89)$$

11.10 PLANE WAVES AND POLARIZATION

A plane wave is a constant frequency wave whose wavefronts are the infinite parallel planes. A plane monochromatic transverse wave propagating in z-direction is given by

$$\begin{aligned}\vec{E}(z,t) &= \vec{E}_0\, e^{i(kz-\omega t)}\\ \vec{H}(z,t) &= \vec{H}_0\, e^{i(kz-\omega t)}\end{aligned} \qquad ...(11.90)$$

The one dimension wave equation in charge free region is

$$\left.\begin{aligned}\frac{\partial^2 E}{\partial z^2} &= \mu_0\epsilon_0 \frac{\partial^2 \vec{E}}{\partial t^2}\\ \text{and} \quad \frac{\partial^2 H}{\partial z^2} &= \mu_0\epsilon_0 \frac{\partial^2 \vec{H}}{\partial t^2}\end{aligned}\right\} \qquad ...(11.91)$$

The equation (11.90) is solution of the wave equations given in (11.91). Of course, there are miscellaneous solutions, in fact, both real and imaginary parts of the equation (11.89) are the solutions of the wave equations. If we write,

$$\vec{E} = \hat{y}\vec{E}_0 \sin(kz - \omega t) \qquad ...(11.92)$$

as a solution of the wave equation (11.90), then $\omega t = 2\pi, 4\pi, 6\pi$ ... and so on. The propagation vector for $z = \lambda$ is given by

$$kz = 2\pi$$

or

$$\boxed{k = \frac{2\pi}{\lambda}} \qquad ...(11.93)$$

where λ is wavelength of the propagative plane wave.

The angular frequency $\omega = 2\pi f$, the propagating wave repeats for every interval $T = \frac{2\pi}{\omega}$.

The phase velocity of a plane wave of a constant phase is

$$kz - \omega t = \text{constant} \qquad ...(11.94)$$

on differentiating it, we get

$$kdz - \omega dt = 0$$

or

$$\boxed{v_p = \frac{dZ}{dt} = \frac{\omega}{k}} \qquad ...(11.95)$$

The electric vector plays a key role in describing the polarization of a wave. A plane wave is polarized with its electric vector $\vec{E}$ in the $\vec{E}_0$ direction (Amplitude), as the wave given in the equation (11.92) is polarised in y-direction. In general, we take both components of the electric field when describing the different polarizing conditions. Generally, the polarizations are of the three types viz,

(a) Linear polarization

(b) Circular polarization

(c) Elliptical polarization.

(a) Linear Polarization. If the electric vector $\vec{E}$ is fixed along a straight line for all space and time co-ordinates, the electromagnetic wave is called linearly polarized. Suppose that a wave is propagating in z direction and is given by the equation

$$\vec{E} = \hat{x}E_0 e^{i(kz - \omega t)} \qquad ...(11.96)$$

This wave is linearly polarized in x-direction as shown in fig. 11.5.

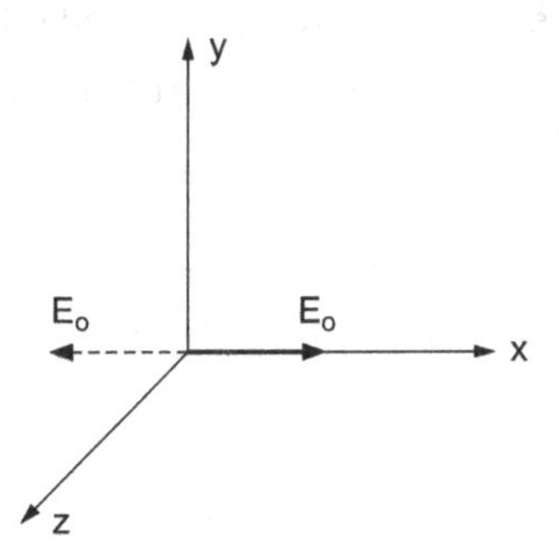

Fig. 11.5. Linearly polarized wave.

(b) Circular Polarization. If the tip of the electric vector $\vec{E}$ traces a circle, the wave is called circularly polarized wave. To describe a circularly polarized wave, we are taking two linearly polarized waves as

$$\vec{E} = \hat{x}e^{i(Kz-\omega t)} + i\hat{y}e^{i(Kz-\omega t)} \quad \text{...(11.97)}$$

Here, $i = \sqrt{-1}$,

Moreover, $$e^{i\pi/2} = \cos\frac{\pi}{2} + i\sin\frac{\pi}{2}$$
$$= i$$

Thus, with phase angle $\phi = \frac{\pi}{2}$, the tip of electric vector $\vec{E}$ rotates in the clockwise direction and for $\phi = -\pi/2$, $\vec{E}$ rotates in the counter clockwise direction.

The equation (11.97) shows a right hand circularly polarized wave as shown in Fig. 11.6.

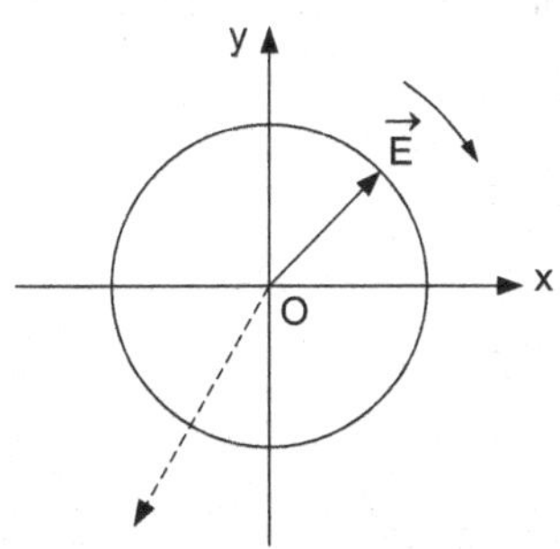

Fig. 11.6. Circularly polarized wave

(c) Elliptical Polarization. The linear polarization and the circular polarization are the special cases of the elliptical polarization. For linear polarization, phase difference, $\phi = 0$ and for circular polarization, $\phi = \pm\pi/2$. If the amplitudes of the two waves are not equal, the polarized wave is called elliptical polarized wave.

An elliptical polarized wave may be constructed by two linearly polarized components as given below.

$$\vec{E} = (E_1\hat{x} + iE_2\hat{y})\, e^{i(Kz - \omega t)} \quad ...(11.98)$$

where $i = \sqrt{-1}$ which decides the phase of the wave.

The angle of the field is given by

$$\theta(z, t) = \tan^{-1}\left(\frac{E_2}{E_1}\right) \quad ...(11.99)$$

11.11 ELECTROMAGNETIC WAVES IN CONDUCTING MEDIA

Consider an electromagnetic wave propagating in the conducting medium in the absence of an external charge.

For conducting medium, there is only ohmic current.

Then, $$\vec{J} = \sigma\vec{E},\ \rho = 0,\ \vec{B} = \mu\vec{H},\ \vec{D} = \in\vec{E} \quad ...(11.100)$$

The electromagnetic wave equations for the conducting medium are given by the equations (11.63) and (11.64) as

$$\nabla^2\vec{E} - \in\mu\frac{\partial^2\vec{E}}{\partial t^2} - \mu\sigma\frac{\partial\vec{E}}{\partial t} = 0 \quad ...(11.101)$$

and $$\nabla^2\vec{H} - \in\mu\frac{\partial^2\vec{H}}{\partial t^2} - \mu\sigma\frac{\partial\vec{H}}{\partial t} = 0 \quad ...(11.102)$$

Now, Maxwell's equations for conducting medium are as

$$\nabla\cdot\vec{E} = 0 \quad ...(11.103)$$

$$\nabla\cdot\vec{H} = 0 \quad ...(11.104)$$

$$\nabla\times\vec{E} = -\mu\frac{\partial\vec{H}}{\partial t} \quad ...(11.105)$$

and $$\nabla\times\vec{H} = \sigma\vec{E} + \in\frac{\partial\vec{E}}{\partial t} \quad ...(11.106)$$

The plane wave solutions of the equations (11.101) and (11.102) are given by

and
$$\left.\begin{aligned}\vec{E}(z,t) &= E_0 e^{-i(kz-\omega t)} \\ \vec{H}(z,t) &= H_0 e^{-i(kz-\omega t)}\end{aligned}\right\} \qquad ...(10.107)$$

Now, we have

and
$$\left.\begin{aligned}\nabla &\equiv -ik \\ \frac{\partial}{\partial t} &\equiv i\omega\end{aligned}\right\} \qquad ...(10.108)$$

Substituting into Equation (11.101) we get

$$(-ik)^2 \vec{E} - \in \mu\, (i\omega)^2 \vec{E} - \mu\sigma(i\omega)\vec{E} = 0 \qquad ...(11.109)$$

or
$$k^2 = i\omega\mu\sigma - \omega^2 \in \mu \qquad ...(11.110)$$

$\therefore$
$$k = \sqrt{i\mu\omega\,(\sigma + i\in\omega)} \qquad ...(11.111)$$

For good conductors, $\sigma >> \in\omega$, Thus, equation (11.111) reduces to

$$k \approx \sqrt{i\mu\omega\sigma} \qquad ...(11.112)$$

$$= \sqrt{i}\,\sqrt{\mu\omega\sigma} \qquad ...(11.113)$$

Now we evaluate $\sqrt{i}$

$$(i)^{1/2} = (\cos \pi/2 + i \sin \pi/2)^{1/2}$$

$$= (\cos \pi/4 + i \sin \pi/4)$$

$\therefore$
$$\sqrt{(i)} = \frac{(1+i)}{\sqrt{2}} \qquad ...(11.114)$$

Thus, the equation (11.113) takes form

$$k = (1+i)\sqrt{\frac{\mu\omega\sigma}{2}} = \alpha + i\beta \qquad ...(11.115)$$

where
$$\alpha = \beta = \sqrt{\frac{\mu\omega\sigma}{2}}$$

or
$$k = \frac{(1+i)}{\delta} \qquad ...(11.116)$$

where δ is called skin depth or depth of penetration, and is given by

$$\boxed{\delta = \sqrt{\frac{2}{\mu\omega\sigma}}} \qquad ...(11.117)$$

Now substituting the value of k from the equation (11.116) into the equation (11.117), we get

$$E = E_0 e^{-z/\delta} e^{-i(z/\delta - \omega t)} \qquad ...(11.118)$$

In the equation (11.118), the term $e^{-z/\delta}$ is known as dumping factor while $E_0 e^{-i(z/\delta - \omega t)}$ represents the wave propagating into conductor. Thus, amplitude of the wave is

$$E = E_0 e^{-z/\delta} \qquad ...(11.119)$$

At $z = \delta$, we have

$$E = E_0 e^{-1} = \frac{E_0}{e} \qquad ...(11.120)$$

It means that the actual amplitude of the wave decreases to $\left(\frac{1}{e}\right)$ times when the wave penetrates the conductor to a distance δ.

Moreover, the current density is

$$J = \sigma E \qquad ...(11.121)$$

$$= \sigma E_0 e^{-kz}$$

or $$J = J_0 e^{-kz} \qquad ...(11.122)$$

where $$J_0 = \sigma E_0$$

$\therefore$ $$\boxed{J = J_0 e^{-z/\delta}} \qquad ...(11.123)$$

The current density decays exponentially with the skin depth δ.

The phase velocity of the wave in good conductor is given by

$$v_p = \frac{\omega}{\beta}$$

or $$\boxed{v_p = \omega\delta} \qquad \therefore \ \beta = \frac{1}{\delta} \qquad ...(11.124)$$

The average power dissipated per unit area is given by

$$\langle S \rangle = \frac{|E||H|}{2} \qquad ...(11.125)$$

$$E = E_0 e^{-z/\delta} e^{-i(z/\delta - \omega t)}$$

The value of H can be obtained using the equation given as

$$\nabla \times E = -\mu \frac{\partial H}{\partial t} \qquad ...(11.126)$$

or $$\frac{\partial H}{\partial t} = -\frac{1}{\mu}\frac{\partial E}{\partial z} \qquad ...(11.127)$$

Thus we get

$$H = \frac{\sigma \delta E_0}{2} e^{-z/\delta} e^{-i(z/\delta - \omega t - \pi/4)}$$

Now,

$$\boxed{\langle S \rangle = \frac{\sigma \delta E_0^2}{4} e^{-2z/\delta}} \quad ...(11.128)$$

11.12 SCALAR AND VECTOR POTENTIALS

In time varying field, the potential ϕ can't be written as

$$\vec{E} = -\text{grad } \phi = -\nabla\phi$$

otherwise $\nabla \times \vec{E} = \nabla \times (-\nabla\phi) = 0$

Now, we write the Maxwell's equation for electric field

$$\nabla \times \vec{E} = -\frac{\partial \vec{B}}{\partial t} \quad ...(11.129)$$

This equation predicts that the curl of the electric field is not equal to zero but it is equal to time rate of change of the magnetic field. The static electric field is conservative but time varying field is not conservative we write, now, Maxwell's equation as

$$\nabla \cdot \vec{B} = 0 \quad ...(11.130)$$

This equation (11.130) enforces to write the magnetic field as a curl of a vector quantity $\vec{A}$ as

$$\vec{B} = \nabla \times \vec{A} \quad ...(11.131)$$

Substituting the value of $\vec{B}$ from the equation (11.131) into equation (11.129), we get

$$\nabla \times \vec{E} = \frac{\partial}{\partial t}(\nabla \times \vec{A}) \quad ...(11.132)$$

or

$$\nabla \times \left(\vec{E} + \frac{\partial \vec{A}}{\partial t} \right) = 0 \quad ...(11.133)$$

Since the curl of sum of two vector quantity is equal to zero, this may be written as the negative gradient of a scalar potential ϕ, Thus,

$$\vec{E} + \frac{\partial \vec{A}}{\partial t} = -\nabla\phi \quad ...(11.134)$$

or

$$\vec{E} = -\nabla\phi - \frac{\partial \vec{A}}{\partial t} \quad ...(11.135)$$

Hence, we get

$$\boxed{\vec{B} = \nabla \times \vec{A}} \quad ...(11.136)$$

and

$$\boxed{\vec{B} = -\nabla\phi - \frac{\partial \vec{A}}{\partial t}} \quad ...(11.137)$$

where ϕ is called as scalar potential and $\vec{A}$ is called as vector potential.

The equation (11.137) consists of two parts viz

(a) $-\nabla\phi$ which is due to the electric charge distribution.

(b) $-\frac{\partial A}{\partial t}$ is due to the time dependent current density $\vec{J}$.

11.13 NON-HOMOGENEOUS WAVE EQUATIONS FOR VECTOR AND SCALAR POTENTIALS

Most of the problems of the electromagnetics are solved with Maxwell's equations. We already mentioned the homogeneous wave equation for free space or non conducting media. A wave equation may be put forward with the scalar and vector potentials.

Starting with Maxwell's equation given as

$$\nabla \times \vec{H} = \vec{J} + \frac{\partial \vec{D}}{\partial t} \quad ...(11.138)$$

$\therefore$

$$\vec{B} = \mu_0 \vec{H}, \ \vec{D} = \epsilon_0 \vec{E}$$

$$\nabla \times \vec{B} = \mu_0 \vec{J} + \mu_0 \epsilon_0 \frac{\partial \vec{E}}{\partial t} \quad ...(11.139)$$

Substituting the values of $\vec{B}$ and $\vec{E}$ from the equations (11.136) and (11.137), we obtain

$$\nabla \times (\nabla \times \vec{A}) = \mu_0 \vec{J} + \mu_0 \epsilon_0 \frac{\partial}{\partial t}\left(-\nabla\phi - \frac{\partial \vec{A}}{\partial t}\right) \quad ...(11.140)$$

using vector identity

$$\nabla \times (\nabla \times \vec{A}) = \nabla(\nabla \cdot \vec{A}) - \nabla^2 \vec{A}$$

then, we have

$$\nabla(\nabla \cdot \vec{A}) - \nabla^2 \vec{A} = \mu_0 \vec{J} + \epsilon_0 \mu_0 \nabla\left(\frac{\partial \phi}{\partial t^2}\right) - \epsilon_0 \mu_0 \frac{\partial^2 \vec{A}}{\partial t^2} \quad ...(11.141)$$

rearranging the terms, we get

$$\nabla^2 \vec{A} - \epsilon_0 \mu_0 \frac{\partial^2 \vec{A}}{\partial t^2} - \nabla\left(\nabla \cdot \vec{A} + \epsilon_0 \mu_0 \frac{\partial \phi}{\partial t}\right) = -\mu_0 \vec{J} \quad ...(11.142)$$

Now, taking Maxwell's equation

$$\nabla \cdot \vec{D} = \rho \quad ...(11.143)$$

or

$$\epsilon_0 \nabla \cdot \vec{E} = \rho \quad ...(11.144)$$

Substituting for $\vec{E}$ from the equation (11.137), we get

$$\nabla \cdot \left(-\nabla \phi - \frac{\partial \vec{A}}{\partial t}\right) = \frac{\rho}{\epsilon_0} \quad ...(11.145)$$

or

$$\nabla^2 \phi + \nabla \cdot \frac{\partial \vec{A}}{\partial t} = \frac{-\rho}{\epsilon_0} \quad ...(11.146)$$

$\therefore$

$$\nabla^2 \phi + \nabla \cdot \frac{\partial \vec{A}}{\partial t} + \epsilon_0 \mu_0 \frac{\partial^2 \phi}{\partial t^2} - \epsilon_0 \mu_0 \frac{\partial^2 \phi}{\partial t^2} = \frac{\rho}{\epsilon_0} \quad ...(11.147)$$

Here, we have added and subtracted the term $\epsilon_0 \mu_0 \frac{\partial^2 \phi}{\partial t^2}$.

rearranging the terms in the equation (11.147), we get

$$\nabla^2 \phi + \epsilon_0 \mu_0 \frac{\partial^2 \phi}{\partial t^2} + \frac{\partial}{\partial t}\left(\nabla \cdot \vec{A} + \epsilon_0 \mu_0 \frac{\partial \phi}{\partial t}\right) = \frac{-\rho}{\epsilon_0} \quad ...(11.148)$$

Now, we put the term in bracket is equal to zero, we get

$$\boxed{\nabla \cdot \vec{A} + \epsilon_0 \mu_0 \frac{\partial \phi}{\partial t} = 0} \quad ...(11.149)$$

The equation (11.149) is Lorentz condition. Thus, the equations (11.142) and (11.148) take the form as

$$\nabla^2 \cdot \vec{A} - \epsilon_0\mu_0 \frac{\partial^2 A}{\partial t^2} = -\mu \vec{J} \quad ...(11.150)$$

and $$\nabla^2\phi - \epsilon_0\mu_0 \frac{\partial^2\phi}{\partial t^2} = \frac{-\rho}{\epsilon_0} \quad ..(11.151)$$

The equation (11.150) and (11.151) are non-homogeneous equations for vector potential $\vec{A}$ and scalar potential ϕ. From these equation it is clear that the source of *em* field is present. In time varying field, if the curl of a vector $\vec{E}$ is zero, then it is represented by the equation (11.137), and if $\nabla \cdot \vec{B} = 0$, then $\vec{B}$ is represented by the equation (11.136).

Moreover, we take Ampere's equation as

$$\nabla \times \vec{B} = \mu_0 \vec{J} \quad ...(11.152)$$

then, $$\nabla \times (\nabla \times \vec{A}) = \mu_0 \vec{J} \quad ...(11.153)$$

or $$\nabla (\nabla \cdot \vec{A}) - \nabla^2 \vec{A} = \mu_0 \vec{J} \quad ...(11.154)$$

there is only possibility to choose

$$\nabla \cdot \vec{A} = 0 \quad ...(11.155)$$

Hence, $$\nabla^2 \vec{A} = -\mu_0 \vec{J} \quad ...(11.156)$$

This is known as Poisson's equation in vector potential $\vec{A}$. In views of the equations (11.150) and (11.151), the solution will be in form of-

$$\phi = \int_V \frac{\rho dV}{4\pi \epsilon_0 r} \quad ...(11.157)$$

and $$A = \int_V \frac{\mu_0 J\, dV}{4\pi r} = \oint \frac{\mu_0 i\, dl}{4\pi r} \quad ...(11.158)$$

Example 11.1. A parallel plate capacitor consists of two plates with area A separated by distance d. If the ac source $v = v_0 \cos \omega t$ is applied between two plates of the capacitor, calculate displacement current.

Solution: $\therefore$ $E = v/d$, then,

displacement current $i_d = A\dfrac{\partial D}{\partial t}$

or $$i_d = \epsilon_0 A \frac{\partial E}{\partial t}$$

$\Rightarrow$ $$i_d = \frac{\epsilon_0 A}{d}\frac{\partial v}{\partial t} = C\frac{\partial v}{\partial t}$$

where $$C = \frac{\epsilon_0 A}{d}$$

$$i_d = C\frac{d}{dt}(v_0 \cos \omega t)$$

$\therefore$ $$i_d = -C\, v_0\, \omega \sin \omega t.$$

Example 11.2. Deduce Poissons's equation from Maxwell's first equation.
Solution: Maxwell's First equation is given by

$$\nabla \cdot \vec{D} = \rho$$

Since $$\vec{D} = \epsilon_0 \vec{E}\text{, then}$$

$$\nabla \cdot \vec{E} = \frac{\rho}{\epsilon_0}$$

or $$\nabla \cdot \left(-\nabla\phi - \frac{\partial \vec{A}}{\partial t}\right) = \frac{\rho}{\epsilon_0}$$

or $$-\nabla^2\phi - \frac{\partial(\nabla \cdot A)}{\partial t} = \frac{\rho}{\epsilon_0}$$

But $$\nabla \cdot \vec{A} = 0$$

Thus, $$\nabla^2\phi = -\frac{\rho}{\epsilon_0}$$

Example 11.3. A monochromatic plane polarized electromagnetic wave travelling in free space is given by

$$\vec{E} = \hat{y}\, E_0 \sin(\omega t - kz)$$

Find magnetic field intensity $\vec{H}$ and poynting vector $\vec{S}$.

Solution: By Maxwell's equation,

$$\frac{-\partial \vec{B}}{\partial t} = \nabla \times \vec{E}$$

$$= \begin{vmatrix} \hat{x} & \hat{y} & \hat{z} \\ \frac{\partial}{\partial x} & \frac{\partial}{\partial y} & \frac{\partial}{\partial z} \\ 0 & E & 0 \end{vmatrix}$$

$$= -\hat{x}\frac{\partial E}{\partial z}$$

$$\frac{-\partial \vec{B}}{\partial t} = -\hat{x}\, E_0 k \cos(\omega t - kx)$$

on integrating it w.r.to t, we get

$$\vec{B} = \hat{x}\, E_0\left(\frac{k}{\omega}\right)\sin(\omega t - kz)$$

or

$$\vec{B} = \hat{x}\frac{E_0}{c}\sin(\omega t - kz)$$

$$\vec{H} = \hat{x}\left(\frac{\mu_0 E_0}{c}\right)\sin(\omega t - kz)$$

or

$$\vec{H} = \hat{x}\, H_0 \sin(\omega t - kz)$$

where $H_0 = \left(\frac{\mu_0 E_0}{c}\right)$ and $c = \left(\frac{\omega}{k}\right)$

Poynting vector

$$\vec{S} = \vec{E} \times \vec{H}$$

or

$$\vec{S} = E_0 H_0 \sin^2(\omega t - kx)(-\hat{z})$$

Example 11.4. The vector potential $\vec{A}$ is given by

$$\vec{A} = \oint \frac{\mu_0\, i\, \vec{dl}}{4\pi r}$$

Find the curl of $\vec{A}$.

Solution: $\therefore$ $$\vec{A} = \frac{\mu_0 i}{4\pi}\oint \frac{\vec{dl}}{r}$$

Now

$$\nabla \times \vec{A} = \frac{\mu_0 i}{4\pi}\nabla \times \oint \frac{\vec{dl}}{r}$$

$$= \frac{\mu_0 i}{4\pi}\oint \frac{\nabla \times \vec{dl}}{r} + \frac{\mu_0 i}{4\pi}\oint \nabla\left(\frac{1}{r}\right) \times \vec{dl}$$

The curl of source term $\vec{dl}$ is zero, $\vec{dl}$ is current source element. Now,

$$\nabla \times \vec{A} = \frac{\mu_0 i}{4\pi}\oint \left(\frac{-\hat{r}}{r^2}\right) \times \vec{dl}$$

$$= \frac{\mu_0 i}{4\pi}\oint \frac{\vec{dl} \times \hat{r}}{r^2}$$

$$\boxed{\nabla \times \vec{A} = \vec{B}}$$

Since, $$\nabla\left(\frac{1}{r}\right) = -\frac{\hat{r}}{r^2}$$

and $$\vec{B} = \oint \frac{\mu_0 i\, \vec{dl} \times \hat{r}}{4\pi r^2}$$

Example 11.5. Show that $\vec{E}$ and $\vec{B}$ are not affected under gauge transformation as $\vec{A}$ and ϕ are given by

$$\vec{A}' = \vec{A} + \nabla\psi$$

$$\phi' = \phi - \frac{\partial \psi}{\partial t}$$

where ψ is any arbitrary scalar.

Solution: Given

$$\vec{A}' = \vec{A} + \nabla\psi$$

$$\phi' = \phi - \frac{\partial \psi}{\partial t}$$

Since,
$$\vec{E} = -\nabla\phi' - \frac{\partial \vec{A}'}{\partial t}$$
$$= -\nabla\left(\phi - \frac{\partial\psi}{\partial t}\right) - \frac{\partial}{\partial t}(\vec{A} + \nabla\psi)$$
$$= -\nabla\phi - \frac{\partial \vec{A}}{\partial t} + \frac{\partial(\nabla\psi)}{\partial t} - \frac{\partial(\nabla\psi)}{\partial t}$$

or
$$\vec{E} = -\nabla\phi - \frac{\partial \vec{A}}{\partial t}$$

Now,
$$\vec{B} = \nabla \times \vec{A}'$$
$$= \nabla \times (\vec{A} + \nabla\psi)$$
$$= \nabla \times \vec{A} + \nabla \times (\nabla\psi)$$

Since curl grad of scalar is zero, then
$$\vec{B} = \nabla \times \vec{A}$$

Example 11.6 A plane polarized wave is travelling in free space and is given by
$$\vec{E} = \hat{y}\, E_0 \sin(\omega t - kx)$$

If $E_0 = 0.02$ V/m and $f = 9$ MHz, calculate E, H, and poynting vector S and $<S>$.

Solution: $\therefore$
$$\omega = 2\pi f = 2\pi \times 10^9 \text{ rad/s}$$

and
$$\lambda = \frac{c}{f} = \frac{3\times10^8 \text{ m/s}}{1\times10^9 \text{ Hz}} = 0.3 \text{ m}$$

Then
$$E = E_0 \sin 2\pi\left(ft - \frac{x}{\lambda}\right)$$
$$E = 0.2 \sin 2\pi(10^9 - 3.33x) \text{ V/m}$$

Now
$$H_0 = \sqrt{\frac{\epsilon_0}{\mu_0}}\, E_0 = \frac{E_0}{Z} = \frac{0.2 \text{ V/m}}{377\Omega}$$

or
$$H_0 = 5.3 \times 10^{-4} \text{ A/m}$$

$\therefore$ $$H = H_0 \sin 2\pi\,(10^9 - 3.33x)\text{ A/m}$$

or $$H = 5.3 \times 10^{-4} \sin 2\pi(10^9 - 3.33x)\text{ A/m}$$

Poynting vector

$$S = |E||H|$$
$$= 1.06 \times 10^{-4} \sin^2 (109 - 3.33x)\text{ W/m}^2$$
$$< S > = \frac{1}{2} \times 1.06 \times 10^{-4}\text{ W/m}^2$$

or $$< S > = 0.53 \times 10^{-4} = 5.3 \times 10^{-5}\text{ W/m}^2$$

EXERCISES

11.1. Write the four equations of Maxwell in differential form.

11.2. Describe the displacement current and show that for a parallel plate capacitor the displacement current is given by

$$i_d = C\frac{dv}{dt}$$

11.3. Write Maxwell's equations in integral form.

11.4. State and prove Poynting theorem and explain the physical significance of the terms in it.

11.5. Derive the wave equation for a monochromatic electromagnetic wave travelling in free space.

11.6. Derive the wave equation for a wave propagating in a conducting medium and for a plane wave show that

$$k^2 = i\,\omega\,\mu\,\sigma\, - \omega^2 \in \mu$$

11.7. Describe linear, circular and elliptical polarizations.

11.8. An electromagnetic wave is propagating in conducting medium, obtain an expression for depth of penetration and show that for current density

$$J = J_0\, e^{-z/\delta}$$

11.9. Obtain the expression for wave equation for vector and scalar potentials.

11.10. Write down the Lorentz condition and explain its physical significance.

11.11. Show that the vector potential may be written as

$$A = \oint \frac{\mu_0 i\, dl}{4\pi r}$$

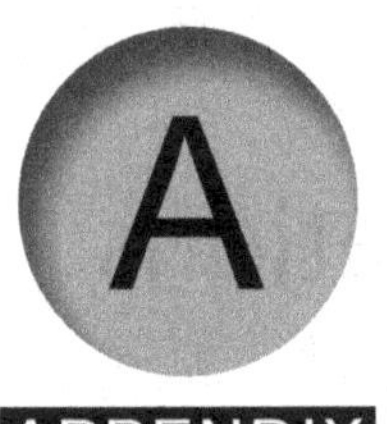

APPENDIX

Physical Constants

Electron charge	e	-1.6×10^{-19} coulomb
Electron Mass	m_e	9.1×10^{-31} kg
Electron charge to mass ratio	e/m_e	1.76×10^{11} coulomb/kg
Permittivity constant	$\in_0$	8.85×10^{-12} coulomb^2/N-m^2
Permeability constant	μ_0	$1.26 \times 10^{-6} = 4\pi \times 10^{-7}$ Wb/A-m
Speed of Light	c	3×10^8 m/s
Proton Mass	m_p	1.67×10^{-27} kg
Electron Volt	eV	1.6×10^{-19} joule
Magnetic Moment of electron	μ_e	9.29×10^{-18} joule-m^2/Wb
magnetic Induction	B	1 Wb/m^2 = 10^4 gauss

APPENDIX

B Trigonometrical Relations

$$\sin(90° - \theta) = \cos\theta$$
$$\cos(90° - \theta) = \sin\theta$$
$$\sin(180° - \theta) = \sin\theta$$
$$\cos(180° - \theta) = -\cos\theta$$
$$\sin(A \pm B) = \sin A \cos B \pm \cos A \sin B$$
$$\cos(A \pm B) = \cos A \cos B \mp \sin A \sin B$$
$$\cos^2\theta + \sin^2\theta = 1$$
$$\sin 2\theta = 2\sin\theta\cos\theta$$
$$\cos 2\theta = 2\cos^2\theta - 1 = 1 - 2\sin^2\theta = \cos^2\theta - \sin^2\theta$$
$$\sin A \pm \sin B = 2\sin\left(\frac{A \pm B}{2}\right)\cdot\cos\left(\frac{A \mp B}{2}\right)$$
$$\cos A + \cos B = 2\cos\left(\frac{A + B}{2}\right)\cdot\cos\left(\frac{A - B}{2}\right)$$
$$\cos A - \cos B = 2\sin\left(\frac{A + B}{2}\right)\cdot\sin\left(\frac{A - B}{2}\right)$$
$$\sin\theta = \theta - \frac{\theta^3}{3!} + \frac{\theta^5}{5!} \ldots\ldots\ldots$$
$$\cos\theta = 1 - \frac{\theta^2}{2!} + \frac{\theta^4}{4!} \ldots\ldots\ldots$$

Angle	0°	30°	45°	60°	90°	120°
sin	0	1/2	$1/\sqrt{2}$	$\sqrt{3}/2$	1	$\sqrt{3}/2$
cos	1	$\sqrt{3}/2$	1/2	0	0	– 1/2
tan	0	$1/\sqrt{3}$	1	$\sqrt{3}$	∞	$-\sqrt{3}$

$$\sin\theta = \frac{e^{i\theta} - e^{-i\theta}}{2i}$$
$$\cos\theta = \frac{e^{i\theta} + e^{-i\theta}}{2}$$

Algebraic Relations

Circle	–	Circumference	–	$2\pi r$
		Area	–	πr^2
		Length of arc	–	$\frac{\theta}{360} \times 2\pi r$
		Area of Sector	–	$\frac{\theta}{360} \times \pi r^2$
Sphere	–	Surface Area	–	$4\pi r^2$
		Volume	–	$\frac{4}{3}\pi r^3$
Cone	–	Surface Area	–	πrh
		Volume	–	$\frac{1}{3}\pi r^3 h$

$$e = 2.718$$
$$e^0 = 1$$
$$e^{-\infty} = 0$$
$$\text{Log}_e\, e = 1$$
$$\text{Log}_e\, 1 = 0$$
$$\text{Log}_e\, xy = \text{Log}_e\, x + \text{Log}_e\, y$$
$$\text{Log}_e \left(\frac{x}{y}\right) = \text{Log}_e\, x - \text{Log}_e\, y$$
$$\text{Log}_e\, x^a = a\, \text{Log}_e\, x$$
$$\text{Log}_e\, x = 2.303\, \text{Log}_{10}\, x$$
$$\text{Log}_e\, (1 + x) = x - \frac{x^2}{2} + \frac{x^3}{3} + \frac{x^4}{4} + \ldots$$
$$1 \text{ radian} = 57.3°$$

$$1° = 0.01745 \text{ radians}$$

$$e^x = 1 + x + \frac{x^2}{2!} + \frac{x^3}{3!} + \frac{x^4}{4!} + \ldots$$

Quadratic Equation

$$ax^2 + bx + c = 0$$

Roots of equation are $x = \dfrac{-b \pm \sqrt{b^2 - 4ac}}{2a}, \; a \neq 0$

Binomial Expansion

$$(1 + x)^n = 1 + nx + \frac{n(n-1)}{2!}x^2 + \frac{n(n-1)(n-2)}{3!}x^3 + \ldots$$

$$(a + x)^n = a^n\left(1 + \frac{x}{a}\right)^n = a^n\left[1 + n\left(\frac{x}{a}\right) + \frac{n(n-1)}{2!}\left(\frac{x}{a}\right)^2 + \ldots\ldots\right.$$

$${}^n c_r = \frac{n!}{r!(n-r)!}$$

Arithmetic Progression

Sum of n terms $S_n = \dfrac{n}{2}[2a + (n-1)d]$

Geometric Progression

Sum of n terms $S_n = \dfrac{a(1 - r^n)}{1 - r}$

Sum of infinite terms $S_\infty = \dfrac{a}{1 - r}$

APPENDIX

Vector Identities

$$\vec{A}\cdot\vec{B} = |\vec{A}||\vec{B}|\cos(\vec{A},\vec{B})$$

$$\vec{A}\times\vec{B} = |\vec{A}||\vec{B}|\sin(\vec{A},\vec{B})$$

$$\vec{A}\times(\vec{B}\times\vec{C}) = (\vec{A}\cdot\vec{C})\vec{B} - (\vec{A}\cdot\vec{B})\vec{C}$$

$$\nabla\cdot\nabla\times\vec{A} = 0$$

$$\nabla\times\nabla\phi = 0$$

$$\nabla\cdot(\phi\vec{A}) = \phi\nabla\cdot\vec{A} + \vec{A}\cdot\nabla\phi$$

$$\nabla\cdot(\vec{A}\times\vec{B}) = \vec{B}\cdot\nabla\times\vec{A} - \vec{A}\cdot\nabla\times\vec{B}$$

$$\nabla\times(\vec{A}\times\vec{B}) = \vec{A}(\nabla\cdot\vec{B}) - \vec{B}(\nabla\cdot\vec{A}) + (\vec{B}\cdot\nabla)\vec{A} + (\vec{A}\cdot\nabla)\vec{B}$$

$$\nabla\times(\phi\vec{A}) = \phi\nabla\times\vec{A} + \nabla\phi\times\vec{A}$$

$$\nabla\left(\frac{1}{r}\right) = \frac{-\hat{r}}{r^2} = -\frac{\vec{r}}{r^3}$$

INDEX

G

H

I

K

L

M

N

O

P

Q

R

S

T

U

V

W

■ ■ ■

Printed in the United States
by Baker & Taylor Publisher Services